NANNOFOSSILS AND THEIR APPLICATIONS

NANNOFOSSILS AND THEIR APPLICATIONS

**Proceedings of the
International Nannofossil Association Conference, London 1987**

Editors:

JASON A. CRUX, B.Sc., Ph.D.
Senior Biostratigrapher
British Petroleum Research Centre, Sunbury-on-Thames, UK

and

SHIRLEY E. VAN HECK
Geologist
Shell Internationale Petroleum
Maatschappij, The Netherlands

Published by
ELLIS HORWOOD LIMITED
Publishers · Chichester

for

THE BRITISH MICROPALAEONTOLOGICAL SOCIETY

Halsted Press: a division of
JOHN WILEY & SONS
New York · Chichester · Brisbane · Toronto

First published in 1989
ELLIS HORWOOD LIMITED
Market Cross House, Cooper Street,
Chichester, West Sussex, PO19 1EB, England
*The publisher's colophon is reproduced from James
Gillison's drawing of the ancient Market Cross, Chichester.*

Distributors:

Australia and New Zealand:
JACARANDA WILEY LIMITED
GPO Box 859, Brisbane, Queensland 4001,
Australia

Canada:
JOHN WILEY & SONS CANADA LIMITED
22 Worcester Road, Rexdale, Ontario, Canada

Europe and Africa:
JOHN WILEY & SONS LIMITED
Baffins Lane, Chichester, West Sussex, England

North and South America and the rest of the world:
Halsted Press: a division of
JOHN WILEY & SONS
605 Third Avenue, New York, NY 10158, USA

South-East Asia
JOHN WILEY & SONS (SEA) PTE LIMITED
37 Jalan Pemimpin # 05–04
Block B, Union Industrial Building, Singapore 2057

Indian Subcontinent
WILEY EASTERN LIMITED
4835/24 Ansari Road
Daryaganj, New Delhi 110002, India

**1989 J.A. Crux and S. E. van Heck/
Ellis Horwood Limited**

British Library Cataloguing in Publication Data
Nannofossils and their applications.
1. Nannofossils
I. Crux, Jason A. (Jason Alexander), *1955–*
II. Heck, Shirley E. van (Shirley Elizabeth van), *1952–*
III. Series
561 .93

Library of Congress Card No. 88–32070

ISBN 0–7458–0237–0 (Ellis Horwood Limited
ISBN 0–470–21203–9 (Halsted Press)

Printed in Great Britain by Unwin Bros., Woking

Contents

Contributors

J. L. Applegate,
Department of Geology, Florida State University, Tallahassee, FL 32306, USA.

M.-P. Aubry,
Woods Hole Oceanographic Institution, Woods Hole, MA 02543, USA.

J. A. Bergen,
Department of Geology, Florida State University, Tallahassee, FL 32306, USA.
now at: Amoco Research Centre, Tulsa, OK 74135, USA.

P. R. Bown,
University College London, Micropalaeontology Unit, Gower Street, London WC1E 6BT, UK.

M. K. E. Cooper,
SSI Ltd., Chancellor Court, 20 Priestly Road, Guildford, Surrey GV2 5YZ, UK.

J. M. Covington,
Department of Geology, Florida State University, Tallahassee, FL 32306, USA.
now at: Texaco USA, Southern Exploration Division, New Orleans, LA 70160, USA.

J. A. Crux,
Stratigraphy Branch, BP Research Centre, Chertsey Road, Sunbury-on-Thames, TW16 7LN
Middlesex, UK.

J.-A. Flores,
Departamento de Paleontologia, Universidad de Salamanca, 3700B Salamanca, Spain.

L. Gallagher,
Micropalaeontology Unit, University College London, Gower Street, London WC1E 6BT, UK.

G. Gard,
Department of Geology, University of Stockholm, S-10691 Stockholm, Sweden.

M. H. Girgis,
Robertson Research International Ltd., Llandudno, Gwynedd, LL30 1SA, UK.

S. E. van Heck,
Shell EXPRO UK, UEE/313, Shell-Mex House, Strand, London WC2R 0DX, UK.
now at: SIPM, EPX/36, P.O. Box 162, 2501 An Den Haag, The Netherlands.

S. Moshkovitz,
Geological Survey of Israel, 30 Malkei Israel Street, Jerusalem, Israel.

J. Mutterlose,
Institut für Geologie und Paläontologie, Universität Hannover, Callinstr. 30, 3000 Hannover, West Germany.

K. Osmond,
Department of Geology, Florida State University, Tallahassee, FL 32306, USA.

K. Perch-Nielsen,
17 Orchard Rise, Richmond, Surrey, UK.

F.-J. Sierro,
Area of Palaeontology, University of Salamanca, 3700B Salamanca, Spain.

V. Toker,
Ankara Universitesi, Fen Fakultesi Jeoloji, Muhendisligi Bolumu, 06100 Ankara, Turkey.

O. Varol,
Robertson Research International, Ty'n-y-Coed, Llanrhos, Llandudno, Gwynedd LL30 1SA, UK.

S. W. Wise, Jr.,
Department of Geology, Florida State University, Tallahassee, FL 32306, USA.

J. R. Young,
Geology Department, Imperial College, London SW7 2BP, UK.
now at: Biochemisch Laboratorium, RUL, Wassenaarseweg 64, 2333 AL Leiden, The Netherlands.

The organisers of the 1987 London Meeting of the International Nannoplankton Association and the editors of these Proceedings wish to thank the following sponsors for their financial support:

    ARCO BRITISH LTD London
    BP RESEARCH CENTRE Sunbury-on-Thames
    BRITOIL PLC Glasgow
    The BRITISH MICROPALAEONTOLOGICAL SOCIETY London
    CHEVRON EXPLORATION NORTH SEA LTD London
    CLYDE PETROLEUM PLC Ledbury
    ESSO EXPLORATION AND PRODUCTION UK LTD Esher
    ROBERTSON RESEARCH INTERNATIONAL LTD Llandudno
    SHELL UK EXPLORATION AND PRODUCTION London

The editors also wish to thank the many referees who reviewed the papers in this volume.

Preface

The present volume is the second that has been produced after a meeting of the INA, the International Nannoplankton Association. The meeting in London, which was attended by 50 participants from Europe, Africa, America and Asia was held from the 19–21st of August 1987. It was followed by a one day excursion to the Upper Cretaceous and Palaeogene sediments cropping out at the South coast of England. Having called the INA meeting in Vienna (1985) the third formal INA Meeting, the meeting in London can be called the fourth, if we consider the workshop held in 1986 in Woods Hole as an informal gathering.

The organising committee consisted of Shirley van Heck and Jason Crux, who also edited this volume, and Alan Lord, who was our host at University College. They were assisted by students and staff of the Micropalaeontology Unit who not only organised the viewing of the slides and the coffee, but also arranged three excellent buffet lunches. The conference party was held onboard a ship cruising up and down the Thames.

The scientific program started with an invited lecture by P. Westbroek on "Coccolithophorids and the integration between earth and life sciences" followed by presentations on living and Neogene coccoliths. The second day was dedicated to various aspects of Cretaceous, Jurassic and Triassic calcareous nannofossils. On the third day, talks included discussions on phylogenetic strategies and on geological processes and events and nannofossil distribution. The many posters on both coccolithophorids and silicoflagellates furnished ample food for discussions. The beginnings of a computer program on systematics of coccoliths (so far mainly Jurassic) were presented by H.J. and K.M. Dockerill, while J. Young demonstrated his program for computer-drawn coccoliths and discoasters. Most abstracts were published in a special edition of the INA Newsletter, 9/2 and a late one in 9/3.

The fieldtrip was organised by Jason Crux who was helped by Ray Milbourne, Nigel Robinson and David Ward.

Three localities were visited where Middle and Upper Albian was sampled at Copt Point near Folkestone, the classical locality treated in detail in the monographs of Black. The Turonian and Coniacian chalk was studied at the Langdon Stairs in the Cliffs of Dover and the basal part of the Thanetian stratotype at Pegwell Bay near Ramsgate. The fine day was concluded with a fine dinner in Canterbury.

INA thanks the organising committee and all those involved in a direct or indirect way in the organisation of the London Meeting, the Excursion and these Proceedings.

London,   Katharina Perch-Nielsen
September 1988   President I.N.A.

Introduction

This volume contains the proceedings of the International Nannoplankton Association London Meeting 1987. Of the 31 papers and posters presented at the meeting, 13 are included in this book. One additional paper is included which was scheduled for the meeting but which could not be presented because at the last moment the author was unable to attend.

The book has been divided into three parts, each focussing on a different aspect of nannoplankton studies. Some papers focus on more than one aspect, and these have been grouped according to that aspect that seemed most prominent to the editors. Within each part the papers are ordered in stratigraphical sequence from young to old. Part I deals with pure morphology and systematics, containing papers on coccolith structure and growth (Young), phylogenetically based taxonomy (Aubry), taxonomy, structure and evolution of *Reticulofenestra* (Gallagher), crystallography and optical properties (Moshkovitz and Osmond), and conical nannofossils (Bown and Cooper). Part II focusses on ecological factors such as palaeobiogeography with papers on variation in assemblages during the last glacial cycle (Gard), temperature-controlled migration in the Aptian (Mutterlose), palaeogeography and biostratigraphy in the Lower Cretaceous (Crux), a comparison of North Atlantic assemblages in the Lower Cretaceous (Applegate, Bergen, Covington and Wise) and nannofossil provincialism in the Late Jurassic–Early Cretaceous (Cooper). Part III contains papers that show stratigraphical applications, such as Tortonian–Messinian stratigraphy (Flores and Sierro), Palaeocene stratigraphy (Varol), Palaeocene–Eocene stratigraphy of Turkey (Toker) and stratigraphical and palaeoenvironmental importance of *Arkhangelskiella* (Girgis).

In general the editors have aimed to restrict the number of references with each paper, and the reader is referred to the contributions of Perch-Nielsen in Bolli, Saunders and Perch-Nielsen, 1985, *Plankton Stratigraphy*, Cambridge University Press, Cambridge, and to the references in the Loeblich and Tappan index and bibliography (1966–1972) and INA Newsletter (1979–1988).

It has been attempted to avoid nomenclaturally incorrect (invalid/illegitimate) names, but exceptions have been made for names that are in general use (such as *Discoaster*) or those for which no correct alternative was available.

To avoid lengthy and cumbersome citations in the text, the book contains a taxonomic index containing the full citation for the taxa. Names of taxa occurring only as part of a zonal name have not been included in this index, nor have informal names such as 'discoasters' or 'nannoconids' been included. The page numbers (213) and plate numbers (13.1) have been included separately, and entries referring to introductions of new names, descriptions of taxa and pictures of holotypes are printed in bold.

Apart from the taxonomical index this volume contains a general subject index.

Part I

Morphology, Systematics and Evolution

The peculiar geometry of a rhombohedral crystal lends itself to the development of a great variety of elaborate patterns when groups of such crystals are forced to fit into a circular or elliptical framework, and the oblique direction of the optic axis leads to very distinctive interference effects in polarized light.

Black 1963

1

Observations on heterococcolith rim structure and its relationship to developmental processes

Jeremy R. Young

Qualitative and quantitative observations are made on the structure and geometry of heterococcolith rims, using primarily published electron micrographs. This information is combined with a brief review of coccolith development during coccolithogenesis. A three-stage model for heterococcolith development is outlined: (1) formation of an organic base-plate; (2) uniform crystal nucleation around the edge of the base-plate, producing a proto-coccolith ring; (3) uniform element growth from this proto-coccolith ring. The application of this model to reticulofenestrid coccoliths and the effects of variable development at each stage are discussed. Its application to various other groups is outlined and its limitations noted.

1.1 INTRODUCTION

Since the application of electron microscopy to nannoplankton research it has been possible to observe the fine structure of coccoliths. This has become a standard descriptive technique with invaluable results for the development of taxonomy. I have used some of the accumulated data in a slightly different way here, investigating certain geometrical aspects of coccolith morphology, and combining the results with information from studies of the development of living coccoliths. These approaches have been used to develop an enhanced understanding of the structure of coccoliths. A prime objective of this is to identify those aspects of the structure of coccoliths which are the result of architectural control, and so to separate other aspects which are of taxonomic significance, or which may be of adaptive significance.

A more direct impetus was provided by an attempt to write a computer program to generate, mathematically, illustrations of coccoliths. This is a useful end in itself, since accurate illustration is an essential means of communication in palaeontology, and diagrams provide a means of synthesising and interpreting data from electron microscope studies. Also, writing the program provided both a need for a model of coccolith geometry and a means of testing the predictions made by such a model. The program I have developed has enabled me to produce diagrams of a number of species. Similar illustrations have, of course, been produced by conventional techniques (e.g. Grün et al. 1974, Theodoridis 1984). However, computer-generated diagrams have the advantage that they can easily be redrawn, modified, and experimented with, although they are not necessarily faster to produce initially. Also there is potential for further development of the technique, to produce perspective illustrations, and to model birefringence patterns.

This paper deals with development of a model for heterococcolith development and geometry, and its application to Neogene coccoliths, particularly reticulofenestrids. This model forms the theoretical basis for the computer program, details of which are given in Young (1987b, c).

1.2. SOME GEOMETRICAL ASPECTS OF COCCOLITH MORPHOLOGY

Circular coccoliths, and other nannofossils with radial symmetry, are geometrically simple, can easily be mathematically modelled, and do not readily yield much information on their mode of development. The same is not true for elliptical coccoliths, since the variable curvature of their rims needs to be accommodated by variation in the shape, size or spacing of the elements. Geometrical investigation of this effect is essential for modelling, and can provide clues to the controls on coccolith morphology.

The simplest way of mathematically deriving an elliptical form from a circular one is via a simple linear stretch. This type of transformation is illustrated in Figs. 1.1B and 1.1C. However, the resultant elliptical coccolith is not realistic, it differs from real coccoliths in a number of important ways, as discussed below.

The following terminology should be noted. (1) Placolith coccoliths are considered here as being divisible into two parts; the *central area,* which is enclosed by the tube (in many specimens the central area is an open aperture), and the *rim,* which consists of the tube, distal shield and outer part of the proximal shield (Fig. 1.2). (2) The term *element* is used in the conventional loose sense for the individual components of cycles. In particular cases these may be distinguished as rays, plates, etc. Where elements from different cycles appear to be united these larger units are referred to here as *crystal units.* (3) For convenience of reference, and computing, coccoliths are described as elliptical; in fact they probably have less regular oval shapes.

(a) Rim width and ellipticity variation

On the deformed coccolith the uniform stretching means that all circles become true ellipses of similar elongation (axial ratio = 1.3 in

Fig. 1.C). As a consequence the rim width varies. It is greater at the ends of the coccolith (i.e. along the long axis) than at the sides, by a factor equivalent to the elongation.

Examination of electron micrographs of real coccoliths shows that a quite different relationship is the normal case. Characteristically the rim width is constant, and as a result the ellipticity varies, decreasing outwards. Thus the central area of most coccoliths is more strongly elliptical than their outer edge. As a result in many 'circular' coccoliths the central area is often distinctly elliptical (e.g. *Calcidiscus leptoporus, Cyclicargolithus floridanus*).

This rim width constancy and ellipticity variation can be seen qualitatively by observing electron micrographs, and is apparent even with light microscopy. It can be demonstrated by measuring the length and width of the inner and outer margins of the rims of individual specimens. Fig. 1.3 gives data of this type, from my own and published micrographs. Rim width and axial ratio variation for sixty placolith coccoliths are plotted on two graphs. As these show, rim width is constant (Fig. 1.3A), and so the axial ratio of the coccolith perimeter is consistently lower than that of the central area (Fig. 1.3B). The correlation of rim widths is remarkably good, strongly suggesting that rim width constancy is a basic feature of coccolith geometry. It implies that rim elements are of similar length all round the coccolith.

(b) Element orientation

If the distorted coccolith is compared with the circular coccolith from which it was derived (Figs. 1.1B and 1.1C), it can be seen that the angular separation of the elements is increased along the sides of the distorted coccolith and decreased at the ends. Thus the elements fan strongly at the sides and only slightly at the ends. This again is the reverse of the normally observed relationship on real coccoliths. As shown by the tracings, Figs. 1.1D–1.1H, maximum fanning occurs around the ends, where the curvature is greatest, whilst the elements along the sides are often sub-parallel.

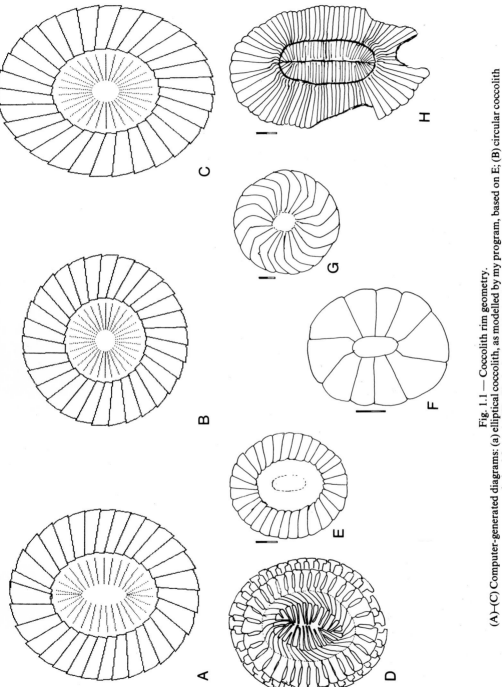

Fig. 1.1 — Coccolith rim geometry.

(A)–(C) Computer-generated diagrams: (a) elliptical coccolith, as modelled by my program, based on E; (B) circular coccolith derived from A; (C) elliptical 'coccolith' formed by stretching B.

(D)–(H) Tracings of coccoliths (scale bars, 1 μm; sources Roth 1970, Bukry 1971, 1974, Perch-Nielsen 1971, 1977): (D) *Emiliania huxleyi*, proximal; (E) *Coccolithus pelagicus*; (F) *C. crater*; (G) *Calcidiscus leptoporus*; (H) *Ellipsolithus macellus*, proximal.

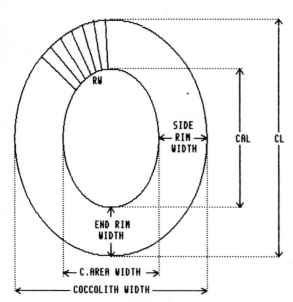

Fig. 1.2 — Biometric parameters.
Key to the features plotted in Figs. 3, 7 and 9. Abbreviations;
CAL, central area length; CL, coccolith length; RW, ray
width (averaged from 3 to 6 rays).

Emiliania huxleyi, Fig. 1.1D) these tend to be bunched at the ends, instead of converging uniformly on the centre. Both these effects suggest that the orientation of elements is more closely related to the orientation of the rim (i.e. the local tangent), than to radial directions from the centre of the coccolith.

This aspect of the geometry of coccoliths is hard to measure reliably, without high quality enlarged micrographs. It can, however, be qualitatively observed in virtually any micrographs of Mesozoic or Tertiary coccoliths.

(c) Element spacing

In the distorted coccolith the element spacing is directly related to the angular divergence, and so is maximum along the edges and minimum at the ends. This again is the reverse of the generally observed relationship, since usually elements are broader at the ends of elliptical coccoliths than along their sides (Figs. 1.1D–1.1H). Closer examination shows that this is predominantly an effect of the outer margin. By contrast, around the inner margin of coccolith rims element spacing usually appears to be more or less uniform. The variation in the width of the ends of the elements is thus primarily a product of the variation in angular divergence, or ray fanning.

The fanning effect is shown particularly clearly by elongated coccoliths such as *Ellipsolithus macellus* (Fig. 1.1H). A related effect is shown by central area elements; in real coccoliths (e.g.

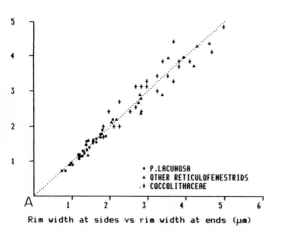

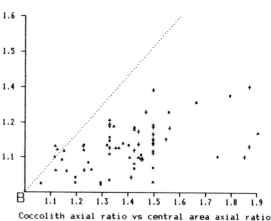

A   Rim width at sides vs rim width at ends (µm)

B   Coccolith axial ratio vs central area axial ratio

Fig. 1.3 — Biometric data illustrating rim width constancy.
(A) Scatter plot of rim width variation in 60 coccoliths. (B) Plot of axial ratios (length:width), illustrating that the axial ratio of the central area is consistently higher than that of the coccolith perimeter. Dotted lines indicate equal abscissa and ordinate values.

(d) Discussion

This succession of mismatches between the coccolith morphology derived by stretching a circular coccolith and that of real coccoliths makes it plain that coccolith morphology is not controlled in this manner. The rim width data suggest that the distortion implicit in the derivation of an elliptical coccolith from a circular one is primarily confined to the central area, whilst the rim is only indirectly affected. Support for this concept is provided by the, tentative, evidence that the orientation and spacing of elements is constant around the inner margin of the rim. Thus it is attractive to consider coccoliths as being formed from elements of constant basic form with coccolith structure determined by the form of the elements, their orientation and positions. Analogous structures can be created with a pack of playing cards.

1.3 COCCOLITH DEVELOPMENT DURING COCCOLITHOGENESIS

An alternative approach to looking at the final form of coccoliths is to examine how they actually develop. Coccolithogenesis in hetero-coccolith producing phases is an intracellular process closely associated with organic scale formation. The process has been followed in detail in three species, *Coccolithus pelagicus* (Parke and Adams 1960, Manton and Leedale 1969), *Emiliania huxleyi* (Wilbur and Watabe 1962, Klaveness 1976, Westbroek *et al.* 1984), and *Pleurochrysis carterae* (Manton and Leedale 1969, Outka and Williams 1971, van der Wal *et al.* 1983).

Although there are important differences in coccolith structure and cell organisation in these three species the sequence of coccolith development is similar in all three. So a general pattern can be suggested, as shown in Fig. 1.4

In each case an organic base-plate scale has been observed to form before calcification starts (Fig. 1.4A). In the case of *E. huxleyi* this was not identified until the study of Westbroek *et al.* (1984), although it was illustrated by Wilbur and Watabe (1962). In the other species base-plate formation prior to calcification is well documented.

Initial calcification occurs around the rim of this base-plate, producing a ring of simple elements (Fig. 1.4B). This ring of elements forms the basis of subsequent coccolith development and can conveniently be referred to as the *proto-coccolith ring*. Good examples are illustrated in Klaveness (1976) and Leadbeater and Morton (1973); in them the elements appear to be uniformly spaced around the ring. Several workers have suggested that the base-plate is important at least in providing a frame of reference for the nucleation sites (Manton and Leedale 1969, Outka and Williams 1971, Westbroek *et al.* 1984).

During subsequent coccolith growth (Figs. 1.4C and 1.4D) the elements remain attached to the base-plate, so it is unlikely that they move relative to each other. Hence the element spacings and orientations determined by the initial nucleation should be retained during coccolith growth.

In *E. huxleyi*, growth occurs in upward, outward and inward directions so that the proto-coccolith ring approximates to the inner margin of the proximal shield (Wilbur and Watabe 1962, Westbroek *et al.* 1984). In *Umbilicosphaera sibogae* var. *foliosa*, for which this stage of coccolithogenesis has been described by Inouye and Pienaar (1984), growth occurs in a very similar manner. *Pleurochrysis carterae* is also similar, although there is little or no outward growth—the coccolith remains a ring-shaped cricolith rather than developing into a placolith (Outka and Williams 1971, van der Wal *et al.* 1983).

In *Coccolithus pelagicus* the coccolith development stages were not so readily discernible. It appears likely, however, from the illustrations of Manton and Leedale (1969) that in this species too growth occurs outward and upward from the proto-coccolith ring (their Figs. 22–28), so that it corresponds to the base of the tube, and that in the mature coccolith the base-plate is attached to the inside edge of the proximal shield (their Figs. 16, 17 and 31).

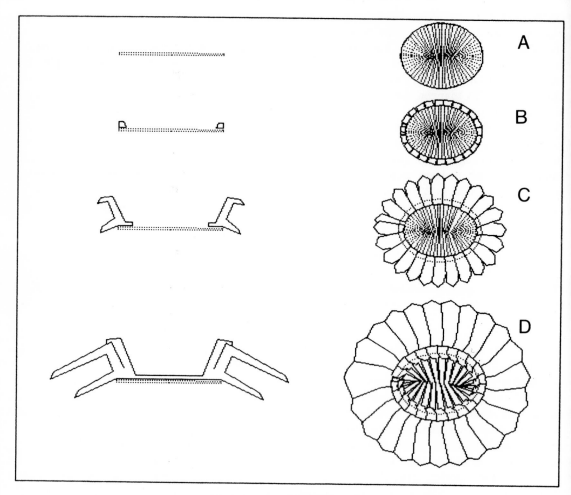

Fig. 1.4 — Coccolith development.
Series of diagrams showing intracellular coccolith development. Organic base-plate scale shown by dotted lines. Left — sections, right — plan views (distal side): (A) base-plate only; (B) base-plate and proto-coccolith ring; (C) partially formed coccolith; (D) final form.

1.4 A BASIC MODEL FOR HETERO-COCCOLITH DEVELOPMENT

The mode of development from one, or a very few, proto-coccolith rings, and the clear succession of stages which occurs during coccolithogenesis, provide a useful sequential framework for the interpretation of final coccolith morphology, as summarised below.

(1) Formation of the base-plate scale: this is a critical precursor to coccolith development since the base-plate acts as a template for crystallite nucleation. Its shape and size is thus one of the main influences on the ultimate form of the coccolith—it is responsible for the ellipticity of elliptical coccoliths.

(2) Crystal nucleation, around the base-plate: it is at this stage that the number, spacing, and orientation of the elements is fixed. The limited observations from coccolith geometry, and the few available illustrations of actual proto-coccolith rings, suggest that, at least to a first approximation, the elements are uniformly

spaced at this stage, with a constant orientation relative to the base-plate margin.

(3) Element growth: this occurs in various directions but leaves the proto-coccolith ring on the proximal surface of the coccolith. Rim cycles are formed by outward growth of elements, and so have different structures from central area features which are formed by inward growth of elements. The constant width of coccolith rims suggests that elements develop to uniform lengths during this process. All the elements of a cycle also tend to have similar shapes, apart from width variation due to differences in curvature of the ring. The asymmetry introduced by elliptical base-plates appears to be compensated for by variable lateral expansion of elements or by their overlapping to varying degrees. The basic form, initial spacing, and orientation of the crystal units do not seem to be affected by the variable curvature.

1.5 RETICULOFENESTRID STRUCTURE

To test the developmental model it needs to be applied to real coccoliths. This involves digression into taxonomy, so only one group is discussed in detail here: the reticulofenestrids, which form a coherent, abundant, and well-known group, with an interesting range of structures. In addition, coccolithogenesis has been extensively studied in the principal living species, *Emiliania huxleyi*.

The term reticulofenestrid is used here for coccoliths with similar rim structure to *Reticulofenestra*. This includes all the Late Eocene to Recent Noelaerhabdaceae, but excludes the earlier genera *Toweius* and *Prinsius*, which have a significantly different rim structure (see e.g. Perch-Nielsen 1985). Representative reticulofenestrid species are illustrated in Fig. 1.5.

Basic reticulofenestrid structure is discussed first, to show how it can be related to growth from a proto-coccolith ring. Then each of the three stages of coccolith development suggested above is looked at in turn to see how it affects aspects of morphology and variation in the group.

(a) Basic structure

Most reticulofenestrids consist on the proximal side of a shield constructed of a single cycle of rays and a grill flooring the central opening (examples in Figs. 1.5D and 1.5G). On the distal side two cycles of elements are visible, an outer cycle forming the distal shield and a discrete inner cycle of smaller 'cover plates' overlying the distal shield. A central tube connects the proximal and distal shields, and this consists of two cycles: an outer tube cycle with clockwise imbrication of the elements, and an inner tube cycle—termed the wall—with anticlockwise imbrication (i.e. the tops of the elements are offset in an anticlockwise sense from the bottoms of the elements). Thus a total of six cycles can be identified (proximal and distal shield, outer tube, wall, cover plate, and grill cycles). However, closer examination shows that the elements of the various cycles are connected.

The cover plates arise from the elements of the wall, which in turn are connected to the rays of the proximal shield, and so to the grill. These relationships are readily apparent in suitable micrographs and have been noted by many authors (e.g. Hay *et al*. 1966, Perch-Nielsen 1971, Edwards 1973, Romein 1979).

Similarly, the distal shield rays continue into the outer tube cycle, and this again merges with the proximal shield. These connections are only visible in fortuitously oriented specimens, and so are not well established. They are, however, clearly shown in the micrographs of Perch-Nielsen (1971, Plate 24.3, and noted p. 31), Bramlette and Wilcoxon (1967, Plate 1.3), and Steinmetz and Stradner (1984, Plate 26.4).

Thus all the various apparently distinct elements are interconnected at the base of the tube (Fig. 1.6). As such the conventionally recognised 'elements' can be seen to be rather arbitrary parts of larger units. These units have a remarkably elaborate four-part structure, but presumably are single calcite crystals, and so are referred to here as *crystal units*. An interesting aspect of the structure, shown in Fig. 1.6, is that the opposite imbrication directions of the wall and outer tube result in the cover plates and

8

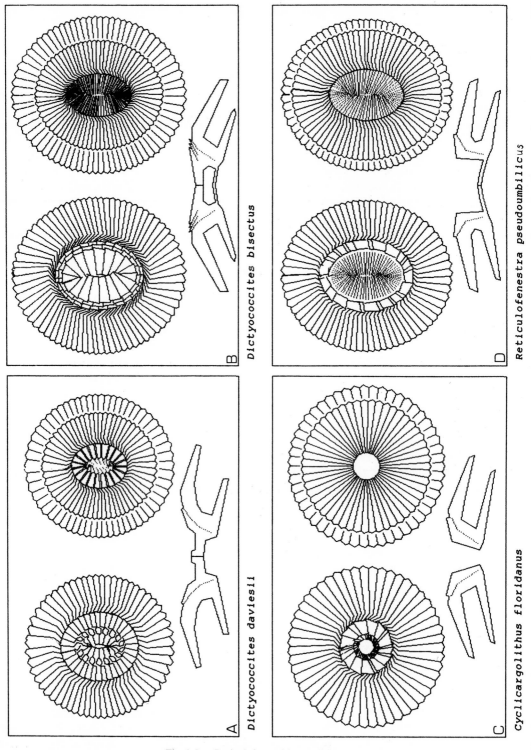

Fig. 1.5 — Reticulofenestrid coccoliths.
Computer-generated diagrams of eight species. Note the following. (1) Each diagram set consists of distal view (left), proximal view (right) and cross-section (bottom). (2) Diagrams are not drawn to a common scale, and the cross-sections are at larger

Dictyococcites daviesii

Dictyococcites bisectus

Cyclicargolithus floridanus

Reticulofenestra pseudoumbilicus

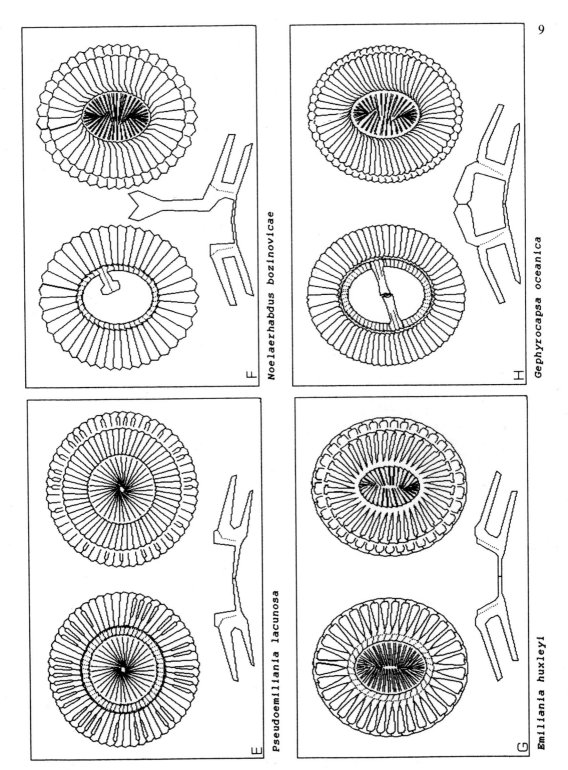

Noelaerhabdus bozinovicae

Gephyrocapsa oceanica

Pseudoemiliania lacunosa

Emiliania huxleyi

Fig 1.5 (cont)
scales than the plan views. (3) Central area details and cross-sections are schematic. (4) All diagrams are based on information from many sources rather than single specimens.

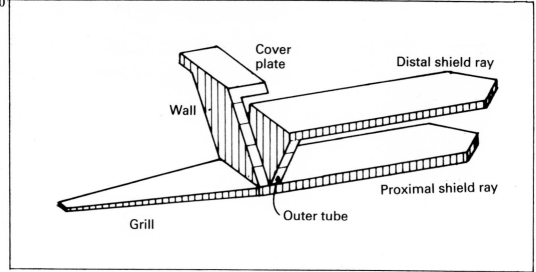

Fig. 1.6 — Reticulofenestrid crystal unit structure.
Simplified drawing of a single reticulofenestrid crystal unit, showing its complex multi-element construction. For clarity all elements are shown as of similar thickness; in fact the column elements are rather thicker. Also, kinks in the rays, grill complexities, etc. are omitted.

distal shield rays of single crystal units being offset. As a result, although the distal shield and wall are formed from the same cycles of crystal units, they do not fuse during diagenesis.

In the particular case of *Emiliania huxleyi* this structure, with the various elements being parts of larger crystal units, was suggested by the observations on coccolithogenesis of Wilbur and Watabe (1962) and subsequently demonstrated by X-ray diffraction examination of individual crystal units by Watabe (1968).

Light microscopy also suggests this inter-connection of elements—all the components of the coccoliths show similar optical behaviour in both plan and side view. The central grill is rarely visible in cross-polarised light but this is almost certainly a result of its extreme thinness rather than different optical orientation.

This structural interpretation, with all the elements traceable to a common origin at the base of the tube (Fig. 1.6), provides strong support for the concept of their growing from a proto-coccolith ring corresponding to the base of the tube, which is a critical prediction of the model outlined above. Conversely the model provides an explanation for the structure of the crystal units.

(b) Effect of proto-coccolith size and ellipticity
An obvious consequence of the growth model is that one of the primary controls on coccolith morphology is the size and shape of the proto-coccolith ring, which in turn is related to the base-plate size and shape. Usually shape, or ellipticity, seems to show a low degree of variation within species, and size rather more. An interesting example of the effects of combined variation in these factors is provided by the *Pseudoemiliania lacunosa–Reticulofenestra doronicoides* group (the name *Pseudoemiliania lacunosa* is adopted here since it has nearly universal currency; it may, however, not be strictly valid—van Heck pers. commun. 1988).

This group is usually regarded as consisting of two or more species, with *R. doronicoides* being a typical small elliptical reticulofenestrid, and *P. lacunosa* a slightly larger circular to sub-circular form with a broad central area, numerous elements and slitting between the rays of the distal shield (Fig. 1.5E). There is, however, a continuous range of morphotypes between these end-members, and it is probably more valid to regard the entire group as a single species. This interpretation was based initially on my own, mainly qualitative, light microscope observations, but the range and style of variation is also clear from published illustrations and descriptions (e.g. McIntyre *et al.* 1967, Samtleben 1979, Nishida 1979, Pujos 1985). Similar interpretations of the taxonomy of the group have been made by McIntyre *et al.* (1967) and Samtleben (1979).

The large number of morphological variables make the group ideal for biometric investigation. I have conducted a crude study of this type, using about forty electron micrographs of *P. lacunosa*, mainly from the literature. Graphs from this data set are given in Fig. 1.7, and a synthetic series of diagrams based on it is given in Fig. 1.8. Unfortunately it is not possible to identify precisely the proto-coccolith ring location, particularly since distal views have had to be used to get a reasonably sized data set. However, since element growth is predominantly outward, the edge of the central area provides a practical approximation. Variation in central area size and shape in *P. lacunosa* can thus be used to investigate the effect of variation in proto-coccolith size and shape.

As shown in Fig. 1.7A central area length and width are strongly correlated but not directly proportional; central area ellipticity decreases with increasing size (see also Figs. 1.7B and 1.8) As a result central area width, which is related to both increasing central area size and decreasing ellipticity, shows greater variation (0.5–3.0 μm) than either coccolith length (3–7 μm) or central area length (1–3 μm). It proved the most discriminating parameter to plot other data against.

The parameters which are most nearly independent of central area width are rim width (Fig. 1.7C) and element spacing around the central area edge (Fig. 1.8B). The variation in these parameters is low and apparently random, or related to other controls. This is of interest since it suggests that variation can occur in one phase of the coccolith development process with little effect on the other phases.

An inevitable consequence of the low rim width variation is that central area size increases relative to coccolith size as the central area width increases (Fig. 1.7D). Similarly, since initial element spacing is approximately constant, the number of rays increases with central area width (Fig. 1.7E).

The number of slits between rays increases very markedly (from 0 to 50), with central area width. This is partially an effect of the increasing number of rays, but in addition the percentage of rays with slits increases (Fig. 1.7F). The cause of this is not clear, but slitting and the change to broader more open coccolith shapes might both be mechanisms for reducing calcite usage, and so could be parallel responses to a single external factor. The distribution of slits around the coccoliths appears to be random, as modelled in Fig. 1.8.

Summary: a number of superficially independent changes in coccolith morphology can be seen to be related to the single control of increase in base-plate size, with faster increase in width than length. This is demonstrated in the diagrams of Fig. 1.8; the only parameters changed in the program were central area width, ellipticity, and degree of slitting.

Interestingly there are analogous cases of correlation in these features, except for slitting, in other groups, notably *Reticulofenestra pseudo-umbilicus–rotaria,* and *Calcidiscus leptoporus–macintyrei*; variation in the *Coccolithus pelagicus* group also shows some such features. More generally variation in base-plate size and shape appears to be an important process in producing interspecific and intraspecific variation.

(c) Nucleation related features
The second phase of coccolith development is crystal nucleation, when the spacing and orientation of crystal units are determined.

It is readily apparent from electron micrographs that there is little variation in the orientation of the elements between the various reticulofenestrid species. Also, between crossed-nicols all reticulofenestrids show birefringence in plan view, and have similar interference figures. This suggests that the crystallographic orientation of the crystal units is constant.

More surprisingly crystal unit spacing also appears to be remarkably constant within the

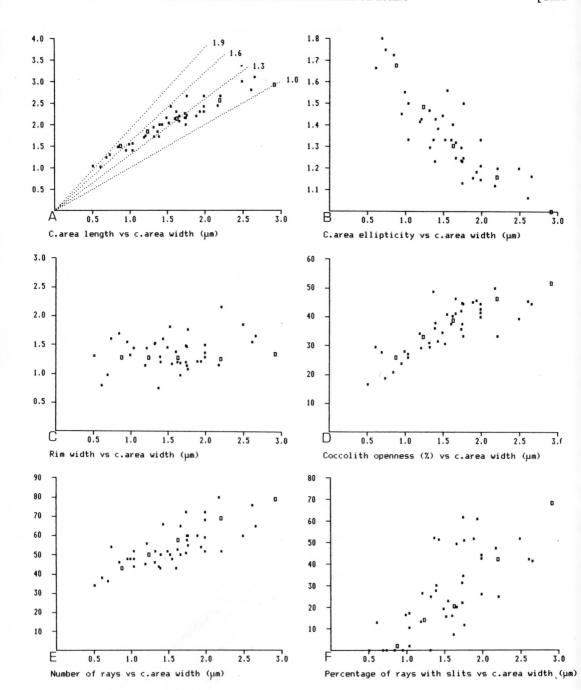

Fig. 1.7 — *Pseudoemiliania lacunosa* biometric data.
Data from measurements of photographs of *P. lacunosa* (s.l.) specimens, illustrating variation in various parameters with aperture width. Open symbols represent the computer-generated specimens of Fig. 1.8. Dotted lines in A show axial ratio.
Coccolith openness (D) 100 × central area length/coccolith length.

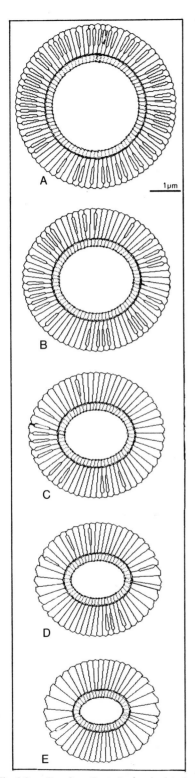

A

1μm

B

C

D

E

Fig. 1.8 — *Pseudoemiliania lacunosa* variation. Computer-generated series of coccoliths, illustrating correlated variation in aperture shape, aperture width, number of rays, and degree of slitting. Diagrams are to constant scale.

group as a whole. This can be demonstrated by direct measurement of the width of the rays at the inner margin of the proximal shield (assuming again that this approximates to the proto-coccolith ring). Measurements of ray width in a range of species, mainly from published micrographs, are given in Fig. 1.9A. The ray widths only show a two-fold variation (0.1–0.2μm), in comparison with a ten-fold variation in coccolith length. A similar pattern is shown by the results from within the single species *P. lacunosa* (Fig.1.9B). The accuracy of the measurements is not high—with errors likely from inaccurate quotation of magnifications, specimen tilting, and choice of location to measure elements. These problems should, however, increase rather than decrease the variability.

Qualitatively the low variability in crystal unit spacing is reflected in the common observation that reticulofenestrids have very narrow rays, rarely resolvable by light microscopy. The only exceptions, such as *Cyclicargolithus floridanus*, have broad rims and narrow central areas (Fig. 1.5C).

Summary: The low variation in both crystallographic orientation and crystal unit spacing suggests that the nucleation processes were relatively constant during evolution of the reticulofenestrids. As such they are of importance as distinctive higher level taxonomic features of the group.

(d) Element growth variation
Apart from the slitting mentioned above, reticulofenestrid crystal units are remarkably similar in all species. Constant features include the four-part structure (Fig. 1.6), the imbrication directions, and kinks in both the proximal and the distal shield rays (Fig. 1.5). Most reticulofenestrid coccoliths also have pointed ray tips, but this is probably related to the crystallography of the elements, and may in part be of diagenetic origin. The grill and wall cycles show most variation in mode and degree of development, and so are important in subdivision of the group.

In *Cyclicargolithus floridanus* and other Early Miocene and Oligocene reticulofenestrids the

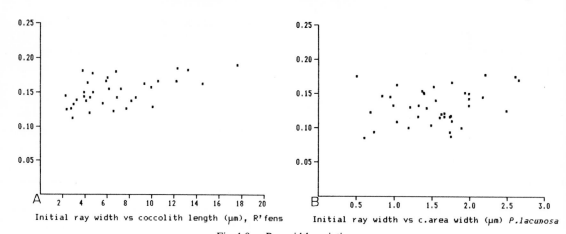

Fig. 1.9 — Ray width variation.
Plots illustrating the low, and random, variation of initial ray width in (A) reticulofenestrids in general and (B) the *P. lacunosa* group.

wall elements, or a limited number of them, overlap onto the distal shield as cover plates (Fig. 1.5C). In most later reticulofenestrids there are no discrete cover plates; instead, a raised collar occurs around the central area (Fig. 1.5E). This is formed jointly by the wall cycle elements and the distal shield elements, which butt against them. A third type of development is for the wall to terminate flush with, or slightly below, the edge of the distal shield, without development of a collar or cover plates (Fig. 1.5G). This morphology is most characteristic of *Emiliania huxleyi*, but is also seen in some *Gephyrocapsa* specimens.

In all three cases the wall cycle can be thickened inwards to close the central area. Forms with this development are often placed in a separate genus, *Dictyococcites*. However, specimens of *Emiliania huxleyi* with closed central areas have been illustrated by several authors (e.g. Black 1968, Heimdal and Gaarder 1981), and this type of variation seems more likely to be of ecological than taxonomical significance. Probable exceptions are *Dictyococcites bisectus* (Fig. 1.5B) in which the central area is closed by a plug of inward-growing cover plates, and *Dictyococcites daviesii* in which a limited number of wall elements contribute to an expanded grill.

The general variability of wall cycle development is continued by two more radical structures. In *Gephyrocapsa* wall elements on opposite sides

of the distal shield become elongated and arch over to form a bridge (Fig. 1.5H). The bridges always show the same diagonal orientation (NE–SW in distal view), and frequently the elements fail to meet precisely, with an anticlockwise offset. Etched specimens show that each half of the bridge is formed from two to five wall elements. *Noelaerhabdus* (a Late Miocene Paratethyan genus, Jerkovič 1970) has a single asymmetrically placed spine, similarly formed from a few wall elements (Fig. 1.5F).

The development of different structures from the wall is thus a consistent theme in reticulofenestrid variation. These new structures are thus produced by modification of the crystal units rather than by production of new crystal units. Moreover, the basic crystal unit structure is not changed; instead the final phase of its development is modified. In this respect the different structures fit well with the model proposed above. In contrast, the spine of *Noelaerhabdus* and the bridge of *Gephyrocapsa* show that crystal units in one part of the cycle can develop quite different forms to those in other parts of the cycle. This is an important modification of the general pattern of similar development of all crystal units. Here final coccolith form appears to be regulating element development, rather than vice versa. Some degree of interplay is nonetheless evident, in the offset of the bridge halves in *Gephyrocapsa,* and in the eccentricity of the spine of *Noelaerhabdus.* Also malformed *Gephyrocapsa* specimens with three or more half bridges grown from different parts of the wall sometimes occur (e.g. Okada and McIntyre 1977, Plate 3). In all these respects the final structure is distorted from symmetrical perfection by the developmental process.

Summary: It appears that basic crystal unit structure is constant in reticulofenestrids. Major modifications are mainly confined to addition of extra features and are usually taken as generic level features. Intraspecific variation can produce changes in the degree of development, causing variations in rim width, in degree of slitting, and in central area closure.

1.6 EXTENSION OF THE MODEL TO OTHER GROUPS

The following notes briefly outline the extent to which the three-phase–proto-coccolith ring model can be applied in other coccolith families. Only Neogene groups are discussed since I know these best.

(a) Coccolithaceae

The Coccolithaceae, like the reticulofenestrids, are placoliths. Coccolithogenesis has been studied in two species, *Coccolithus pelagicus* and *Umbilicosphaera sibogae,* and, as discussed above, coccolith development in them is very similar to that in *E. huxleyi.* The family is more heterogeneous than the reticulofenestrids. It contains at least three discrete groups; *Coccolithus* (plus *Chiasmolithus, Cruciplacolithus,* and *Clausicoccus*), *Umbilicosphaera,* and *Calcidiscus.* Within each of these groups there is a constant rim structure analogous to that of the reticulofenestrids, and similar variation patterns in such features as ellipticity, ray number and central area width can be recognised.

However, all the Coccolithaceae have at least two discrete cycles of crystal units, since the proximal and distal shields show contrasting optical behaviour, and the proximal shield is usually bicyclic. My interpretation of the structure is that there are two discrete rim cycles (Fig. 1.10A). An upper rim cycle (non-birefringent) forms the distal shield, outer tube, and upper part of the proximal shield. The lower rim cycle (birefringent) forms the lower part of the proximal shield. This obviously requires two separate sets of crystal nuclei. The plates of Manton and Leedale (1969, Figs. 28 and 31) provide striking evidence for this. They show the edge of the base-plate curving up during proto-coccolith formation, and separate crystals developing on either side of it, thus forming a double proto-coccolith ring.

In the *Coccolithus* group, further complexity is provided by the presence of a wall (i.e. inner tube elements, Fig. 1.10A). This is birefringent like the proximal shield, but is composed of many cycles of small elements. It is difficult to envisage these as all being connected to the proximal shield; instead it seems likely that they too require separate nucleation, which considerably increases the amount of crystal nucleation. Central area structures, such as the cross in *Cruciplacolithus,* are formed from similar elements and so unlike the central area structures in reticulofenestrids do not show any relation to the rim geometry.

(b) Helicosphaeraceae

The coccoliths, or helicoliths, of *Helicosphaera* consist of three main parts; proximal plate, flange and blanket (cf. Theodoridis 1984; see Fig. 1.7). *Helicosphaera* has not been grown in culture, so there is no information on cocco-lithogenesis in the genus. Norris (1971) has, however, illustrated a base-plate scale on *H. carteri,* entirely covering the proximal plate, and Gaarder (1970) recorded similar observations. Hence this is a likely position for the proto-coccolith ring. Strong support for this is provided by the form of the elements of the proximal plate, which show ray bunching and wedging inward, suggesting inward growth. The proximal plate can thus be interpreted as a normal cycle, but grown inwards, like the grills in the reticulofenestrids. The flange is composed of the same number of elements as the proximal plate, and they have similar optical orientations (Young 1987b), so the elements of these two units are probably two components of a single cycle of crystal units. The flange, however, is a strongly modified cycle, with apparently considerable variation in crystal unit structure and length, producing the spiral effect. This is quite different from the uniform rim development of placolith coccoliths, and indicates a significant modification of the coccolith development process. It has a direct functional significance; the specialised rim form allows unusually regular and close interlocking of the coccoliths on the coccosphere (Young 1987a).

The proximal plate (and so proto-coccolith ring) shape is also irregular. It is often somewhat tapered or rhomboidal rather than truly

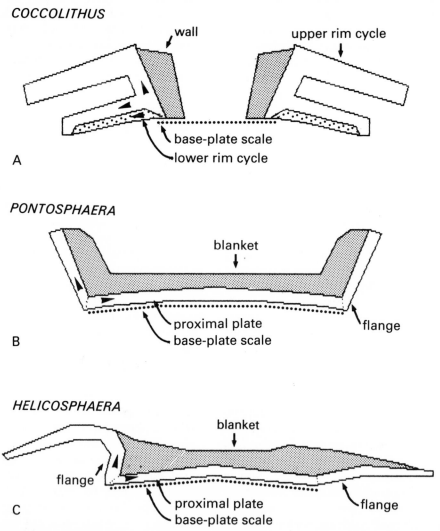

Fig. 1.10 — Schematic cross-sections through typical Coccolithaceae, Pontosphaeraceae and Helicosphaeraceae. Densely stippled parts are formed of numerous irregular cycles of concentrically arranged elements, which are birefringent in plan view. The other parts are cycles of larger elements, suggested to have developed from proto-coccolith rings located at the edges of the base-plate scales; arrow-heads indicate postulated directions of crystal growth. Except for the lower rim cycle of *Coccolithus* (light stipple) these cycles are not birefringent in plan view.

elliptical. Nucleation processes, as in the other groups, appear more constant, with a characteristic orientation and spacing of the elements

A separate problem is provided by the blanket. The elements of this unit have different optical orientations from those of the flange and proximal plate. However, they are floored by the proximal plate, and so must have nucleated on this rather than the organic base-plate.

(c) Pontosphaeraceae

The Pontosphaeraceae have a structure analogous to that of the Helicosphaeraceae, with proximal plate, flange and an inner–upper blanket of smaller elements (Fig. 1.10C). Gaarder (1970) has illustrated a base-plate scale entirely covering the proximal plate on an ordinary coccolith of *Scyphosphaera apsteinii*. The flanges in both the Helicosphaeraceae and the Pontosphaeraceae display anticlockwise element imbrication, but otherwise have very different structures and forms. The relationship between the two families is by no means certain, but in terms of the coccolith development they appear to be similar.

There are also distinct similarities between the blanket elements in these two groups and the wall elements of the Coccolithaceae. In all cases they are small, sub-vertically oriented, lath-like, elements arranged in numerous irregular concentric cycles. They show strong birefringence in plan view, with an oblique extinction cross.

These elements thus differ from the rim cycles discussed above in that they are not arranged in regular cycles, did not nucleate on the edge of the base-plate, and have very simple structure. It would appear that they are the product of a different type, or fourth phase, of development, characterised by formation of numerous small crystal units, with less precise distribution in cycles, and less precise regulation of growth. Structures formed from these elements tend to be rather variable, and they are widely used for species level taxonomy.

(d) Holococcoliths

The diagnostic feature of holococcoliths is that they are formed of numerous minute identical elements. These holococcolith elements do not interconnect to form larger crystal units and are not arranged in uniform cycles; instead, elements show similar orientations over large zones, forming irregular pseudo-crystals. Also, few elements are in contact with the base-plate. Plainly the concept of development from a proto-coccolith ring is not applicable to holococcoliths. A quite different developmental model is needed to interpret their morphology. This is also suggested by evidence that they form by extra-cellular calcification (Manton and Leedale 1963, Rowson *et al.* 1986), and they show entirely different styles of morphological variation to heterococcoliths.

(e) Other groups

The model of development from a proto-coccolith ring is probably applicable to Syracosphaeraceae, Calciosoleniaceae, Zygodiscaceae and most Mesozoic heterococcoliths, but I have not examined them in detail. Discoasters and spheno-liths can readily be envisaged as developing in an analogous manner, but with little or no base-plate development.

Groups with structures which appear to reflect different development processes include Rhabdo-sphaeraceae—spirally arranged cycles of small elements (but rim is analogous), Braarudo-sphaeraceae—layered ultrastructure, and Cerato-lithaceae and Triquetrorhabdulaceae—completely different shape and structure.

1.7 SUMMARY

(1) Examination of coccolith morphology suggests that the three-phase pattern of heterococcolith cycle development observed in studies of coccolithogenesis is of general application. The phases are base-plate formation, crystal nucleation, and element growth. The first two stages produce a proto-coccolith ring, the third develops it into a coccolith.

(2) Both nucleation and growth processes are usually uniform around cycles, except where

elements interfere with each other. Departures from this pattern, such as flange development in *Helicosphaera* and bridge formation in *Gephyrocapsa*, probably require functional explanations.

(3) Structural complexity in heterococcoliths is primarily a result of elaborate crystal unit structure, rather than production of numerous separate cycles of crystal units. This can be a valuable perspective for elucidating coccolith structure and looking for homologous structures within families.

(4) Intraspecific variation can occur, and evolution operate, during any of the developmental phases, with varying effects on final form. Some aspects, however, are more stable than others. Most stable are nucleation-related features—element spacing, orientation and number of cycles. Crystal unit structure is only slightly less stable, but element length and degree of calcification are both, and independently, liable to intraspecific variation. Base-plate—and so proto-coccolith ring—diameter is similarly variable, and in some species is accompanied by shape variation.

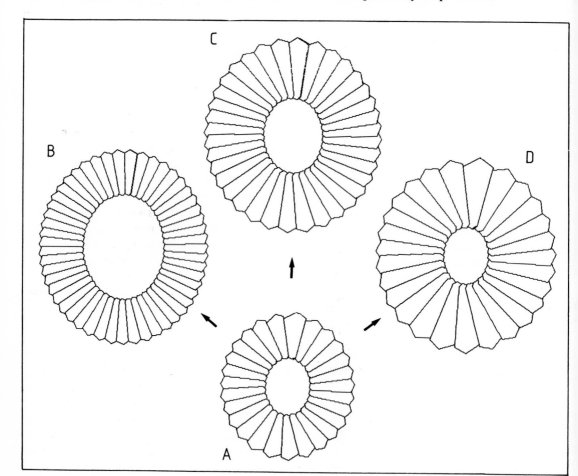

Fig. 1.11 — Allometric coccolith growth.
Three large coccoliths (B)–(D) derived from a smaller one (A) without variation in initial ray spacing, and consequently showing variation in form: (A), (B) increase in proto-coccolith ring width only (e.g. *P. lacunosa*); (A)–(C) balanced increase in rim and ring width (e.g. *R. pseudoumbilicus*); (A)–(D) increase in rim width only (e.g. some *E. huxleyi* specimens).

(5) An important result of the stability of crystal unit spacing is that larger coccoliths inevitably differ from smaller ones of the same species—whether the size variation is caused by variation in proto-coccolith ring diameter, rim width, or a combination of the two (Fig. 1.11),

(6) This model is applicable to the development of typical rim, flange, and proximal plate cycles in Neogene heterococcoliths. It does not appear to be directly applicable to some central area structures, to blanket elements, or to holococcoliths.

(7) The model is directly applicable to the problem of mathematically describing coccoliths, and forms the theoretical basis of a computer program used to produce illustrations of coccoliths.

1.8 ACKNOWLEDGEMENTS

This work comes rather directly from my doctoral thesis—I am very grateful for the support of my mother, in particular, during the course of writing up. Conversations with Paul Bown, Liam Gallagher, and Rob Haddock aided in the development of the ideas discussed here. Katharina Perch-Nielsen and Shirley van Heck made valuable comments on the manuscript.

1.9 REFERENCES

Black, M. 1968. Taxonomic problems in the study of coccoliths. *Palaeontology*, **11**, 793–813.

Bramlette, M.N. and Wilcoxon, J.A. 1967. Middle Tertiary calcareous nannoplankton of the Cipero Section, Trinidad, West Indies. *Tulane Stud. Geol. Paleontol.*, **5**, 93–131, 220.

Bukry, D. 1971. Coccolith stratigraphy, Leg 6 DSDP. *Init. Rep. DSDP*, **6**, 965–1004.

Bukry, D. 1974. Coccoliths as paleosalinity indicators—evidence from the Black Sea. In *The Black Sea—Geology, Chemistry, and Biology. Mem. Am. Assoc. Pet. Geol.*, **20**, 353–363.

Edwards, A.R. 1973. Key species of New Zealand calcareous nannofossils. *N.Z. J. Geol. Geophys.*, **16**, 68–89.

Gaarder, K.R. 1970. Three new taxa of coccolithineae. *Nytt Mag. Bottanik*, **17**, 113–126.

Grün, W., Prins, B. and Zweili, F. 1974. Coccolithophoriden aus dem Lias epsilon von Holzmaden (Deutschland). *N. Jb. Paläont., Abh.*, **147**, 294–328.

Hay, W.W., Mohler, H.P. and Wade, M.E. 1966. Calcareous nannofossils from the Nal'Chik (NW Caucasus). *Eclogae geol. Helv.*, **55**, 497–517.

Heimdal, B.R. and Gaarder, K.R. 1981. Coccolithophorids from the northern part of the eastern Atlantic (II) Heterococcoliths. *"Meteor" Forschungs Ergebnisse*, **D(33)**, 37–69.

Inouye, I. and Pienaar, R.N. 1984. New observations on the coccolithophorid *Umbilicosphaera sibogae* var. *foliosa* (Prymnesiophyceae) with reference to cell covering, cell structure, and flagellar apparatus. *Br. phycol. J.*, **19**, 357–369.

Jerković, M.L. 1970. *Noelaerhabdus* nov. gen. type d'une nouvelle famille de Coccolithophoridées fossiles: Noelaerhabdaceae du Miocène supérieur de Yougoslavie. *C.R. Acad. Sci. Paris*, **270D**, 468–470.

Klaveness, D. 1976. *Emiliania huxleyi* (Lohmann) Hay & Mohler III. Mineral deposition and the origin of the matrix during coccolith formation. *Protistologica*, **12**, 217–224.

Leadbeater, B.S.C. and Morton, C. 1973. Ultrastructural observations on the external morphology of some members of the Haptophyceae from the coast of Jugoslavia. *Nova Hedwigia*, **24**, 207–233.

Manton, I. and Leedale, G.F. 1963. Observations on the micro-anatomy of *Crystallolithus hyalinus* Gaarder & Markali. *Arch. Mikrobiol.*, **47**, 115–136.

Manton, I. and Leedale, G.F. 1969. Observations on the microanatomy of *Coccolithus pelagicus* and *Cricosphaera carterae*, with special reference to the origin and nature of coccoliths and scales. *J. Mar. Biol. Ass. UK*, **49**, 1–16.

McIntyre, A., Bé, A.W.H. and Preiksitas, R. 1967. Coccoliths and the Pliocene–Pleistocene boundary. *Progr. Oceanogr.*, **4**, 3–25.

Nishida, S. 1979. Restudies of the calcareous nannoplankton biostratigraphy of Tonohama Group, Shikoku, Japan. *Bull. Nara Univ. Education*, **28**, 97–110.

Norris, R.E. 1971. Extant calcareous nannoplankton from the Indian Ocean. In A. Farinacci. (Ed.), *Procs. II Planktonic Conf., Roma 1970*, **2**, Tecnoscienza, Rome, 899–910.

Okada, H. and McIntyre, A. 1977. Modern coccolithophoridae from the Pacific and North Atlantic Oceans. *Micropaleontology*, **23**, 1–55.

Outka, D.E. and Williams, D.C. 1971. Sequential coccolith morphogenesis in *Hymenomonas carterae*. *J. Protozool.*, **18**, 285–297.

Parke, M. and Adams, I. 1960. The motile *(Crystallolithus hyalinus* Gaarder & Markali) and non-motile phases in the life history of *Coccolithus pelagicus* (Wallich) Schiller. *J. Mar. Biol. Ass. UK*, **39**, 263–274.

Perch-Nielsen, K. 1971. Elektronenmikroscopische Untersuchungen an Coccolithen und Verwandten Formen aus dem Eozän von Dänemark. *Biol. Skr. Danske Vidensk. Selsk.*, **18**, 1–76.

Perch-Nielsen, K. 1977. Albian to Pleistocene calcareous nannofossils from the western South Atlantic, DSDP Leg 39. *Init. Rep. DSDP*, **39**, 699–825.

Perch-Nielsen, K. 1985. Cenozoic calcareous nannofossils. In H.M. Bolli, J. B. Saunders and K. Perch-Nielsen (Eds.), *Plankton Stratigraphy*, Cambridge University Press, Cambridge, 427–554.

Pujos, A. 1985. Quaternary nannofossils from the Goban Spur, eastern North Atlantic Ocean DSDP Holes 548–549A. *Init. Rep. DSDP*, **80/II**, 797–792.

Romein, A.J.T. 1979. Lineages in early Palaeogene calcareous nannoplankton. *Utrecht. Micropaleont. Bull.*, **22**, 1–231.

Roth, P.H. 1970. Oligocene calcareous nannoplankton biostratigraphy. *Eclogae geol. Helv.*, **63**, 799–881.

Rowson, J.D., Leadbeater, B.S.C. and Green, J.C. 1986. Calcium carbonate deposition in the motile *(Crystallolithus)* phase of *Coccolithus pelagicus* (Prymnesiophyceae). *Br. phycol. J.*, **21**, 359–370.

Samtleben, C. 1979. Pliocene to Pleistocene coccolith assemblages from the Sierra Leone Rise, Site 366, Leg 41. *Init. Rep. DSDP*, **41**, 913–931.

Steinmetz, J. and Stradner, H. 1984. Cenozoic calcareous nannofossils from DSDP Leg 75, Southeast Atlantic Ocean. *Init. Rep. DSDP*, **75**, 671–753.

Theodoridis, S. 1984. Calcareous nannofossil biostratigraphy of the Miocene and revision of the helicoliths and discoasters. *Utrecht Micropaleont. Bull.*, **32**, 1–271.

van der Wal, P., de Jong, E.W., Westbroek, P., de Bruin, W.C. and Mulder-Stapel, A.A. 1983. Polysaccharide localization, coccolith formation, and golgi dynamics in the coccolithophorid *Hymenomonas carterae*. *J. Ultrastruct. Res.*, **85**, 139–158.

Westbroek, P., de Jong, E.W., van der Wal, P., Borman, A.H., de Vrind, J.P.M., Kok, D., de Bruijn, W.C. and Parker, S.B. 1984. Mechanism of calcification in the marine alga *Emiliania huxleyi*. *Phil. Trans. R. Soc.*, **B304**, 435–444.

Watabe, N. 1968. Crystallographic analysis of the coccolith of *Coccolithus huxleyi*. *Calcified Tissue Res.*, **1**, 114–121.

Wilbur, K.M. and Watabe, N. 1962. Experimental studies on calcification in molluscs and the alga *Coccolithus huxleyi*. *Ann. New York Acad. Sci.*, **103**, 82–112.

Young, J.R. 1987a. Possible functional interpretations of coccolith morphology. *Abh. geol. B.-A.*, **39**, 305–313.

Young, J.R. 1987b. Neogene calcareous nannofossils from the Makran of Pakistan and the Indian Ocean. *Thesis, Imperial College, University of London*, 1–288 (unpublished).

Young, J.R. 1987c. Basic techniques for computer illustration of nannoplankton. *INA Newsletter*, **9**, 116–119.

2

Phylogenetically based calcareous nannofossil taxonomy: implications for the interpretation of geological events

Marie-Pierre Aubry

As in other fossil groups, a phylogenetically based taxonomy is a fundamental requirement for deciphering evolutionary patterns among calcareous nannoplankton. There are, however, difficulties associated with recognising phylogenetically related taxonomic groups, in particular since the palaeontological concept of calcareous nannoplankton species is so far removed from the biological concept of calcareous nannoplankton species. There are also difficulties associated with parallel and convergent evolution. Based on a discussion of the probable phylogenetic relationships between five genera (*Zygodiscus*, *Lophodolithus*, *Helicosphaera*, *Pontosphaera* and *Neocrepidolithus*), the status of a number of Cenozoic families is discussed, and a reinterpretation of the relationships between Mesozoic and Cenozoic calcareous nannoflora is presented.

2.1 INTRODUCTION

Calcareous nannofossils have become an important tool for interpreting the geological record. Integrated with biostratigraphical schemes based on other marine microfossils, and with magnetic and isotope stratigraphies, calcareous nannofossil biostratigraphy aids in improving the stratigraphical and chronological resolution needed for deciphering subtle geo-

logical changes which occurred during short intervals of time. At the same time, they provide crucial evidence for interpreting main geological events and for evaluating the validity of current interpretations of those events. For instance calcareous nannofossils provide important evidence for characterising the terminal Cretaceous event(s). The nature of the evidence has repeatedly changed, however, since Bramlette and Martini (1964) first pointed to the major calcareous nannoplankton renewal across this boundary. Improved biostratigraphical resolution near the Mesozoic–Cenozoic boundary has been gained as more extended and complete sections were studied but also as calcareous nannoplankton taxonomy improved; and the description of the changes that calcareous nannoplankton underwent at the boundary has been further refined. Increasing evidence led to different interpretations of the effects of the event(s) on the Late Cretaceous calcareous nannoplankton and in turn to various interpretations of the event(s) itself (themselves) (see review in Aubry in press a). It would appear that successive interpretations based on calcareous nannofossil studies largely reflect the changing state of calcareous nannofossil taxonomy. In other words, the nature and the quality of the evidence that calcareous nannofossils provide are mainly dependent on the reliability of the taxonomy of the group. Indeed, validity of interpretations of geological events based on palaeontological evidence largely depends on the accuracy of taxonomic schemes.

During the last two decades, major efforts have been devoted to clarifying calcareous nannofossil taxonomy, to deciphering phylogenetic relationships between genera and to establishing a phylogenetically sound taxonomy. Despite these efforts, large discrepancies remain between classification schemes. No agreement has yet been reached concerning the content of many genera (see reviews in Perch-Nielsen 1985a, 1985b; and discussion in Aubry in press b), lineages within genera and phylogenetic links between genera (see discussion concerning the genus *Helicosphaera* in Aubry in press c, for example). In addition, the generic contents of families differ widely among authors and relationships between Mesozoic and Cenozoic families and genera remain to be established.

The present ambiguities of calcareous nannofossil taxonomy reflect, in fact, difficulties which are inherent to the group and which, for the most part, were not recognised until recently. The object of this paper is to evaluate critically the likelihood that calcareous nannofossil taxonomy based on coccolith morphology and structure alone might reflect the biological complexity known in living Coccolithophorales and true phylogenetic relationships between species. I will first review the aspects of the biology of living Coccolithophorales which are relevant to discussing taxonomy. I will then discuss some of the problems derived from convergent evolution. Lastly, the incidences of parallel and convergent evolution on taxonomical schemes will be stressed and I will plead in favour of a 'dynamic taxonomy', i.e. a plea for erecting taxonomical categories which reflect highly probable phylogenetic relationships among calcareous nannofossils. As in other groups (Steineck and Fleisher 1978), it is only on the basis of a phylogenetically based taxonomy that it is possible to estimate the rates of diversification and extinction among the calcareous nannofossils and to decipher global evolutionary patterns.

The discussion will centre mainly on the study of the relationships between five genera: *Zygodiscus*, *Lophodolithus*, *Helicosphaera*, *Pontosphaera* and *Neocrepidolithus*.

2.2 CALCAREOUS NANNOFOSSILS AND CALCAREOUS NANNOPLANKTON

The calcareous nannofossils represent a very heterogeneous grouping of diverse remains which have in common the following features: very small size (generally smaller than 50 μm), calcitic composition, and probable marine phytoplankton origin. Some groups (e.g. schizosphaerellids, calpionellids, pithonellids) have a restricted geographical and stratigraphical occurrence and their biological affinities remain controversial. The large majority of calcareous nannofossils correspond, however, to a more uniform group commonly referred to as 'coccoliths' although it would be best to distinguish between coccoliths s.s. and nannoliths. Coccoliths (s.s.) and nannoliths are small calcitic particles with a complex morphology and structure, usually smaller than 20 μm, which occur together, often in great abundance, in marine deposits. Fossil coccoliths refer to those particles with a general morphology and structure equivalent to that known in the calcitic secretions which surround cells of the living Coccolithophorales (Haptophytes). Placoliths, cribriliths, lopadoliths are coccoliths. Nannoliths refer to those fossil particles with a general morphology and structure reminiscent of that of coccoliths but without modern equivalents or with modern equivalents which are produced by obscure groups other than the Coccolithophorales. Sphenoliths, ceratoliths, discoasters are nannoliths.

To establish a meaningful taxonomy, it is necessary that the palaeontological concept of species be as close as possible to the biological concept of species. This can be achieved satisfactorily in most palaeontological groups, including planktonic foraminifera and radiolaria, whose taxonomy is guided by knowledge of the biology of modern representatives. Such is not the case, however, with fossil Coccolithophorales for which a species corresponds only to a certain conceptualised morphotype.

The complicated biology of living Coccolithophorales indicates that morphology and structure of the coccoliths does not accurately reflect the biological content of the species in this group.

(1) Coccolithophorales are characterised by the external envelope of coccoliths, the coccosphere, with which the cell surrounds itself. Polymorphism and dithecatism commonly occur. *Scyphosphaera apsteinii* offers a well-known example of dimorphism with its equatorial ring of lopadoliths which contrasts with the low cribriliths which cover the remainder of the cell (for example see Nishida 1979, Plate 11, Figs. 1(a) and 1(b)). Many other forms of polymorphism (e.g. in Nishida 1979, Plate 10, Figs. 1–3; Heimdal and Gaarder 1981, Plate 6, Figs. 28–31, Plate 2, Figs. 5–8) and forms of dithecatism (e.g. in Nishida 1979, Plate 8, Fig. 1(a); Heimdal and Gaarder 1981, Plate 5, Figs. 23–26) have been illustrated showing coccospheres bearing adjacent coccoliths of different morphologies and unrelated structures.

(2) Even when coccospheres are formed of a single coccolith type, ecological morphotypes may be very distinct. Wide morphological intraspecific variability of the placoliths of *Emiliania huxleyi* has been related to water temperatures (Okada and Honjo 1973, Okada and McIntyre 1977).

(3) More deceptive is the production of coccoliths of different types at different stages of the life-cycle of some species. Except for the mechanisms related to the calcification of the coccoliths, very little is known of the biology of most Coccolithophorales. The only data available on their life-cycle from laboratory cultures have shown, in some species, a succession of several phases characterised by the production of cells of different types, the

coccolith-bearing cells secreting coccoliths of different morphologies and structures during different phases (Parke 1971, Lefort 1971, Leadbeater 1971, Klaveness 1972a). Unless the natural sequential production of coccospheres with coccoliths of different morphologies and unrelated structures is recognised, the creation of superficial taxa may result (e.g. *Coccolithus pelagicus* and *Crystallolithus hyalinus* which in fact correspond to a single biological species (Parke and Adams 1960)).

From the facts described above, it appears that a biological definition of species of Coccolithophorales is difficult to establish and it is clear that coccolith morphologies and structures alone are inadequate for the creation of a species-level taxonomy bearing a biological meaning. This, however, conflicts with the species concept of Coccolithophorales in palaeontology. Specific taxonomy of fossil Coccolithophorales is solely based on the morphology and structure of coccoliths which occur predominantly as isolated individuals in sediments. It is exceptional when entire coccospheres are fossilised and thus the characteristics (size, shape, presence or absence of apical opening, polymorphism, dithecatism, which all appear to be characteristic within modern genera) of the coccospheres of most fossil species may never be known. Except in very rare cases, the palaeontological record does not allow matching different coccoliths which were part of the same coccospheres, nor does it enable us to associate different coccoliths which were produced during successive phases of the life-cycle of an extinct biological species. As a result the species concept in fossil Coccolithophorales as currently practised is essentially unsatisfactory and to a large extent artificial. The palaeontological meaning of species is usually very restricted with respect to the biological meaning. The palaeontological meaning of genera is equally distorted. For instance, modern species assigned to the extant genus *Scyphosphaera* bear both cribriliths and lopadoliths, while modern species of the genus *Pontosphaera* carry only cribriliths. However, because coccospheres are rarely preserved in sediments, the palaeontological meaning of *Scyphosphaera* is restricted to lopadoliths whereas that of *Pontosphaera* is broadened to include all cribriliths (see discussion in Aubry in press b). In groups biologically and genetically as complex as the Coccolithophorales, where the relationships between the various types of cells produced in cultures are not yet understood (Klaveness 1972a), where sexual reproduction is only suspected (Rayns 1962), and where genetic differentiation between clones as indicated by comparative reproductive rates is not necessarily reflected in coccolith morphology (Brand 1982), delineation of fossil species can only be very approximate.

Few questions have been answered as yet concerning genetic differentiation in the living Coccolithophorales (Brand 1981, 1982) and virtually nothing is known about morphological variations of coccoliths in response to genetic differentiation. Morphological variations shown by the placoliths of *Emiliania huxleyi* are often assumed to be ecophenotypic (Okada and Honjo 1973, Okada and McIntyre 1977) although Winter (1985) questions this interpretation and Brand (1982) suggests that those variations could result from genetic differentiation on large spatial scales. The fact is that until more is understood concerning the genetic variability in living Coccolithophorales and its relationship to coccolith morphology, it will be difficult to separate phenotypic variations resulting from palaeoenvironmental changes (comparable with those described by Lohmann and Malmgren (1983) in *Globorotalia truncatulinoides*) from phenotypic variations reflecting genetic differentiation, particularly in groups as complex and morphologically variable as the *Reticulofenestra* group.

In addition, there are the occurrences of dimorphic coccospheres which bear coccoliths of species otherwise apparently well characterised and, in monospecific cultures, the spontaneous appearance of coccospheres formed of coccoliths of another supposedly well-characterised species (Lefort 1975). Dimorphic coccospheres have

been found in living plankton (Kamptner 1941, Nishida 1979, Winter *et al.* 1979), in cultures (Lefort 1971, Gayral and Fresnel-Morange 1971) and occasionally preserved in fossil assemblages (Clochiatti 1971, Gard 1987). These and the spontaneous phenomena described by Lefort (1975) suggest close phylogenetic relationships between the species involved (Lefort 1971, 1975, Winter 1985) or, perhaps, the present inaccuracy of the biological concept of Coccolithophorales species. 'Combined cocco-spheres' were suggested by Kamptner (1941) to result from hybridisation, a means of speciation suggested to exist in some microfossil groups (e.g. radiolaria, see review in Lazarus 1983). However, it might be difficult to test hybrid-isation as a current means of speciation in fossil Coccolithophorales because of the restrictions imposed by the palaeontological record.

In summary, the palaeontological concept of Coccolithophorales species is far removed from the biological concept. The fact that isolated coccoliths rather than whole coccospheres are usually preserved in sediments adds to the difficulties resulting from the complicated life-cycle known in some living Coccolithophorales species. Aspects of the biology of living Coccolithophorales warn about the limitations of a species-level taxonomy based on coccolith morphology and structure and, thereby, about the risks in establishing phylogenetic links based on these grounds. The risks of confusion are even larger when dealing with nannoliths because their original position on or in the cell remains unknown in most cases. Norris (1971) has shown that modern ceratoliths occur within a cell. However, modern arrangements are not necessarily reliable models for the fossil. The dodecahedral shell of *Braarudosphaera bigelowii* is well known but very different from that of the Mesozoic species in the genus as beautifully exemplified by Lambert (1986). Arrangements of discoasters, ortholiths, sphenoliths, and other nannoliths without modern representatives are even more speculative. That the palaeontological concept of species might be largely unrealistic has no bearing on biostratigraphical studies and to a lesser degree on biogeographical studies. In contrast, it might be consequential for evolutionary studies, for testing the modalities and rates of speciation in fossil calcareous nannoplankton, for delineating lineages, and, consequently, for establishing a phylogenetically based taxonomy. Thus caution must be taken not to define taxa, particularly species, on too narrow morphological and structural grounds.

Yet we observe evolutionary trends with sequential evolution of the morphology and structure of calcareous nannofossils. Such trends have been described in Mesozoic genera, particularly in *Broinsonia* (Lauer 1975, Verbeek 1977, Crux 1982), *Discorhabdus* (Crux 1987), *Micula* (Roth and Bowdler 1979, Crux 1982), *Stephanolithion* (Rood and Barnard 1972), *Stradnerlithus* (Rood and Barnard 1972) and in Cenozoic genera such as *Chiasmolithus* (Gartner 1970, Romein 1979), *Discoaster* (Bukry 1971), *Cruciplacolithus* (Romein 1979), *Triquetrorhabdulus* (Biolzi *et al.* 1981) and *Helicosphaera* (Haq 1973, Aubry in press c).

Coccoliths of two kinds are produced in living Coccolithophorales. Holococcoliths are formed of tiny rhombs and are precipitated outside the cell membrane (Parke and Adams 1960, Manton and Leedale 1963, Klaveness 1973). They are rarely preserved in the sediments. Secretion of the commonly fossilised heterococcoliths occurs within the cell under the close control of the Golgi apparatus or the reticulum body (Wilbur and Watabe 1963, Klaveness 1972b). Heterococcolith formation is thus genetically controlled (probably more closely genetically controlled than holococcolith formation), and changes in morphology and structure of the heterococcoliths must reflect at least large-scale genetic changes. The evolutionary trends exhibited by groups of coccoliths indicate that the notion of genus applied to calcareous nannofossils is not strictly morphological, but warrants its phylogenetic character. This statement contradicts the opinion expressed by Noël (1984) that for calcareous nannofossils the genus is an entirely artificial taxonomical category, which is nothing but a tool to ease

determinations. Although rarely stated, this attitude, which led Deflandre (1952, 1958), Deflandre and Deflandre-Rigaud (1948, 1970) and Kamptner (1955) to introduce Croneis's (1938) utilitarian classification, has been implicitly adopted by many workers. However, a major effort has been pursued in recent years to base genera on distinctive structural characters, resulting in a somewhat paradoxical situation: for calcareous nannofossils, the genus, which corresponds to a grouping of forms which share the same basic morphological and structural characteristics and appear to be phylogenetically linked, is often a less superficial taxon than the species itself which is essentially restricted to a morphotype. The genus *Helicosphaera* is a good example of a phylogenetic rather than morphological genus. It is a sum of lineages even though parts of some of these lineages cannot be reconstructed with any degree of confidence (Aubry in press c). In contrast, in present palaeontological understanding, the genus *Scyphosphaera*, which includes only lopadoliths, is a morphological genus. A phylogenetically based taxonomy is the only source of information on evolutionary patterns. Only from such a scheme is it possible to assess the nature and effects of radiations and mass extinctions. The study of evolutionary trends enables us to distinguish innovations experienced by a genus. For instance, innovations in the genus *Helicosphaera* do not occur sequentially but rather multiple innovations appear at particular stratigraphical horizons, a pattern which leads to parallel evolution. It is likely that the study of evolutionary trends within families will also provide valuable information. As a forerunner, Black (1971) entertained the idea that family relationships could 'sort themselves out of phylogenetic lineages' and discussed phylogenetic continuity in some taxa from the Recent back to the early Cenozoic and Cretaceous. I will show below that phylogenetically based families can be recognised through careful analytical comparison of the structural similarities between genera. There are problems however, associated with the establishment of phylogenetic relationships. One difficulty concerns the origination of Coccolithophorales species (morphospecies) which is complicated by a certain 'provincialism' exhibited by at least some species. Unlike some other microfossils (e.g. planktonic foraminifera, radiolaria), Cocco-lithophorales are not restricted to oceanic waters and fossil assemblages (when well preserved) reach higher diversity in epicontinental deposits than in oceanic oozes (Aubry 1983, and unpublished data). It has been reported that some species first occur earlier in epicontinental than in oceanic deposits (e.g. *Reticulofenestra umbilicus, Lanternithus minutus, Discoaster bifax*, (Aubry 1983); see discussion below). Diachroneity of species extinction is more difficult to establish because of the recurrent problem of reworking. However, the example offered by *Gephyrocapsa protohuxleyi*, assumed to have become extinct about 75 thousand years ago but then reported alive in the Gulf of Elat (Winter *et al.* 1978) casts doubt about the synchroneity of some calcareous nannofossil extinction datum levels. The consequences for phylogenetic studies are that the direct ancestor of a given species or group of species might be difficult to identify, particularly if evolution proceeds in isolated areas such as restricted basins of epicontinental seas where strati-graphical sections are usually discontinuous. An even greater difficulty in establishing phylo-genetic relationships results from convergent evolution as discussed below.

2.3 EVOLUTION AND CLASSIFICATION: TOWARDS A DYNAMIC TAXONOMY

I have shown elsewhere that the structural units which compose coccoliths placed in the six genera *Zygodiscus, Lophodolithus, Helico-sphaera, Neocrepidolithus, Pontosphaera*, and *Scyphosphaera* are homologous (Aubry in press c) (Fig. 2.1). These coccoliths consist of (1) a basal plate composed of radiating, wedge-shaped elements, surrounded by (2) a flange formed of strongly overlapping, distally flaring elements,

and (3) a distal cover built of strongly overlapping mainly tangentially arranged elements, which lines both the basal plate (entirely) and the flange (entirely or partly) (in *Scyphosphaera*, the distal cover develops upwards beyond the flange). In addition, in some coccoliths, there is (4) a transverse bar which spans the central opening delineated by the basal plate. In discussing the origin of *Helicosphaera* I have also shown that close phylogenetic relationships exist between this genus and the five genera cited above (Aubry in press c). There are two distinct groups: the *Zygodiscus–Lophodolithus–Helicosphaera* group is char-acterised by an elliptical basal plate which delineates a central opening spanned in many cases, by a transverse bar; the *Neocrepidolithus–Pontosphaera–Scyphosphaera* group is char-acterised by a more compact basal plate and the absence of a transverse bar. Close phylogenetic links between genera in this latter group are generally assumed (see discussion in Aubry in press b, c). A direct phylogenetic link between *Zygodiscus* and *Lophodolithus* is also accepted (see discussion in Aubry in press b, c). The possibility that *Zygodiscus* might be the ancestor of *Helicosphaera* or that both genera share a common ancestor has been discussed in Aubry (in press c). Except for *Scyphosphaera* and *Lophodolithus* which seem to have directly originated from *Pontosphaera* and *Zygodiscus* respectively, the direct or indirect nature of phylogenetic links between the remaining genera cannot be established as yet. In particular, the earliest appearances of representatives of *Pontosphaera* and *Zygodiscus* are unclear (Fig. 2.2). Coccoliths described as *Prolatipatella multicarinata* and similar to *Pontosphaera plana* are reported from the Late Maastrichtian from Texas (Gartner 1968), Egypt (Shafik and Stradner 1971), Austria (von Priewalder 1973) and Tunisia (Perch-Nielsen 1979a). *Zygodiscus pomerolii* was described from the Late Maastrichtian of Madagascar (Perch-Nielsen 1973). Despite a close morphological and structural resemblance between these Late Maastrichtian forms and Cenozoic coccoliths assigned to *Pontosphaera* and *Zygodiscus,* the latter have been regarded as unrelated to the former because of apparent discontinuous occurrences between the Late Maastrichtian and the Late Palaeocene (Zone NP5 for *Zygodiscus,* Zone NP9 for *Pontosphaera*: Perch-Nielsen 1985b). If Late Cretaceous forms were unrelated to the Late Palaeocene ones, convergent evolution in the Late Cretaceous and Late Palaeocene would be striking. No other comparable example of convergent evolution among calcareous nannofossils is known as yet.

There are, however, some difficulties in accepting taxonomic–phylogenetic separation between these Late Cretaceous and Late Palaeocene forms. First, convergent evolution reaching such a perfect degree of repetition

appears suspicious. Second, coccoliths assigned to the genera under discussion are more common and diversified in shallow water than in deep-sea deposits. Because shallow-water sections are hiatus prone, the stratigraphical distribution of many species in these genera, in particular in *Pontosphaera,* is poorly known. Also, these coccoliths are mostly represented in warmer seas and less commonly occur at high latitudes. Third, stratigraphical discontinuity between the oldest, rare and geographically restricted representatives and the younger, more common ones is known in some genera, in particular in *Helicosphaera.* The stratigraphically oldest reported occurrence of *Helicosphaera* is in the lower Upper Palaeocene (Zone NP5) of Iran (Haq 1971, 1973); the next younger one is in the

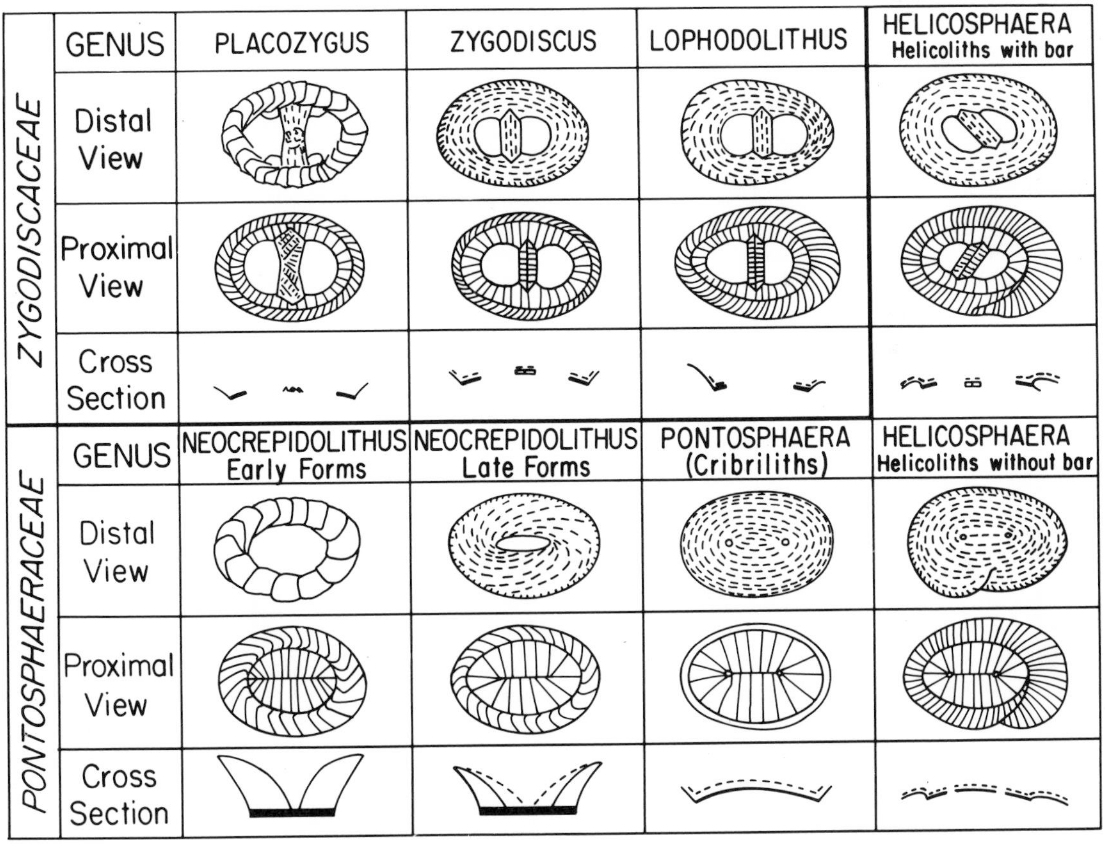

Fig. 2.1 — Interpretation and comparison of the structures of coccoliths in *Placozygus, Zygodiscus, Lophodolithus, Helicosphaera* (helicoliths), *Neocrepidolithus* and *Pontosphaera* (cribriliths). See text for further explanation. For the structure of coccoliths in *Scyphosphaera* (lopadoliths) refer to Aubry, in press b.

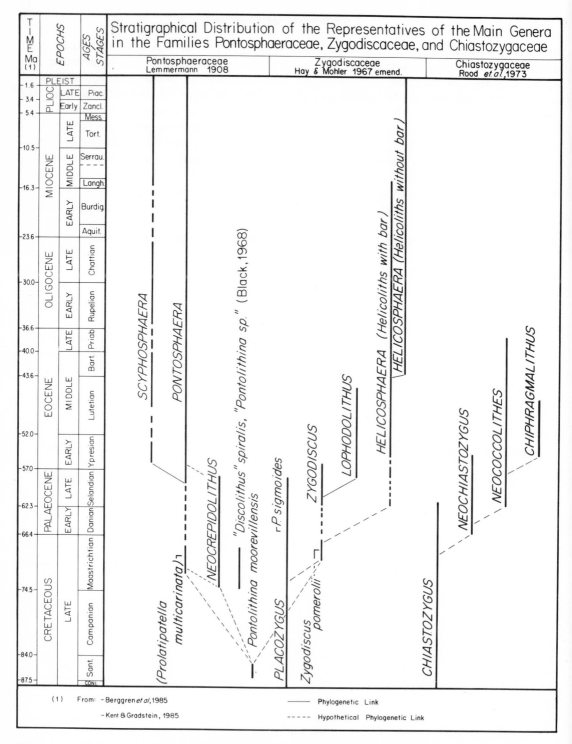

Fig. 2.2 — Global stratigraphical range of representatives of the Cenozoic genera in the families Pontosphaeraceae, Zygodiscaceae and Chiastozygaceae and suggested evolutionary links. Numerical ages from Kent and Gradstein (1985) for the Mesozoic and from Berggren *et al.* (1985) for the Cenozoic.

uppermost Palaeocene (Zone NP9) at Site 354 on the Ceara Rise in the South Atlantic Ocean (Perch-Nielsen 1977); the stratigraphically oldest described species occurs in the upper Lower Eocene (Zone NP12) from which level species of *Helicosphaera* become common and widely distributed geographically. Similarly, the stratigraphical record of *Scyphosphaera* which extends from Lower Eocene (Zone NP12) to Recent is discontinuous. No species of this genus has yet been reported from the Lower Miocene below the Langhian. Its diversity in the Oligocene has long been assumed to be extremely low until Martini (1980) found lopadolith-rich Upper Oligocene sediments at Site 448A in the Philippine Sea. It may be of some significance that the Late Miocene–Early Pliocene high specific diversity in *Scyphosphaera* is known from only three shallow water deposits. Finally, the number of sections through Lower Palaeocene shallow-water deposits from tropical and subtropical areas is limited. As a result, the Lower Palaeocene stratigraphical record of some genera may be obscured. Assessment of the levels at which representatives of the genera *Zygodiscus* and *Pontosphaera* first occurred is difficult although critical in our search for evolutionary links as well as in our documentation of parallel evolution. Although an origin of *Pontosphaera* from *Neocrepidolithus* is favoured, Perch-Nielsen (1981a, b) indicated also the possibility that *Zygodiscus* might be ancestral to *Pontosphaera*, through reduction of the bar. If this hypothetical link is disproven, parallel evolution in *Pontosphaera* and *Zygodiscus* can be illustrated. *Placozygus sigmoides*, a species which ranges across the Cretaceous–Tertiary boundary, has been suggested as ancestral to *Zygodiscus* (Perch-Nielsen 1981a, b). Like early forms of *Neocrepidolithus* with which it occurs in the Late Cretaceous and earliest Palaeocene, *P. sigmoides* has no distal cover (Fig. 2.1). Its bar is close to that of *Zygodiscus* in forming a bevel in the basal plate and in being of different structure on the distal and proximal sides, but it has a primary helicoidal structure and may bear a distal stem. Finally, poorly documented forms illustrated by

Black (1968) as *Pontolithina moorevillensis* and *Pontolithina* sp. respectively from the Santonian and Maastrichtian of Alabama, and by Pienaar (1968) as *Discolithus spiralis* from the Maastrichtian of South Africa, show definite structural features equivalent to structural units of cribriliths and coccoliths in *Neocrepidolithus* and *Zygodiscus* (cf. Black 1968, Plate 149, Figs. 4 and 5, Pienaar 1968, Plate 70, Fig. 3, Plate 71, Fig. 4, Pienaar 1969, Plate 3, Fig. 3). Forms assigned to *Pontolithina* were suggested as possible ancestors of several Tertiary lineages, in particular of the *Pontosphaera* lineage (Black 1968, pp. 801, 806). In the present state of our knowledge, tracing direct evolutionary links between these two forms and *Placozygus sigmoides*, *Neocrepidolithus*, *Zygodiscus*, and *Pontosphaera* may still appear speculative but it is probable that additional information on such links is to be found in Upper Cretaceous rocks.

Despite these phylogenetic uncertainties, the structural homologies described between the seven genera under discussion are an incentive to grouping them in two closely related families.

Ideally, classification at the family level should reflect evolutionary links between genera included in the family. Comparison between various classification schemes presented in recent years for Mesozoic and Cenozoic calcareous nannofossils show that such is not the case. Great disagreements concerning the content of Mesozoic and Cenozoic families (cf. Hay 1977, Tappan 1980, Perch-Nielsen 1985a, b) reveal major weaknesses of the present calcareous nannofossil family concepts. Also, there are substantial overlaps in the generic contents of the Mesozoic and Cenozoic families presented by Perch-Nielsen (1985a, b) some genera being transferred from one family into another across the Cretaceous–Tertiary boundary. Thus, and contrary to Perch-Nielsen (1985c), it may be best to restrict the use of calcareous nannofossil families until family concepts are revised and established on firmer grounds, for superficial family concepts may be more misleading than helpful.

The genera *Zygodiscus*, *Lophodolithus*,

Table 2.1—Family assignment of the genera *Placozygus, Zygodiscus, Lophodolithus, Helicosphaera, Neocrepidolithus, Pontosphaera,* and *Scyphosphaera* according to three recent overviews on calcareous nannofossils. *Neocrepidolithus* and *Placozygus* are included in Hay's (1977) and Tappan's (1980) concepts of *Crepidolithus* and *Zygodiscus* respectively. Hay (1977) placed *Crepidolithus* in the family Stephanolithiaceae and Tappan (1980) placed it in the family Crepidolithaceae.

Family	Hay 1977	Tappan 1980	Perch-Nielsen 1985b
Zygodiscaceae Hay & Mohler 1967	*Zygodiscus* (+ several Mesozoic genera)	*Chiastozygus* *Chiphragmalithus* *Cyclolithella* *Isthmolithus* *Nannotetrina* *Neochiastozygus* *Zygodiscus* (+ numerous Mesozoic genera)	*Chiastozygus* *Chiphragmalithus* *Isthmolithus* *Lophodolithus* *Neochiastozygus* *Neococcolithes* *Placozygus* *Zygodiscus* (+ several Mesozoic genera)
Helicosphaeraceae Black 1971	—	*Helicosphaera*	*Helicosphaera*
Scyphosphaeraceae Jafar 1975	—	*Calciopilleus* *Nannocorbis* *Scyphosphaera* (+ Recent genus)	—
Pontosphaeraceae Lemmermann 1908	*Pontosphaera* *Transversopontis†* *Discolithina** *Lophodolithus* *Helicosphaera* *Koczyia** *Crassapontosphaera** *Scyphosphaera* (+ Recent genera)	*Crassapontosphaera** *Discolithina** *Koczyia** *Lophodolithus* *Pontosphaera* *Transversopontis†* (+ Recent genera and Mesozoic genera) *Pontolithina*	*Calciopilleus* *Neocrepidolithus* *Nannocorbis* *Pontosphaera* *Crassapontosphaera** *Discolithina** *Koczyia** *Scyphosphaera* *Transversopontis†*

*Genera regarded as synonymous with *Pontosphaera* by Perch-Nielsen (1985b) and Aubry (in press b).
†Genera regarded as synonymous with *Pontosphaera* by Aubry (in press).

Helicosphaera, Pontosphaera and *Scyphosphaera* have been placed in four different families and their arrangement in these families varies with various authors (Table 2.1). Whereas Tappan (1980) distributed the five genera among the four families, Perch-Nielsen (1985a) rejects the family Scyphosphaeraceae and Hay (1977) retains only two families, Zygodiscaceae and Pontosphaeraceae.

The family Helicosphaeraceae was introduced by Black (1971) and is monogeneric.

The family Scyphosphaeraceae was introduced by Jafar (1975) 'to accommodate the peculiarly constructed genus *Scyphosphaera,* which is characterised by possessing large barrel-shaped lopadoliths at the equatorial region of the cell and in this respect differs markedly from the closely related genus *Pontosphaera* Lohmann, 1902' (Jafar 1975, p. 369). Creating the family Scyphosphaeraceae to distinguish a group of lopadolith- and cribrilith-bearing cells (genus *Scyphosphaera*) from a group of cribrilith-bearing cells (genus *Pontosphaera*) is superfluous as the genus *Scyphosphaera* itself was created to differentiate it from *Pontosphaera* on the same grounds (Lohmann, 1902). Tappan (1980) later included *Nannocorbis, Calciopilleus* and *Lohmannosphaera* in the family. As pointed out by Perch-Nielsen (1985a), only isolated coccoliths of *Nannocorbis* and *Calciopilleus* are known so that it is uncertain whether the coccospheres in these genera were built as those in *Scyphosphaera*. Moreover, similarity in outline should not be equated with homology in structure: coccoliths in *Nannocorbis* show a spiral structure, coccoliths in *Calciopilleus* have transverse ribs. Also, *Lohmannosphaera* is not a synonym of *Scyphosphaera*, from which it is probably very different (Aubry in press b).

The family Zygodiscaceae was introduced by Hay and Mohler (1967) who applied a very broad concept to it. Hay (1977) assigns numerous Mesozoic genera to it and only one Cenozoic genus, *Zygodiscus* (genotype of the family). Tappan (1980) and Perch-Nielsen (1985b) include several other Cenozoic genera in this family, genera erected for coccoliths which despite gross morphological similarity differ profoundly in structure. Coccoliths placed in the genera *Chiastozygus*, *Chiphragmalithus*, *Neococcolithes* and *Neochiastozygus* are undoubtedly closely related (Perch-Nielsen 1981a, b, Aubry in prep.) and questionably might be related to coccoliths of *Isthmolithus* (cf. Wise and Mostajo 1983). However, they differ noticeably in structure from coccoliths assigned to *Zygodiscus*, *Placozygus*, *Lophodolithus* and *Helicosphaera*. Coccoliths in the former group of genera are composed of two (*Chiastozygus*, *Chiphragmalithus*, *Neococcolithes*, *Isthmolithus*) or three (*Neochiastozygus*) structural units. These are a (1) mono-cyclic or (2) bicyclic margin formed of inclined, vertical, strongly imbricated elements which delineate a wide elliptical central area spanned by (3) a central cross-structure. No basal plate is present and the central cross-structure, unlike the bar of coccoliths in the second group, is composed of large imbricated crystals undifferentiated between the proximal and distal sides. Also, at its tips, the central cross-structure widens to obtain direct support from the margin, whereas the bar narrows and forms a bevel in the basal plate. Thus, despite superficial analogies, the structural units of coccoliths assigned to *Chiastozygus*, *Chiphragmalithus*, *Neococcolithes*, *Neochiastozygus* and *Isthmolithus* are not homologous with the structural

units of coccoliths assigned to *Placozygus*, *Zygodiscus*, *Lophodolithus* and *Helicosphaera*. Therefore, the genera in these two groups should be placed in separate families. Perch-Nielsen's (1985a, b) placement of the Cenozoic species of *Chiastozygus* in the family Zygodiscaceae and of the Mesozoic representatives of this same genus in the family Chiastozygaceae (Tappan 1980 considers the latter family synonymous with the former) clearly reflects the fact that, at present, the Cenozoic generic content of the family Zygodiscaceae is unsatisfactory. It appears that the present assignment of Mesozoic genera to this family (Hay 1977, Tappan 1980, Perch-Nielsen 1985a, b) is equally unsatisfactory. As it is not the purpose of this discussion to revise family assignments of Mesozoic genera, we shall only remark that whereas assignment of genera such as *Zeugrhabdotus* to the Zygodiscaceae appears well-founded (the structure of coccoliths in *Zeugrhabdotus* is very close to that of coccoliths in *Zygodiscus*), a placement in this family of other genera as *Tranolithus* or *Barringtonella* cannot be justified except by superficial analogies.

The family Pontosphaeraceae was introduced by Lemmermann (1908). Hay (1977) restricts it to the Cenozoic; Tappan (1980) includes the Late Cretaceous genus *Pontolithina* in it. Ambiguously, Perch-Nielsen assigns the Cenozoic species of the genus *Neocrepidolithus* to it (1985b) and its Mesozoic species to the Mesozoic family Crepidolithaceae (1985a).

Among the Cenozoic genera placed in the family Pontosphaeraceae by Hay (1977) and Tappan (1980), *Crassapontosphaera*, *Discolithina* (nom. subst. for *Discolithus* Huxley 1868, Loeblich and Tappan 1963) and *Koczyia* are regarded as synonymous with *Pontosphaera* by Perch-Nielsen (1985b). The genus *Transversopontis* is also regarded herein as synonymous with *Pontosphaera* as there are no major structural differences between the coccoliths in either genus (Aubry, in press). The discussion presented above supports Perch-Nielsen's (1985b) placement of *Scyphosphaera* and *Neocrepidolithus* together with *Ponto-*

sphaera in the family Pontosphaeraceae. Assignment of *Lophodolithus* (see Hay 1977, Tappan 1980) and *Helicosphaera* (see Hay 1977) to this family is, however, contradicted by evidence which I have presented above for grouping these two genera with the genus *Zygodiscus*. Placement in the Pontosphaeraceae of *Calciopilleus* and *Nannocorbis* (see Perch-Nielsen 1985b) can only be tentative for reasons given above and is not recommended.

2.4 TAXONOMIC REVISION

The discussion presented above supports Hay's (1977) choice of two families, Zygodiscaceae and Pontosphaeraceae, which are retained here to regroup the Cenozoic genera *Placozygus, Zygodiscus, Lophodolithus, Helicosphaera, Pontosphaera, Scyphosphaera* and *Neocrepidolithus* (Table 2.2). It does not, however, support Hay's (1977) determination of the generic content of these families.

> **Family:** Zygodiscaceae Hay and Mohler 1967 emend this paper
> **Type genus:** *Zygodiscus* Bramlette and Sullivan 1961

Emendation: Hay and Mohler's (1967) broad definition of the family allows numerous genera to be included in it. A narrower definition provided by Hay (1977) does not properly account for the structure of the coccoliths of the family type (*Zygodiscus*).

Definition: this family includes mostly elliptical coccoliths usually formed of four structural units: (1) a proximal basal plate composed of radiating elements, surrounded by (2) a flaring flange consisting of strongly imbricated inclined elements, (3) a transverse bar usually aligned with the minor axis of the ellipse but sometimes oblique, formed of two parallel segments, and, in Cenozoic representatives tightly fitted at both ends in the basal plate in which it forms a bevel, (4) a distal cover formed of primary concentrically arranged elements. The distal cover extends over the basal plate and the bar on which its elements are aligned with the minor axis of the ellipse. It covers part or all of the flange.

Morphological variations: the basal plate varies from restricted to a narrow elliptical ring delineating a wide central opening to a large ring delineating a narrow central opening. The flange may be low or high, symmetrically or

Table 2.2—Revision of the Cenozoic generic content of the families Zygodiscaceae, Pontosphaeraceae and Chiastozygaceae. The genus *Pontosphaera* includes the genera *Crassapontosphaera, Koczyia, Discolithina (Discolithus)* and *Transversopontis.* The genus *Prolatipatella* is also regarded as synonymous with *Pontosphaera*. The taxonomic position of *Isthmolithus* is uncertain. Exclusively Mesozoic genera are not considered in this revision.

Family	Generic content	
	Cenozoic	Mesozoic
Zygodiscaceae Hay and Mohler 1967 emend this paper	*Zygodiscus* *Lophodolithus* *Helicosphaera* *Placozygus*	*Zygodiscus* *Placozygus*
Pontosphaeraceae Lemmermann 1908	*Pontosphaera* *Scyphosphaera* *Neocrepidolithus*	*Pontosphaera* (*Prolatipatella*) *Neocrepidolithus* *Pontolithina*
Chiastozygaceae Rood, Hay and Barnard 1973	*Chiastozygus* *Neococcolithes* *Neochiastozygus* *Chiphragmalithus* ? *Isthmolithus*	*Chiastozygus*

asymmetrically developed. Coccoliths are symmetrical or asymmetrical. Evolutionary loss of the bar occurred in some of the youngest representatives of the genera *Lophodolithus* and *Helicosphaera*.

Generic content: the family Zygodiscaceae is represented by four genera in the Cenozoic: *Lophodolithus* and *Helicosphaera* are restricted to that era; *Placozygus* is represented from the Cretaceous to the Late Palaeocene; *Zygodiscus* is primarily represented in the Palaeocene but also in the Late Cretaceous (i.e. *Z. pomerolii*).

Family: Pontosphaeraceae Lemmermann 1908
Type genus: *Pontosphaera* Lohmann 1902
Definition: this family includes elliptical coccoliths which consist of three structural units: (1) a proximal basal plate composed of radiating elements surrounded by (2) a flaring flange composed of strongly imbricated inclined elements; (3) a distal cover formed of mostly concentrically arranged strongly imbricated elements which line both the basal plate and the flange.

Morphological variations: variable development of one or two openings in the basal plate. Height of flange variable. Distal cover may develop beyond the flange.

Generic content: the family Pontosphaeraceae is represented by three Cenozoic genera; *Scyphosphaera* is restricted to the Cenozoic; *Neocrepidolithus* is represented in the Late Cretaceous and the Palaeocene; *Pontosphaera* is mostly a Cenozoic genus which is also represented in the Late Cretaceous (*Prolatipatella* being regarded as synonymous with *Pontosphaera*). The genus *Pontolithina* is the only genus restricted to the Mesozoic and assigned to this family.

Family: Chiastozygaceae Rood, Hay and Barnard 1973
Type genus: *Chiastozygus* Gartner 1968
Definition: 'Coccoliths with an eiffellithalid rim having a central structure consisting of a cross aligned in the equal axes of the ellipse' (Rood *et al.* 1973, p. 370).

Remarks: it is clear that the genera *Chiastozygus, Chiphragmalithus, Neochiastozygus, Neococcolithes* and *Isthmolithus* are not directly related to the family Zygodiscaceae. I agree with Hay's (1977) assignment of *Chiphragmalithus, Chiastozygus* and *Neococcolithes* to the family Chiastozygaceae and I suggest placing *Neochiastozygus* in this family. However, the taxonomic position of *Isthmolithus* remains uncertain.

Generic content: the family Chiastozygaceae includes three Cenozoic genera: *Chiphragmalithus, Neococcolithes* and *Neochiastozygus*; one genus, *Chiastozygus*, is represented in the Mesozoic and in the Early Palaeocene.

As defined, the families Zygodiscaceae, Pontosphaeraceae and Chiastozygaceae include a large number of Cenozoic taxa. For example, the genus *Helicosphaera* (Late Palaeocene to Recent) includes over 40 species, the genus *Scyphosphaera* (Early Eocene to Recent) comprises about 50 species and over 100 species have been described and assigned to the genus *Pontosphaera* (Late Cretaceous to Recent). These three families are represented in the Mesozoic and the Cenozoic, and at least one genus in each is represented in both eras.

2.5 GEOLOGICAL IMPLICATIONS: SURVIVAL AT THE CRETACEOUS–TERTIARY BOUNDARY

Because of the position of the calcareous nannoplankton in the food chain, changes of diversity patterns of calcareous nannofossils through time and particularly around main geological boundaries are of prime interest. The most profound change in the Mesozoic and Cenozoic history of the calcareous nannoflora concerns its Late Cretaceous decline and Early Palaeocene radiation.

The progressive changes which affect the nature and composition of the calcareous nannoflora across the Mesozoic–Cenozoic boundary have been described in great detail (Perch-Nielsen 1969, 1979b, Percival and Fischer

1977, Perch-Nielsen *et al.* 1982). Three groups of species have been distinguished. The first consists of species which form the bulk of Late Cretaceous assemblages. They became extinct at the boundary or shortly above depending on how their presence in Danian sediments is interpreted (*in situ* or reworked; cf. Thierstein and Okada 1979, Thierstein 1982, Perch-Nielsen *et al.* 1982). These species have been referred to as 'vanishing species' (Percival and Fischer, 1977) or 'Cretaceous coccoliths' (Perch-Nielsen *et al.* 1982). They are regarded as having no descendants in the Cenozoic. The second group is formed by a few species which occur both in the Late Cretaceous and in the Early Palaeocene but with a higher frequency in Lower Palaeocene assemblages than in Upper Cretaceous ones. These species, which represent almost an equal number of genera, have been referred to as 'survivors' (Perch-Nielsen *et al.* 1982) or 'persistent species' (Percival and Fischer 1977). They are regarded as the genetic stock from which all the Cenozoic calcareous nannoflora diversified (Perch-Nielsen 1981 a, b, Perch-Nielsen *et al.* 1982). The third group, which characterises the Cenozoic Era, corresponds to the 'Tertiary coccoliths' of Perch-Nielsen *et al.* (1982) or the 'incoming species' of Percival and Fischer (1977). They have arisen from the 'survivors' and have no other apparent ancestry in the Mesozoic. In this respect, *Biantholithus* (*B. sparsus*) is the first typical Cenozoic taxon.

The generally accepted relationships between the Mesozoic and Cenozoic calcareous nannoplankton taxa are illustrated in Fig. 2.3. It is assumed that the Cenozoic taxa evolved from a restricted phylogenetic stock. It is, however, remarkable that none of the surviving genera (about 15 according to Perch-Nielsen 1981a) became successful during the Cenozoic (Table 2.3). The long-ranging genera such as *Braarudosphaera* and *Scapholithus* crossed the boundary without showing any evolutionary change and remained restricted in diversity during the Cenozoic. With the exception of *Markalius* which was still represented in the Oligocene, the other surviving genera, each

represented by a limited number of species, became progressively extinct during the Palaeocene and the Early Eocene. If persistant species are at the origin of the Cenozoic groups, profound evolutionary reorganisations must have occurred during the Palaeocene to produce new structures. For instance, the Early Palaeocene evolution of *Neochiastozygus* from *Chiastozygus* involved a doubling of the margin (Perch-Nielsen 1981a). Similarly, the generally accepted evolution of *Zygodiscus* from *Placozygus* in the early Late Palaeocene and of *Pontosphaera* from *Neocrepidolithus* in the Late Palaeocene would have involved major structural reorganisations (Fig. 2.2 and see Aubry in press c).

It is also accepted that none of the structures characteristic of the Cenozoic nannoplankton genera, i.e. those taxa which were highly successful in the Cenozoic, had already evolved in the Mesozoic. In attempting to establish a phylogenetically based taxonomy, I have discussed above (see also Aubry in press b) the likelihood that the structures characteristic of the genera *Pontosphaera* and *Zygodiscus* had already evolved during the Late Cretaceous, that these genera were already established at this time and that the families Pontosphaeraceae and Zygodiscaceae stemmed in the Late Cretaceous from an as-yet unknown Mesozoic ancestor (Fig. 2.2). In this view, the 'persistent' species are no longer the exclusive phylogenetic link between Mesozoic and Cenozoic genera as illustrated in Fig. 2.4. I suggest that, in fact, several of the structures which became well represented among the Cenozoic nannoflora, and which each serve to differentiate a Cenozoic genus, were being established in the Late Cretaceous. Taxa with the structures characteristic of species assigned to *Pontosphaera*, *Zygodiscus*, as well as *Ellipsolithus* and *Hornibrookina*, among others, occur in the Late Cretaceous.

This reinterpretation of the origin of at least part of the Palaeocene calcareous nannoflora implies that the mechanism of massive extinction at the Mesozoic–Cenozoic boundary was discriminatory. The record shows that some of

Table 2.3—Stratigraphical ranges of the 'survivors'. The 16 genera listed have been regarded as the genera which gave rise to the new Cenozoic nannoflora in the course of the Palaeocene (Perch-Nielsen 1981a, Table 2). This list clearly suggests that none of the genera which became successful in the Cenozoic was already established in the Mesozoic (occurrence of *Pontosphaera* in the Late Cretaceous is questioned by Perch-Nielsen who clearly favours a Late Palaeocene evolution of this genus from *Neocrepidolithus*). Nine (*) of these genera have a long range in the Mesozoic and became extinct during the Palaeocene. Three (**) have a long range in the Mesozoic and have no representative in the Cenozoic; their phylogenetic links with Cenozoic genera are very hypothetical. The genus *Markalius* differs little from *Cyclagelosphaera* and does not represent a major innovation. *Goniolithus fluckigeri* remained an obscure form. *Neocrepidolithus* and *Placozygus* are closely related to *Pontosphaera* and *Zygodiscus* in particular, and it has been suggested that the latter evolved from the former in the Palaeocene. It is herein suggested that *Pontosphaera* and *Zygodiscus* had already become established in the Maastrichtian and that the four genera evolved from a common ancestor. *Pontosphaera* and *Zygodiscus*, and to a lesser extent *Neocrepidolithus* and *Placozygus*, represent major structural innovations and differ noticeably from other Mesozoic genera.

Genus	Stratigraphic range	Remarks
*Biscutum**	Jurassic–Early Palaeocene	
*Braarudosphaera**	Cretaceous–Recent	
*Chiastozygus**	Early Cretaceous–Early Palaeocene	
*Cyclagelosphaera**	Middle Jurassic–Early Palaeocene	
Goniolithus	Maastrichtian–Oligocene	Monospecific; no clear phylogenetic links
*Lapideacassis**	Albian–Early Palaeocene	
Markalius	Maastrichtian–Oligocene	Extremely close to *Cyclagelosphaera*
*Micrantholithus**	Cretaceous–Recent	
*Micula***	Late Cretaceous	Suggested as ancestral to mid-Eocene *Nannotetrina*
Neocrepidolithus	Maastrichtian-Early Palaeocene	Assigned herein to the Pontosphaeraceae
*Octolithus**	Campanian–Early Palaeocene	
Placozygus	Late Cretaceous–Palaeocene	Assigned herein to the Zygodiscaceae
*Repagulum***	Middle and Late Cretaceous	Regarded as ancestral to *Ellipsolithus*
*Scampanella**	Albian–Early Eocene	
*Scapholithus**	Hauterivian (or older)–Recent	
*Sollasites***	Early Jurassic–Maastrichtian	Suggested as ancestral to *Cruciplacolithus* and closely related genera

the large Cenozoic calcareous nannoplankton families originated in the Cretaceous. The paradox of the end-Cretaceous calcareous nannofossil extinction is that the taxa which became extinct (i.e. which did not give rise to Palaeocene lineages) are those which were most abundant and widespread, while those which were spared, and from which at least part of the Cenozoic nannoflora sprang, are those which were extremely rare and localised. This observation is in conflict with the extinction and survival patterns that would result from extraterrestrial causes. Extinctions resulting from an extraterrestrial impact are expected to be at random, and survival is expected to be proportional to the number of individuals in the populations (Stanley 1979, Alvarez 1987, Donovan 1987).

In our effort to understand what processes led to the Mesozoic–Cenozoic boundary renewal, it is certainly of major importance to establish carefully the relationships between the Late Cretaceous and Early Palaeocene taxa and to analyse the nature of the extinctions and subsequent radiations.

2.6 ACKNOWLEDGEMENTS

I am grateful to W. A. Berggren, D. Lazarus and G. P. Lohman (Woods Hole Oceanographic Institution), K. Perch-Nielsen (London) and H. Stradner (Vienna) for constructive criticism on

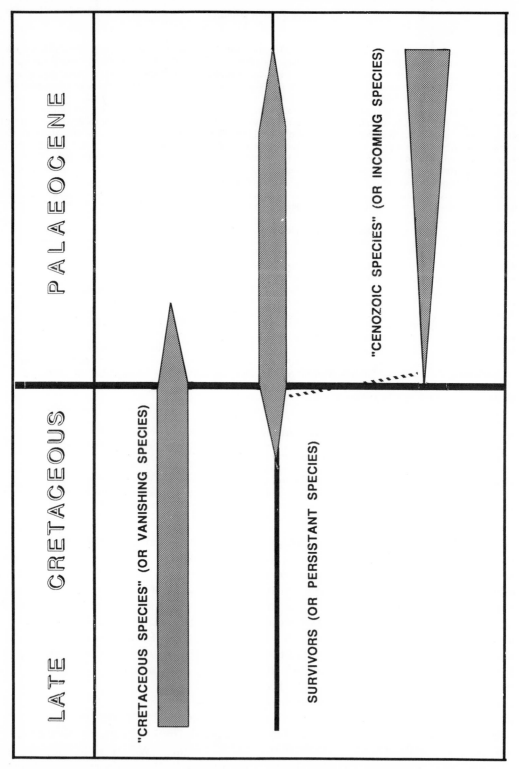

Fig. 2.3 — Generally accepted schematic interpretation of the relationships between Mesozoic and Cenozoic calcareous nannoplankton taxa. Survivors are regarded as the only link between Mesozoic and Cenozoic taxa.

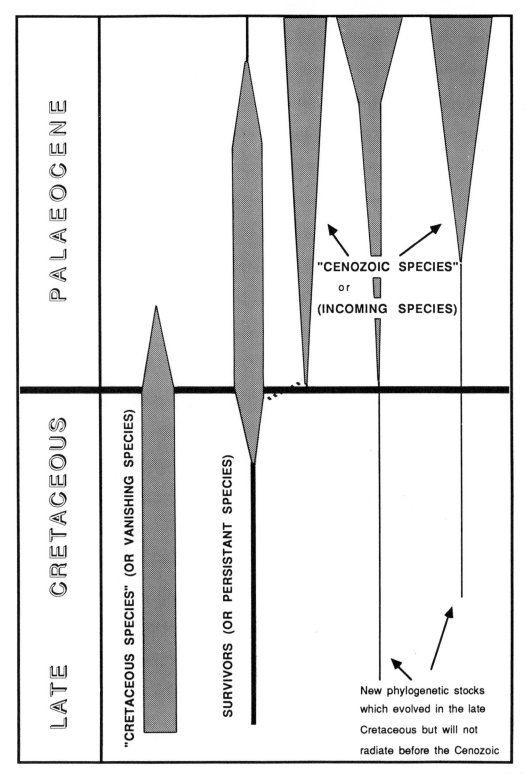

Fig. 2.4 — Revised schematic interpretation of the relationships between Mesozoic and Cenozoic calcareous nannoplankton taxa. The Cenozoic calcareous nannoplankton developed, at least in part, from taxa which evolved during the Late Cretaceous. These taxa exhibit new structures which contrast with those typical of Mesozoic species but are identical to those which characterise a large part of the Cenozoic calcareous nannofloras.

an early draft of this paper, and to A. R. Arnold and W. A. Berggren for inviting me to present the results presented in this paper at the Symposium on Tempo and Evolution held during NAPC IV in Boulder (August 1986).

2.7 REFERENCES

Alvarez, L. W. 1987. Mass extinctions caused by large bodies impacts. *Physics Today*, 24–33.

Aubry, M.-P. 1983. Biostratigraphie du Paléogène épicontinental de l'Europe du Nord-Ouest. Etude fondée sur les nannofossiles calcaires. *Docum. Lab. Géol. Lyon*, **89**, 1–317.

Aubry, M.-P. in press a. The Cretaceous/Tertiary boundary event: paradoxes and dilemma. *Geobios.* (submitted).

Aubry, M.-P. in press b. *Handbook of Cenozoic Calcareous Nannofossils*, **4**, Micropaleontology Press.

Aubry, M.-P. in press c. Phylogeny of the Cenozoic calcareous nannoplankton genus *Helicosphaera*. *Paleobiology.*

Berggren, W. A., Kent, D. V., Flynn, J. J. and van Couvering, J. A. 1985. Cenozoic geochronology. *Bull. Geol. Soc. Amer.*, **96**, 1407–1418.

Biolzi, M., Perch-Nielsen, K. and Ramos, I. 1981. *Triquetrorhabdulus*—an Oligocene/Miocene calcareous nannofossil genus. *INA Newsletter*, **3**, 89–92.

Black, M. 1968. Taxonomic problems in the study of coccoliths. *Palaeontology*, **11**, 793–813.

Black, M. 1971. The systematics of coccoliths in relation to the Palaeontologic record. In: B. M. Funnel and W. R. Riedel (Eds.), *The Micropaleontology of the Oceans*, Cambridge University Press, Cambridge, 611–624.

Bramlette, M. N. and Martini, E. 1964. The great change in calcareous nannoplankton fossils between the Maastrichtian and Danian. *Micropaleontology*, **10**, 291–322.

Bramlette, M. N. and Sullivan, F. R. 1961. Coccolithophorids and related nannoplankton of the early Tertiary in California. *Micropaleontology*, **7**, 129–188.

Brand, L. E. 1981. Genetic variability in reproduction rates in marine phytoplankton populations. *Evolution*, **35**, 1117–1127.

Brand, L. E. 1982. Genetic variability and spatial patterns of genetic differentiation in the reproductive rates of the marine coccolithophores *Emiliania huxleyi* and *Gephyrocapsa oceanica*. *Limnol. Oceanogr.*, **27**, 236–245.

Bukry, D. 1971. *Discoaster* evolutionary trends. *Micropaleontology*, **17**, 43–52.

Clochiatti, M. 1971. Sur l'existence de coccosphères portant des coccolithes de *Gephyrocapsa oceanica* et de *Emiliania huxleyi* (coccolithophoridés). *C.R. Acad. Sci., Paris*, **273(D)**, 318–321.

Croneis, C. 1938. Utilitarian classification for fragmentary fossils. *Jour. Geol.*, **46**, 975–984.

Crux, J. A. 1982. Upper Cretaceous (Cenomanian to Campanian) calcareous nannofossils. In A. R. Lord (Ed.), *A Stratigraphical Index of Calcareous Nannofossils*, Ellis Horwood, Chichester, 81–135.

Crux, J. A. 1987. Concerning dimorphism in early Jurassic coccoliths and the origin of the genus *Discorhabdus* Noël 1965. *Abh. Geol. B.-A.*, **39**, 51–55.

Deflandre, G. 1952. Sous-embranchement des Flagellés (classe des Coccolithophoridés). In J. Piveteau (Ed.), *Traité de Paléontologie*, **1**, Masson, Paris, 107–115.

Deflandre, G. 1958. Exposé et discussion sur la parataxonomie. *C.R. Somm. Soc. Géol. France*, 141–143.

Deflandre, G. and Deflandre-Rigaud, M. 1948. A propos de l'introduction des parataxons dans la nomenclature zoologique. *Bull. Zool. Nomencl.*, **15**, 705–724.

Deflandre, G. and Deflandre-Rigaud, M. 1970. Commentaire sur la systématique et la nomenclature des nannofossiles calcaires. *Cah. Micropaléont.*, **2**, 1–17.

Donovan, S. K. 1987. How sudden is sudden? *Nature*, **328**, 109.

Gard, G. 1987. Observation of a dimorphic coccosphere. *Abh. Geol. B.-A.*, **39**, 85–87.

Gartner, S. 1968. Coccoliths and related calcareous nannofossils from Upper Cretaceous deposits of Texas and Arkansas. *Univ. Kansas Paleontol. Contrib.*, **48**, 1–56.

Gartner, S. 1970. Phylogenetic lineages in the lower Tertiary coccolith "*Chiasmolithus*". N. Amer. Paleont. Conv., September 1969, Proc. G., 930–957.

Gayral, P. and Fresnel-Morange, J. 1971. Résultats préliminaires sur la structure et la biologie de la Coccolithophoracée *Ochrosphaera neapolitana* Schussnig. *C.R. Acad. Sci., Paris*, **273(D)**, 1683–1686.

Haq, B. U. 1971. Paleogene calcareous nannoflora. Part I: The Paleocene of West-Central Persia and the Upper Paleocene–Eocene of West Pakistan. *Stockholm Contr. Geol.*, **25**, 1–56.

Haq, B. U. 1973. Evolutionary trends in the Cenozoic coccolithophoridae genus *Helicopontosphaera*. *Micropaleontology*, **19**, 32–52.

Hay, W. W. 1977. Calcareous nannofossils. In A. T. S. Ramsay (Ed.), *Oceanic Micropaleontology*, **2**, Academic Press, London, 1055–1200.

Hay, W. W. and Mohler, H. P. 1967. Calcareous nannoplankton from Pont Labau, France, and zonation of the Paleocene and lower Eocene. *J. Paleontol.*, **41**, 1505–1541.

Heimdal, B. R. and Gaarder, K. R. 1981. Coccolithophorids from the northern part of the eastern central Atlantic. II. Heterococcoliths. *"Meteor" Forschungs Ergebnisse*, **D/33**, 37–69.

Jafar, S. A. 1975. Some comments on the calcareous nannoplankton genus *Scyphosphaera* and the neotypes of *Scyphosphaera* species from Rotti, Indonesia. *Senckenbergiana Lethaea*, **56**, 365–379.

Kamptner, E. 1941. Die Coccolithineen der Südwestküste von Istrien. *Ann. Naturhist. Mus., Wien*, **51**, 54–149.

Kamptner, E. 1955. Fossile Coccolithineen-Skelettreste aus Insulinde. Eine micropaläontologische Untersuchung. *Verh. Kon. Ned. Akad. Wetensch., Afd. Natuurk. R2*, **50**, 1–87.

Kent, D. V. and Gradstein, F. 1985. A Cretaceous and Jurassic geochronology. *Bull. Geol. Soc. Amer.*, **96**, 1419–1427.

Klaveness, D. 1972a. *Coccolithus huxleyi* (Lohmann) Kamptner. II. The flagellate cell, aberrant cell types, vegetative propagation and life cycles. *J. Bryol. Phycol.*, **7**, 309–318.

Klaveness, D. 1972b. *Coccolithus huxleyi* (Lohmann) Kamptner. I. Morphological investigations on the vegetative cell and the process of coccolith formation. *Protistologica*, **8**, 335–346.

Klaveness, D. 1973. The microanatomy of *Calyptrosphaera sphaeroidea*, with some supplementary observations on the motile stage of *Coccolithus pelagicus*. *Norw. J. Bot.*, **20**, 151–162.

Lambert, B. 1986. La notion d' espèce chez le genre *Braarudosphaera* Deflandre 1947. Mythe et réalité. *Rev. Micropaléont.*, **28**, 255–264.

Lauer, G. 1975. Evolutionary trends in the Arkangelskiellaceae (calcareous nannoplankton) of the upper Cretaceous of central Oman, SE Arabia. *Arch. Sci. Genève*, **28**, 259–262.

Lazarus, D. 1983. Speciation in pelagic Protista and its study in the planktonic microfossil record: a review. *Paleobiology*, **9**, 327–340.

Leadbeater, B. S. G. 1971. Observations on the life history of the Haptophycean alga *Pleurochrysis scherffelii* with special reference to the microanatomy of the different types of motile cells. *Ann. Botany*, **35**, 429–439.

Lefort, F. 1971. Sur l'appartenance à une seule et même espèce de deux Coccolithacées, *Cricosphaera carterae* et *Ochrosphaera verrucosa*. *C.R. Acad. Sci., Paris*, **272(D)**, 2540–2543.

Lefort, F. 1975. Etude de quelques Coccolithophoracées marines rapportées aux genres *Hymenomonas* et *Ochrosphaera*. *Cah. Biol. Mar.*, **16**, 213–229.

Lemmermann, E. 1908. Flagellatae, Chlorophyceae, Coccosphaerales und Silicoflagellatae. In K. Brandt and C. Apstein (Eds.), *Nordisches Plankton. Botanischer Teil*, Lipsius and Tischer, Kiel and Leipzig, 1–40.

Loeblich, A. R. and Tappan, H. 1963. Type fixation and validation of certain calcareous nannoplankton genera. *Proc. Biol. Soc., Washington*, **76**, 191–196.

Loeblich, A. R. and Tappan, H. 1966. Annotated index and bibliography of the Calcareous nannoplankton. I. *Phycologia*, **5**, 81–216.

Lohmann, H. 1902. Die Coccolithophoridae, eine Monographie der Coccolithen bildenden Flagellaten zugleich ein Beitrag zur Kenntnis des Mittelmeerauftriebs. *Arch. Protistenk.*, **1**, 89–165.

Lohman, G. P. and Malmgren, B. A. 1983. Equatorward migration of *Globorotalia truncatulinoides* ecophenotypes through the late Pleistocene: Gradual evolution or ocean change? *Paleobiology*, **9**, 377–389.

Manton, I. and Leedale, G. F. 1963. Observations on the micro-anatomy of *Crystallolithus hyalinus* Gaarder and Markali. *Arch. Mikrobiol.*, **47**, 115–136.

Martini, E. 1980. Oligocene to Recent calcareous nannoplankton from the Philippine Sea, Deep Sea Drilling Project Leg 59. *Init. Rep. DSDP*, **69**, 547–565.

Nishida, S. 1979. Atlas of Pacific nannoplanktons. *News Osaka Micropaleont., Special Paper*, **3**, 1–31.

Noël, D. 1984. Le genre chez les nannofossiles calcaires. *Bull. Soc. Géol. France*, **26**, 583–589.

Norris, R. E. 1971. Extant calcareous nannoplankton from the Indian Ocean. In A. Farinacci (Ed.), *Proc. II Plankt. Conf., Roma, 1970*, **2**, Tecnoscienza, Rome, 899–909.

Okada, H. and Honjo, S. 1973. The distribution of oceanic coccolithophorids in the Pacific. *Deep-Sea Res.*, **20**, 355–374.

Okada, H. and McIntyre, A. 1977. Modern coccolithophores of the Pacific and North Atlantic Oceans. *Micropaleontology*, **23**, 1–55.

Parke, M. 1971. The production of calcareous elements by benthic algae belonging to the class Haptophyceae (Chrysophyta). In A. Farinacci (Ed.), *Proc. II Plankt. Conf., Roma, 1970*, **2**, Tecnoscienza, Rome, 929–937.

Parke, M. and Adams, I. 1960. The motile (*Crystallolithus hyalinus* Gaarder and Markali) and non-motile phases in the life history of *Coccolithus pelagicus* (Wallich) Schiller. *Jour. Mar. Biol. Ass. U.K.*, **39**, 263–274.

Perch-Nielsen, K. 1969. Die Coccolithen einiger Dänischer Maastrichtian und Danian Lokalitäten. *Bull. Geol. Soc. Denmark*, **21**, 51–66.

Perch-Nielsen, K. 1973. Neue Coccolithen aus dem Maastrichtian von Dänemark, Madagascar und Ägypten. *Bull. Geol. Soc. Denmark*, **22**, 306–333.

Perch-Nielsen, K. 1977. Albian to Pleistocene calcareous nannofossils from the western South Atlantic, Deep Sea Drilling Project Leg 39. *Init. Rep. DSDP*, **39**, 699–823.

Perch-Nielsen, K. 1979a. Calcareous nannofossil zonation at the Cretaceous/Tertiary boundary in Denmark. In T. Birkelund and R.G. Bromley (Eds.), *Cretaceous–Tertiary Boundary Events (Copenhagen, Sept. 18–24, 1979), Volume 1: The Maastrichtian and Danian of Denmark*, University of Copenhagen, Copenhagen, 115–135.

Perch-Nielsen, K. 1979b. Calcareous nannofossils in Cretaceous/Tertiary boundary sections in Denmark. In W.K. Christensen and T. Birkelund (Eds.), *Cretaceous–Tertiary Boundary events, Volume 2: Proceedings*, University of Copenhagen, Copenhagen, 120–126.

Perch-Nielsen, K. 1981a. Les coccolithes du Paléocène près de El Kef, Tunisie et leurs ancêtres. *Cah. Micropaléont.*, **3**, 7–23.

Perch-Nielsen, K. 1981b. New Maastrichtian and Paleocene calcareous nannofossils from Africa, Denmark, the U.S.A. and the Atlantic, and some Paleocene lineages. *Ecologae Geol. Helv.*, **74**, 831–863.

Perch-Nielsen, K. 1985a. Mesozoic calcareous nannofossils. In H. M. Bolli, J. B. Saunders and K. Perch-Nielsen (Eds.), *Plankton Stratigraphy*, Cambridge University Press, Cambridge, 329–426.

Perch-Nielsen, K. 1985b. Cenozoic calcareous nannofossils. In H. M. Bolli, J. B. Saunders and K. Perch-Nielsen (Eds.), *Plankton Stratigraphy*, Cambridge University Press, Cambridge, 427–554.

Perch-Nielsen, K. 1985c. Calcareous nannofossil Families and Genera—Remarks about relations and "non-relations." *INA Newsletter*, **7**, 78–79.

Perch-Nielsen, K., McKenzie, J. and He, Q. 1982. Biostratigraphy and isotope stratigraphy and the "catastrophic" extinction of calcareous nannoplankton at the Cretaceous/Tertiary boundary. *Geol. Soc. Amer., Special Paper*, **190**, 353–371.

Percival, S. F. and Fischer, A.G. 1977. Changes in calcareous nannoplankton in the Cretaceous–Tertiary biotic crisis at Zumaya, Spain. *Evolutionary Theory*, **2**, 1–35.

Pienaar, R. N. 1968. Upper Cretaceous coccolithophorids from Zululand, South Africa. *Palaeontology*, **11**, 361–367.

Pienaar, R. N. 1969. Upper Cretaceous calcareous nannoplankton from Zululand, South Africa.

Palaeontologia Afric., **12**, 75–149.

Priewalder, H. von 1973. Die Coccolithophoridenflora des Locus Typicus von *Pseudotextularia elegans* (Rzehak), Reingruberhohe, Niederösterreich (Maastricht). *Jahrb. Geol. B.-A.*, **116**, 3–34.

Rayns, D. G. 1962. Alternations of generations in a Coccolithophorid, *Cricosphaera carterae* (Braarud and Fagerl.) Braarud. *Jour. Mar. Biol. Ass. U.K.*, **42**, 481–484.

Romein, A. J. T. 1979. Lineages in early Paleogene calcareous nannoplankton. *Utrecht Micropaleont. Bull.*, **22**, 1–231.

Rood, A. P. and Barnard, T. 1972. On Jurassic coccoliths: *Stephanolithion, Diadozygus* and related genera. *Eclogae Geol. Helv.*, **65**, 327–342.

Rood, A. P., Hay, W. W. and Barnard, T. 1973. Electron microscope studies of lower and middle Jurassic coccoliths. *Eclogae Geol. Helv.*, **66**, 365–382.

Roth, P. H. and Bowdler, J. L. 1979. Evolution of the calcareous nannofossil genus *Micula* in the late Cretaceous. *Micropaleontology*, **25**, 272–280.

Shafik, S. and Stradner, H. 1971. Nannofossils from the Eastern Desert, Egypt with reference to Maastrichtian nannofossils from the USSR. *Jahrb. Geol. B.-A.*, **17**, 69–104.

Stanley, S. M. 1979. *Macroevolution,* Freeman, San Francisco.

Steineck, P. L. and Fleisher, R. L. 1978. Towards the classical evolutionary reclassification of Cenozoic Globigerinacea (Foraminiferida). *J. Paleontol.*, **52**, 618–635.

Tappan, H. 1980. Haptophyta, Coccolithophores and other calcareous nannoplankton. *The Paleobiology of Plant Protists,* Freeman, San Francisco, Chap. 9, 678–803.

Thierstein, H. R. 1982. Terminal Cretaceous plankton extinctions: A critical assessment. *Geol. Soc. Amer., Special Paper,* **190**, 385–399.

Thierstein, H. R. and Okada, H. 1979. The Cretaceous/Tertiary boundary event in the North Atlantic. *Init. Rep. DSDP,* **43**, 601–616.

Verbeek, J. W. 1977. Calcareous nannoplankton biostratigraphy of middle and upper Cretaceous deposits in Tunisia, southern Spain and France. *Utrecht Micropaleont. Bull.*, **16**, 1–157.

Wilbur, K. M. and Watabe, N. 1963. Experimental studies on calcification in molluscs and the alga *Coccolithus huxleyi. Ann. New York Acad. Sci.*, **109**, 82–112.

Winter, A. 1985. Distribution of living coccolithophores in the California current system, southern California borderland. *Mar. Micropaleont.*, **9**, 385–393.

Winter, A., Reiss, Z. and Luz, B. 1978. Living *Gephyrocapsa protohuxleyi* McIntyre in the Gulf of Elat ('Aqaba). *Mar. Micropaleont.*, **3**, 295–298.

Winter, A., Reiss, Z. and Luz, B. 1979. Distribution of living coccolithophores assemblages in the Gulf of Elat ('Aqaba). *Mar. Micropaleont.*, **4**, 197–223.

Wise, S.W. and Mostajo, E.L. 1983. Correlation of Eocene-Oligocene calcareous nannofossil assemblages from piston cores taken near Deep Sea Drilling sites 511 and 512, Southwest Atlantic Ocean. *Init. Rep. DSDP,* **71**, 1171–1180.

3

Reticulofenestra: A critical review of taxonomy, structure and evolution

Liam Gallagher

A discussion of the generic name *Reticulofenestra,* and its subsequent emendations, is given together with comments on ultrastructure and taxonomy. An emended definition of the genus is produced, and evolutionary links between species are suggested.

'La taxonomie du groupe des *Reticulofenestra* est complexe et confusé . . .'

Aubry 1983

3.1 INTRODUCTION
In 1966 Hay *et al.* proposed the generic name *Reticulofenestra*. Since then numerous authors have used and adapted this description in various ways without formal emendation of the original description. The taxonomic significance of such features as central area grill, tube cycle and rim structure is discussed with respect to previously published observations and the author's personal research on Tertiary nannofloras from the North Sea, Cyprus, Indonesia, New Zealand and southern England. A formal emendation of the genus is made together with a comprehensive classification of key *Reticulofenestra* species. The terminology employed in this classification has been derived from the original type descriptions and supplemented by the author's observations. Scanning electron microscopy (SEM) analysis of reticulofenestrids at various stages of mechanical breakdown has revealed details of the

ultrastructure of these forms, complementing SEM, transmission electron microscopy (TEM) and light microscopy (LM) work previously published. Appreciation of the ultrastructural detail enables comment to be made on possible lines of evolution, both within the genus *Reticulofenestra* and between members of the family Noelaerhabdaceae.

3.2 THE GENERIC NAME RETICULOFENESTRA
In 1966 Hay *et al.*, introduced the generic name *Reticulofenestra* and defined the genus as follows (p. 386): 'Placoliths with a large central opening spanned by a reticulate or lacy net, a tube cycle of tall imbricate wedges is exposed distally, but not proximally, a narrow proximal rim of many thin rays and a wide distal rim of many tubular elements. Type species = *R. caucasica* Hay, Mohler and Wade 1966'.

The most important part of this definition, the criterion by which *Reticulofenestra* species are differentiated from other members of the family Noelaerhabdaceae, is the presence, or evidence of the presence, of a 'lacy net'. A 'tube cycle' is also possessed by the other genera of this family and a 'large central opening' is a very subjective criterion which certainly does not apply to many accepted species of *Reticulofenestra*.

Two years later, in 1968, Stradner (Stradner

and Edwards 1968, p. 19) gave an emended definition of the genus with the important additional observation that the tube cycle could be 'contracted' towards the centre of the ellipse, so forming a 'ceiling-like' cover to the central area. Stradner observed that a form with this structure on the distal side had been described by Hay *et al.*, (1966) as a species of *Syracosphaera*.

Edwards (1973) produced an extremely detailed definition of a species within the genus *Reticulofenestra* (*R. hampdenensis*), documenting in particular the complexity of the rim elements. His emended diagnosis of the genus certainly elaborated on the structure of *Reticulofenestra*, but utilised fine detail only observable in the most meticulously analysed material (see Fig. 3.1) and excluded forms with a 'plugged' central area (e.g. *R. scissura* and *R. callida*). Waghorn (1981), however, reintroduced the idea that *Reticulofenestra* can have a 'plugged' central area via a series of diagrams (see Fig. 3.3), but made no formal emendation of the genus.

Romein (1979) described *Reticulofenestra* in great detail based upon Hay *et al.*'s (1966) definition. He highlighted the characteristics,

(Romein 1979, p. 127) but omitted any detailed reference to central area structure.

In 1980 Backman applied statistical analysis to the study of Neogene reticulofenestrids in order to differentiate between very similar species which were found at the same stratigraphical levels. There was no attempt to emend the original generic definition, but criteria for separating species with similar rim and central area characteristics were established.

Perch-Nielsen (1985b, p. 86) outlined the need to clarify the generic and species concepts of the family Noelaerhabdaceae (= Prinsiaceae; invalid, ICBN Art. 37.1) in order that biostratigraphical, ecological and other inferences could be made with a greater degree of confidence.

Backman and Hermelin (1986) reiterated the problem of species classification and set up rigorous morphometric parameters for their recognition within a still loosely defined genus. Most recently Pujos (1987) set out to clarify the taxonomy of 'small and medium' sized reticulofenestrids, but in isolation from related species and with no clear definition of the genus.

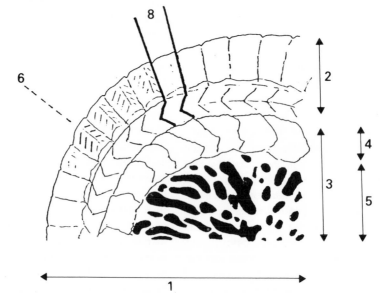

1 - Coccoliths radius
2 - Distal rim
3 - Central area
4 - Tube cycle
5 - Central area grill
6 - Longitudinal ridge
7 - Chevron ridge
8 - √ shaped ridge

Fig. 3.1 — Schematic diagram to show the important details of *Reticulofenestra* species. Re-drawn and modified from Edwards (1973)

3.3 ULTRASTRUCTURE AND TAXONOMY

If a morphological feature is gradational in nature then clearly it cannot be used taxonomically without unequivocal qualitative data. In species placed in *Reticulofenestra* in the literature, tube cycle and central area characteristics can be seen to be variable or gradational. This aspect of the morphology has been repeatedly ignored, and new genera have been erected which might be better considered as *Reticulofenestra* morphovariants, e.g. *Dictyococcites*.

Careful consideration of the ultrastructure of the reticulofenestrid morphotype and of its potential for variation (in terms both of the number of shield elements to the number of central grill elements, and of central grill architecture), provides a working definition from which coherent, unambiguous species descriptions can be derived. The majority of workers have, however, continued to follow the original definition of Hay *et al.* (1966) despite the observations of Stradner (Stradner and Edwards 1968) suggesting the possibility of various shield width to central area grill width ratios and the observations of Waghorn (1981) illustrating the potential for central area grill variation. New genera have been erected when deviations from the type description have been noted.

Perch-Nielsen (1971b, pp. 30–31) for example described several species of *Reticulofenestra*, but also erected the new genus *Cribrocentrum* for forms in which every second element of the shield contributed to the distally projected central grill. This can be interpreted as being synonymous with the 'group 3' *Reticulofenestra* of Stradner (Stradner and Edwards 1968, text Fig. 21) and the 'partially closed' *Reticulofenestra* of Waghorn (1981) and could be included in *Reticulofenestra* on the basis of Edwards' (1973) emended definition. The generic names *Cribrocentrum* and *Dictyococcites* are therefore considered superfluous as they are defined solely by changes in the central area structure. On this basis, these particular changes should only have a species ranking and not a generic one.

Bown (1987) recently demonstrated that consistent rim structures are diagnostic features at family level. The family Noelaerhabdaceae (Jerković 1970) is certainly characterised by a consistent rim structure and therefore represents a clear taxonomic grouping. Changes in the detail of the rim structure are used to define genera within a family, as has been done for the Noelaerhabdaceae, but central area characteristics may occur repeatedly in spatially and temporally disjunct forms and thus are not used for classification above the species level.

The nature of the reticulate grill which covers the central opening is considered here to be a feature reflecting species level. This opinion has been a matter of debate by previous authors. Stradner (Stradner and Edwards 1968) recognised three groups of *Reticulofenestra* based on shield width to central area grill width ratios, but made no taxonomic inferences other than to suggest that the reticulate net may vary within the genus. Similarly Waghorn (1981) illustrated three groups of *Reticulofenestra* based on central area structure, but again no taxonomic conclusions were drawn.

In contrast Haq (1968) and Perch-Nielsen (1971b) described new genera, *Stradnerius* and *Cribrocentrum* respectively, on the basis of deviations from the central area characteristics of the type species of *Reticulofenestra*. Haq (1971) stated that 'plugging' of the central area grill is an extension of crystal growth within the reticulate structure and is not, therefore, of taxonomic value at generic level. Haq (1971, p. 72) also noted that types of reticulate structure previously attributed to diagenetic changes occurred consistently in samples from many different areas with different states of preservation, and so were of value at specific level (see Fig. 3.2 for examples from North Sea study material). If the effects of preservation are not considered, conspecific specimens may easily be interpreted as separate species.

The structure of Noelaerhabdaceae forms in cross-section has been illustrated previously by Romein (1979, Fig. 38) in his evolutionary lineage and by Perch-Nielsen (1985a, Fig. 56 as Prinsiaceae) in an overview of the family (see Fig.

R. scissura

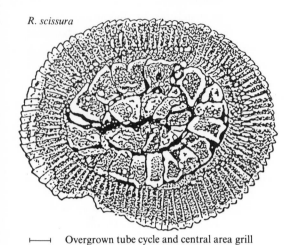

├───┤   Overgrown tube cycle and central area grill

R. foveolata

├─────┤   Overgrown central area grill

R. foveolata

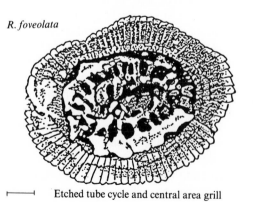

├───┤   Etched tube cycle and central area grill

Fig. 3.2 — Illustration of both 'coarse' and 'closed' central areas in *Reticulofenestra* species, and the effect preservation has on such structures. Scale bars = 1 μm.

3.3). The reconstructions of Romein (1979) are unfortunately inaccurate in a number of important details whilst Perch-Nielsen's diagram is a simple compilation to show terminology for the Noelaerhabdaceae.

Young (1987, Fig. 12) produced schematic cross-sections for many Neogene species, including the various morphotypes of *Reticulofenestra*, and also presented a diagram of a single reticulofenestrid element (Young 1987, Fig. 11) which accurately illustrated the interdependent relationships of the ultrastructural components.

By examining broken and etched specimens in the scanning electron microscope it has been possible to elucidate the nature of the cross-section in reticulofenestrid coccoliths. A gradational change is seen in the structure of the coccoliths from *Prinsius* to *Toweius* to *Reticulofenestra*, the crucial developments of which involved the proximal shield and the tube cycle (see Fig. 3.3). In *Prinsius* and *Toweius* Romein (1979, Fig. 38) considered the 'upper centro-distal cycle' to lie as a 'crown' on top of the 'lower centro-distal cycle', which is an extension of the proximal shield. However, by tracing the tube cycle development through *Prinsius* and *Toweius* to *Reticulofenestra* it is clear that the inner tube cycle (= 'upper centro-distal cycle') is a projection of the proximal shield, whilst the outer tube cycle (= 'lower centro-distal cycle') fuses with the elements of the shields (see Fig. 3.3). The representation of the central area grill as 'doubled' (in *R. dictyoda* and *R. umbilicus*) by Romein (1979, Fig. 38) is an attempt to produce a structure from the 'upper centro-distal cycle' (of *Toweius*) which would otherwise have to be lost. SEM has revealed that the proximal shield is not doubled in *Reticulofenestra*, as it is in *Toweius* and *Prinsius*; micrographs of reticulofenestrids which do show separation of the proximal shield (e.g. Perch-Nielsen 1972, Plate 7, Fig. 5) are considered to be of slightly etched specimens of *Toweius*, not *Reticulofenestra*.

Theodoridis (1984) erected two new species of *Reticulofenestra*, using the original diagnosis of

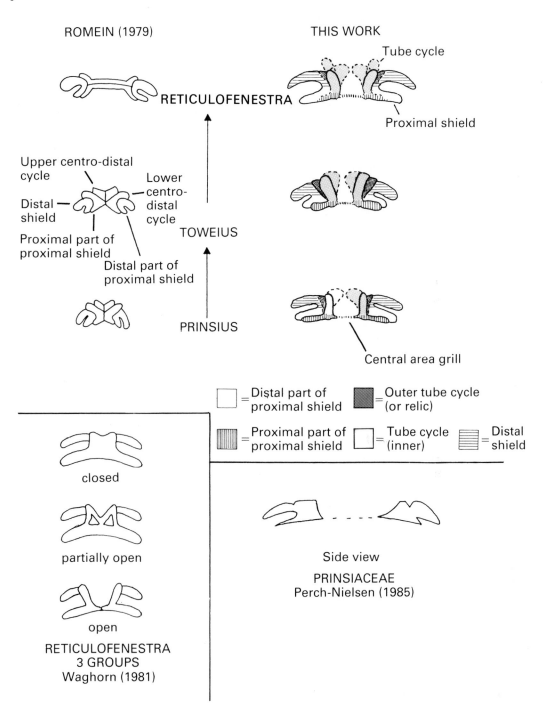

Fig. 3.3 — Schematic diagram illustrating ultrastructural detail of Noelaerhabdaceae forms in cross-section and the development of these in the evolution of the group.

Hay *et al.* (1966) as a basis for identification. He also considered the generic name *Cyclicargolithus* to be a junior synonym of *Reticulofenestra*. The type description of *Cyclicargolithus* clearly states that an open central area must be present, whereas the type description of *Reticulofenestra* (and emendations) states that a central area grill must be present. The lack of a central area structure of any kind is clearly a characteristic worthy of generic status, as opposed to refinements of a central area structure (e.g. *Cribrocentrum*) which are not. Both *R. rotaria* and *R. ampliumbilicus* must, therefore be assigned to *Cyclicargolithus*, together with *R. floridana*.

3.4 SYSTEMATIC CLASSIFICATION

Family: Noelaerhabdaceae Jerkovič 1970
Genus: *Reticulofenestra* Hay, Mohler & Wade 1966 emend.
Type species: *Reticulofenestra umbilicus* (Levin 1965) Martini & Ritzkowski 1968. (Synonymy: *R. caucasica* Hay, Mohler & Wade 1966.)

Emended diagnosis: elliptical to circular placoliths in which the central area contains a grill of variable geometry and a central area tube cycle. Both proximal and distal shields are bright under crossed-polarised light, with strongly curved extinction lines, and the distal shield elements are imbricated.
Description: placoliths, circular to elliptical in outline and concavo-convex in side view, consisting of two appressed shields (structurally a single unit including the tube cycle) and a central opening spanned by a reticulate grill of variable geometry. The proximal shield contains a number of radial to slightly imbricated segments which are extended in a variety of combinations into the central opening to form the central grill and are projected distally to form the tube cycle which lines the margin of the central area and abuts the elements of the distal shield. The distal shield consists of wedge-shaped, imbricated laths separated by oblique sutures. These sutures are sharply kinked

adjacent to the tube cycle. Projection of the tube cycle can give the central area a closed appearance and differential crystal arrangement or preservational effects may give the central grill a 'plugged' appearance (see Fig. 3.3).

A full synonymy for both the genus and key species is listed in Gallagher (in prep.). The following species are here placed in the genus *Reticulofenestra* Hay, Mohler and Wade, 1966 emend.:

Species: *Reticulofenestra martinii* (Hay & Towe, 1962) nov. comb.
Basionym: 1962 *Cyathosphaera martinii* Hay & Towe, p. 510, Plate 4, Figs. ?2,3,4.
Species: *Reticulofenestra productella* (Bukry, 1975) nov. comb.
Basionym: 1975 *Crenalithus productellus* Bukry, p. 690.

See compendium for full discussion of these species and their synonymy.

3.5 EVOLUTION
(a) Introduction

A number of studies have considered aspects of the evolution of reticulofenestrids, either with respect to one species (or more strictly 'morphotype') evolving into another (Backman 1980, Backman and Hermelin 1986) or aspects of the development of the family Noelaerhabdaceae (Romein 1979). However, no attempt has been made to trace the evolutionary links of all *Reticulofenestra* species, to group them on the basis of their ultrastructure, and to relate the structures within the family Noelaerhabdaceae in a taxonomical sense. The problems of classification within this group are notorious and a matter of dispute. It is probably due to this complexity that no previous attempt has been made to establish a comprehensive evolutionary scheme.

Electron micrographs and light micrographs were used, in association with published information, to classify all species considered to be valid and within the emended definition of *Reticulofenestra* and to relate these in an evolutionary framework (Fig. 3.5). Romein

(1979, p. 72) traced the evolution of the family Noelaerhabdaceae from its inception in the earliest Palaeocene from *Prinsius petalosus* (a coccolith of extremely simple construction), through the development of the genus *Toweius* (in which two tube cycles become well established), to the earliest form of *Reticulofenestra* (*R. dictyoda*). Since Romein's (1979) study, however, there have been emendations to the ranges of a number of *Toweius* species (e.g. *T. eminens, T. tovae*) and the recognition of some new forms; thus the early part of the lineage requires modification.

(b) *Prinsius* development

The transition of *P. petalosus* to *P. dimorphosus* was well documented by Romein (1979, p. 71), the gradual reduction in the 'crown' of elements giving rise to a simply constructed coccolith, elliptical in outline with one or two rows of elements in the tube cycle. The reduction of the crown of elements in *P. petalosus* may also have given rise to *P. africanus* and *P. tenuiculus* in the Early Danian with the development of a single row of elements in the tube cycle and a circular outline. The lineage from *P. dimorphosus* through *P. martinii* to *P. bisulcus* can be clearly traced in the Early Palaeocene via more regular arrangement and greater number of shield and tube cycle elements. It is clear from the ultrastructure that the Palaeocene development of the family Noelaerhabdaceae does not involve any relatively 'complex' (in terms of structure) lower Danian *Biscutum* forms such as *B. romeinii*, but may include 'simpler' *Biscutum* species with a similar rim construction to *P. petalosus* and *P. africanus*, or perhaps *Prediscosphaera*-like coccoliths.

(c) *Toweius* development

P. bisulcus gave rise to *T. pertusus* in the late Early Palaeocene which then, by rapid development of the inner tube cycle, gave rise to *T. eminens, T. tovae* and *T. selandianus*. Romein (1979, p. 73) considered *T. pertusus* to be the ancestor of all subsequent species, but the number and diversity of reticulofenestrid species

which develop in the first few million years of their existence may indicate polyphyletic origins (see Figs. 3.4 and 3.5).

(d) Early development of *Reticulofenestra*

T. crassus (= *T. callosus*) is assumed to have developed from *T. pertusus* in the Late Palaeocene by contraction of the inner tube cycle elements and subsequent loss of the distally projected grill. *T. crassus* and *T. pertusus* range into the Early Eocene where they gave rise to *R. dictyoda* and *R. martinii* respectively. The development from *Toweius* to *Reticulofenestra* involved a realignment of the shield crystals' orientation, as seen in the change from a dark to a bright image under cross-polarised light, but the processes involved in this change are poorly understood. Perch-Nielsen (1985a, p. 505) commented on the similarity in appearance of *T. crassus* and *R. dictyoda* in the light microscope. *R. martinii* is seldom referred to in the literature, but its strongly constructed central area grill closely resembles that of *T. pertusus* in proximal view.

The key developments in the transition from *Toweius* to *Reticulofenestra* are the loss of the outer tube cycle (fused with rim elements of the distal shield), an increase in imbrication of the rim elements and an increase in the birefringence of the rim in cross-polarised light.

(e) Middle Eocene development of *Reticulofenestra*

The Middle Eocene was a time of rapid diversification for the reticulofenestrids with approximately ten new species arising within ten million years. *R. dictyoda* gave rise to *R. hillae* by an increase in overall size and central area opening, and to *R. scrippsae* by distal contraction of the tube cycle elements over the central area opening. The first occurrence datum of *R. scrippsae* is uncertain because of taxonomic confusion (see discussion of No. 40 in compendium).

In parallel with these developments, *R. martinii* gave rise to *R. minuta* (there is some dispute as to whether Eocene forms should be called *R. insignita;* see Pujos (1987, p.250)) by a reduction in size and ellipticity. *R. coenura* also developed at this time from *R. martinii* by expansion of the central area opening and reticulation of the central area grill (visible in crossed-polarised light). Subsequent diversification from *R. coenura* included *R. foveolata* by a reduction in overall size and slight thickening of the central area grill at its centre, *R. callida* by strengthening of the tube cycle and 'plugging' of the centre of the central area grill, and *R. reticulata* by evolution of a circular outline and a distinctive reticulate grill structure.

Middle Eocene diversification continued with the separation of *R. umbilicus* from *R. hillae* at an overall placolith size of 14 μ m (Backman and Hermelin 1986) which is recognisable as forming a distinct population at 44.5 Ma *R. scissura* developed from *R. scrippsae* at this time also, by an increase in overall placolith size although no limits have yet been set. *R. daviesii*, a form very similar to *R. callida*, is seen to arise in the Middle Eocene by the addition of pores in the central area 'plug'.

(f) Late Eocene to Pliocene development of *Reticulofenestra*

Ancestors for Late Eocene forms are less easy to recognise. *R. oamaruensis*, an enigmatic species, may have evolved from large forms of *R. coenura* (or *R. onusta*) — its range is very limited and published references are scarce. *R. laevis* is well documented by Roth (1970, p. 850) from the Middle Oligocene of Alabama and appears to have a natural ancestor in *R. reticulata*. However, since their published ranges do not overlap this link is questionable. The range and geographic distribution of *R. gartneri* is poorly understood owing to taxonomic confusion involving *R. hesslandii* (see Pujos 1987) and a scarcity of published information, but it probably descended from *R. hillae* at the end of the Eocene and is a possible ancestor for *R. pseudoumbilicus* in the Middle Miocene.

The progression from *R. perplexa* to *R. productella* has been discussed previously (Backman 1980, Pujos 1987) and, depending on

latitude, the former gives rise to the latter at some time within the Late Miocene. The most likely ancestor of *R. perplexa* is *R. scrippsae* as it possesses the distinctive 'closed' central area. Backman (1980, pp. 44–45) defined the limits of size by which *R. minuta* (< 3 μm) and *R. minutula* (> 3 μ m) should be separated. The change appears to occur at the base of the Miocene although confusion with taxonomy may move this into the Middle Miocene.

The evolutionary lineage described here divides the Noelaerhabdaceae genera *Prinsius, Toweius* and *Reticulofenestra* into five easily differentiated groups (which do not represent taxonomic divisions) see Figs. 3.4 and 3.5.

(1) Open: the tube cycle produces a 'collar' around the margin of the central area, but does not contract towards the centre. The reticulate grill structures are finely constructed in the *Reticulofenestra* forms.

(2) Closed: reticulofenestrids in which the tube cycle is contracted over the central area so as to cover the central area grill. Members of this group have previously been assigned to the genus *Dictyococcites*.

(3) Dark: this group is restricted to small *Prinsius* and *Toweius* species in which the rim elements remain dark in cross-polarised light.

(4) Small: confined to forms of *Reticulofenestra* with an overall placolith size of approximately 3 μ m.

(5) Coarse: representatives from this group consistently possess a coarse central area grill. Members of this group have previously been assigned to the genus *Cribrocentrum*.

3.6 COMPENDIUM OF *RETICULO-FENESTRA* SPECIES

The following set of tables is an attempt to bring together all the commonly referenced species of *Reticulofenestra* in order that cross-references of similar forms may be made and is an advance on the chart produced by Perch-Nielsen (1985b, p. 86) in which simple but inflexible and ambiguous

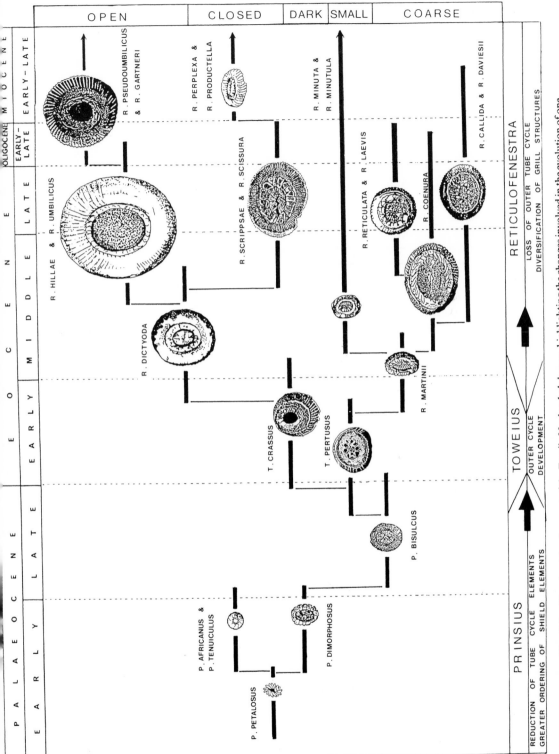

Fig. 3.4 — Structural development of the Family Noelaerhabdaceae, highlighting the changes involved in the evolution of one form to another and the four morphological groups which constitute the genus *Reticulofenestra*.

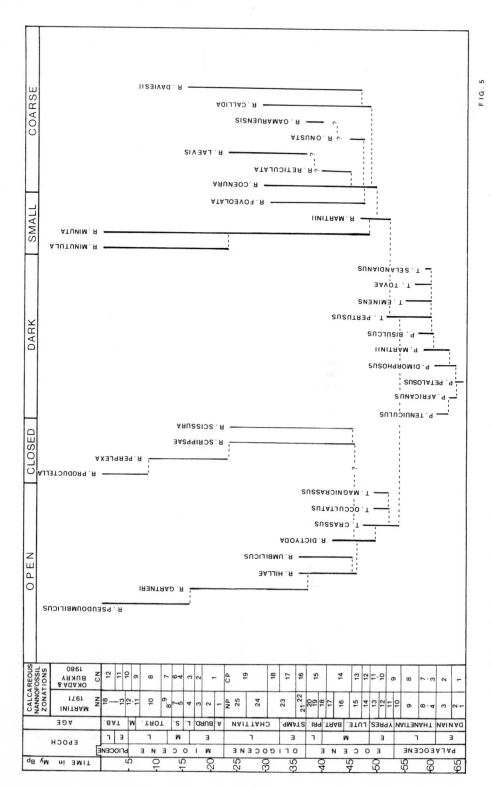

FIG. 5

Fig. 3.5 — Evolutionary lineage showing ancestor-descendant relationships for selected members of the family Noelaerhabdaceae.

structures were listed.

The information presented for each species is taken from its type description, literature sources, and the author's personal observations, then edited and the terminology standardised, in the following format.

Species name
Author(s)
Type locality

Rim structure
Proximal and distal shield characteristics.

Age range
Shape and size
Remarks

Central area structure
Tube cycle and central area grill characteristics.

(1) *R. alabamensis*
Roth 1970
Base of Red Bluff Fmn, Lone Star Cement Qy, Alabama, USA

Distal shield has 35–40 tabular elements which imbricate strongly dextrally, sutures radial at periphery, but bend counterclockwise at the centre. Proximal shield is slightly smaller with a similar arrangement and number of elements.

Early Oligocene (NP21–NP23).
Small, elliptical outline.
Holotype length = 3.2 μm,
average length = 3.5–4.6μm

Slits between bars of central grill along the periphery. The bars join in the centre to form a raised area with irregular perforations. Central area = 50% of total coccolith area. The tube cycle consists of sinistrally imbricate tabular elements.

Remarks: Roth (1970) differentiated this species from *R. foveolata* on the basis that it had a smaller central area grill, an outer cycle of narrow slits and some small pores in the elevated central portion. However, illustrations in Haq (1971) indicate that the two species are very similar in terms of size and central area characteristics. Waghorn (1981) placed *R. clatrata* in synonymy with *R. alabamensis*, even though the former may be up to twice the size of the latter. As the holotype of *R. clatrata* is poorly preserved it is difficult to tell whether this is correct or not.

---

(2). *R. ampliumbilicus*
Theodoridis 1984
Sicily, Italy.

Birefringent proximal and distal shields. No data were presented for the number and arrangement of shield elements, and in the absence of electron micrographs it is not possible to comment on these.

Middle Miocene *H. orientalis* subzone (NN6). Large, subcircular, often circular outline.
Holotype length = 9.0 μm.

Large central opening, apparently lacking a central area grill. No details of the tube cycle were given and cannot be interpreted from the holotype micrographs.

Remarks: this species lacks a central area grill of any kind and is, therefore, incorrectly described as a species of *Reticulofenestra*. It is more likely to belong to the genus *Cyclicargolithus*, but no formal recombination can be attempted until electron micrographs are available.

---

(3) *R. bisecta*
(Hay, Mohler & Wade 1966)
Roth 1970
Nal'chik, Nal 11.

Broad distal shield of 66 elements, which are wedgeshaped and dextrally imbricate in distal view. Sutures inclined slightly clockwise. Proximal shield is smaller with the same number of elements (radial).

Middle Eocene to Oligocene (NP15–NP25).
Elliptical. Holotype length = 8.2 μ m, holotype width = 7.1μ m. Average length = 8.0–11.0μ m, average width = 7.0–9.5μ m.

Plates normally cover distal side of central area, but a central grill is proved by electron micrographs. Every third shield element contributes to the construction of the central grill. The tube cycle has 'multiple' cycles of imbricate tabular elements.

Remarks: this form was first described in the same article as *R. scissura* (Hay, Mohler and Wade 1966) under the generic name *Syracosphaera*. It is now known that *R. scissura* and *S. bisecta* are the same species originally illustrated in different views. *R. scissura* was first chosen over *S. bisecta* and all forms subsequently described as *R. bisecta* must be re-named *R. scissura*.

---

(4) *R. callida*
(Perch-Nielsen 1971)
Bybell 1975
Orby, Denmark.

Distal shield is composed of 50–70 elements; each element forms a point at the margin. The sutures of the distal shield are sharply bent clockwise adjacent to the tube cycle. The proximal shield is slightly smaller than the distal shield, but has the same number of elements.

Early Eocene to Middle Oligocene (NP13–NP23).
Elliptical (rounded).
Holotype length = 7.0 μ m.

Diagnostic knob in centre of central area with slender radiating rods connecting it to the collar. Each tube cycle element merges into a central area grill element. (only every second element according to Perch-Nielsen (1971b)).

Remarks: a similar species to *R. daviesii*, but readily distinguished in the electron microscope (less easily in the light microscope) by a complete lack of pores in the central area knob.

---

(5) *R. clatrata*
Müller 1970
Bohrung Doberg 10,0m, Germany.

Both proximal and distal shields have 30–40 elements. The poor quality of the holotype micrograph prevents comment being made on the imbrication of the elements and/or the inclination of the sutures.

Oligocene (NP23–NP24).
Elliptical.
Length = 8.0–10.0 μ m.

Large elliptical central area, wide central area grill which has 18–25 rods meeting in the middle to form an irregular grill. The rods emanate from a small inner ring of 36 elements (= tube cycle).

Remarks: may prove to be a junior synonym of *R. alabamensis* (see (1)) and, therefore, a junior synonym of *R. foveolata*.

---

(6) *R. coenura*
(Reinhardt 1966) Roth 1970
Dolgen 1/63, Germany.

50–60 elements in the distal shield. Proximal shield is smaller by the ratio 10:8, but has the same number of elements. The elements slope down to the periphery which has a serrate edge.

Early Eocene to Late Oligocene (NP13–NP25).
Elliptical.
Diameter = 5.0–10.0 μ m.

Central area is 40–50% of the total coccolith area. There are numerous small perforations in the central area grill which also has a groove along its axis.

Remarks: *R. coenura* is a species distinguished by its relatively large central area grill which is coarsely reticulate. Perch-Nielsen (1971b) provided excellent electron micrographs of this species and other species which it most resembles (e.g. *R. foveolata*), enabling the relevant distinctions to be clearly made.

---

(7) *R. danica*
(Black 1967) Roth 1970
Grey Marls, Grundfor,
nr Aarhus, Denmark.

40–60 slender and smoothly curved elements in each shield. Sutures only faintly visible.

Middle Eocene to Oligocene (NP15–NP25).
Broadly elliptical. Holotype distal shield = 7.2 μm × 6.0 μm. Holotype proximal shield = 6.0 μm × 5.0 μm. Central opening = 2.8 μm × 1.8 μm.

Central opening surrounded by a cycle of imbricate, wedgeshaped elements (tube cycle) only visible on the distal side. Central area grill bars are straight and join along the major axis of the ellipse.

Remarks: described as having a larger central area and straighter non-anastomosing central area grill bars than *R. scissura*. However, only the proximal view is represented in the type description and it is considered here that *R. danica* should be placed in synonymy with *R. scissura*.

---

(8) *R. daviesii*
(Haq 1968) Haq 1971
NW Germany

Numerous laminar elements in both proximal and distal shields which are slightly imbricate. Sutures inclined sharply clockwise adjacent to the tube cycle.

Eocene to Miocene (NP14–NN4?).
Elliptical.
Holotype length = 7.6 μm,
Holotype width = 6.4 μm.

10–15 pores merge to form a reticulate structure with numerous smaller pores. The central area is 40–50% of the total coccolith area. Septa form from every second or third shield element. The tube cycle consists of a number of inclined elements which may stand above the level of the distal shield. There is a perforate 'knob' in the centre of the central area grill.

Remarks: very similar to *R. callida* in its general ultrastructure, but differentiated by the presence of many small pores in the central area knob.

---

(9) *R. dictyoda*
(Deflandre & Fert 1954)
Stradner, in Stradner &
Edwards 1968
Donzacq, France.

Wide shields with 40–60 wedgeshaped elements. The distal shield is 25–30% larger than the proximal, both shields have the same number of elements.

Early to Middle Eocene (NP12–NP16).

Central area relatively small (25% of total coccolith area) contains a grill which has long holes at the margin and irregular ones in the

Elliptical outline.
Length = 4.0–10.0 μ m,
width = 3.9–9.0 μ m,
central grill = 1.2–4.0 μ.

centre. Rods are derived from every second proximal shield element. The tube cycle is well developed with many inclined elements.

Remarks: Stradner and Edwards (1968, pp. 19–20) described this species in detail, placing it in the genus *Reticulofenestra*. However, their illustrations are of species other than *R. dictyoda;* hence see Haq (1971) for illustrations.

---

(10) *R. falcata*
(Gartner & Smith 1967)
Roth 1970
Yazoo Fmn, Louisiana, USA.

The larger distal shield has approximately 50 elements which are terminated in a rounded point. Viewed proximally the sutures of the distal shield are slightly curved, inclined clockwise and imbricate sinistrally.

Middle Eocene–Oligocene
(NP15–NP25). Slightly
elliptical. Maximum diameter
= 7.5 μ m.

The central opening is elliptical and contains the remnants of a grill. The tube cycle is an elliptical cylinder having knob-like protrusions where it joins the proximal shield.

Remarks: the electron micrographs shown by Gartner and Smith (1967, Plate 1, Figs. 5 and 6) are inconclusive, but strongly resemble *R. scissura;* thus *R. falcata* is superfluous and is considered a junior synonym of *R. scissura*.

---

(11) *R. floridana*
(Roth & Hay in Hay *et al.*
1967) Theodoridis 1984
JOIDES 5, Blake Plateau,
554ft 10in below top.

Placolith with 40–41 dextrally imbricate elements separated by sutures having a slight clockwise inclination on the distal surface, but straight on the proximal surface. Both proximal and distal shield have the same number of elements.

Middle Eocene to early Middle
Miocene (NP14–NN6).
Circular to broadly elliptical.
Holotype diameter = 3.6 μ m.
Paratype diameter = 4.1–
5.0 μ m.

The tube cycle consists of a series of stout, uneven blocks. The central opening is approximately 33% of the total coccolith area and lacks a grill of any kind.

Remarks: Theodoridis (1984, p. 85) assigned this form to the genus *Reticulofenestra* rather than *Cyclicargolithus*. However, as no central area grill of any style or geometry has ever been found it is considered to belong to the genus *Cyclicargolithus*.

---

(12) *R. foveolata*
(Reinhardt 1966) Roth 1970
Bohrung Bad Doberan, 5/61.

Distal shield of 34–36 elements; the proximal shield is 10% smaller, but has the same number of elements. Perch-Nielsen (1971b) quotes 50–70 elements in the shields.

Middle Eocene to Late

Central area is 60% of the total diameter. The septa enclose long

Oligocene (NP14–NP23).
Small, elliptical.
Length = 2.5–6.0 μm,
width = 2.0–2.3 μm.

pores. An outer ring of slit-like perforations surrounds an area pierced by 18–26 small, circular holes. Tube cycle of flat rectangular crystals.

Remarks: Roth (1970, p. 848) differentiated this species from *R. insignita* by its narrower rim and fewer elements and by its more elongate pores in a relatively larger central area grill. Comparison of the type figures of *R. foveolata, R. insignita* and *R. alabamensis* suggests strong similarity between the characteristics listed above (differences attributable to preservational effects?) and it is suggested that the two latter species be placed, therefore, in synonymy with the former as it was the first to be described.

(13) *R. gabrielae*
Roth 1970
13ft above base of Red Bluff
Fmn, Lone Star Cement Qy,
Alabama, USA

18–22 wedgeshaped, dextrally imbricate elements in the distal shield. Elements are separated by radial suture lines. Detail of the proximal shield is not known.

Oligocene (NP21–NP25).
Very small.
Length = 1.8 μm.

Relatively large central area grill with approximately 50 pores arranged regularly more or less parallel to the ellipse. No wedges present around central area grill (= no tube cycle present).

Remarks: the fact that the tube cycle is absent is clearly a preservational trait and there is no valid reason to erect a new genus. *R. gabrielae* is similar to *R. foveolata,* but has fewer elements, more circular perforations, and is much smaller.

(14) *R. gartneri*
Roth & Hay in Hay *et al.*
1967 JOIDES 5, Blake Plateau,
554ft 10in below top.

Distal shield has 150 fine, thin elements which are inclined counter-clockwise in distal view. The proximal shield is slightly smaller than the distal shield with the same number of elements.

Early Oligocene to Early
Miocene (NP21–NN4).
Broadly elliptical, smooth
outline. Length = 5.8 μm.

Central grill perforated by approximately 120 small holes which are circular to oblong in shape. There are no marginal slits. The tube cycle consists of a ring of imbricate elements.

Remarks: *R. gartneri* lacks the slits around the margin of the central area as seen in *R. caucasica* and *R. insignita* according to Roth and Hay in Hay *et al.* (1967) and has a wider outline and more pores in the relatively smaller central area. The electron micrograph figured as the holotype resembles *R. coenura,* but has a more finely perforate central area grill. Pujos (1987) noted that this form is latitudinally restricted and, presumably for this reason, rarely quoted in the literature. It is considered to be a possible ancestor of *R. pseudoumbilicus* and may be synonymous with *R. tenuistriata* (Varol pers. commun.).

(15) *R. gelida*
(Geitzenauer 1972)
Backman 1978
Pacific Subantarctic deep sea
core, Eltanin 25–11, 520–640 cm.

Numerous (68–74) slightly imbricated elements in the distal shield. Between crossed-nicols the entire shield is strongly birefringent. Extinction cross clearly visible to the edge of the shields.

Middle Miocene–Pliocene.
(NN4–NN18).
Oval shaped.
Size = 6.0–12.0μ m.

Central area usually filled with a wall of irregularly shaped calcite crystals (= tube cycle). Central opening circular to slit like. No bridge in central area.

Remarks: clearly differentiated from *C. pelagicus* as outlined by Geitzenauer (1972), but not so definitely from *R. pseudoumbilicus* as discussed by Backman (1980). Only in high latitudes and in placoliths exceeding 140 μ m^2 in area can a difference be detected, and only then in the size of the central opening. Pujos (1987) considered *R. gelida* to be a subspecies of *R. pseudoumbilicus*. However, this is unnecessary splitting and it is recommended herein that *R. gelida* be considered a junior synonym of *R. pseudoumbilicus*.

---

(16) *R. hampdenensis*
Edwards 1973
Hampden Fmn, above basal
Greensand, Hampden Beach,
New Zealand.

28–55 strongly sinistrally imbricate elements in both shields. Proximal shield has parallel sutures and a serrate margin, whereas the distal shield is made up of narrow wedgeshaped elements with counterclockwise inclined sutures.

Early Eocene–Oligocene
(NP13–NP24).
Oval to subcircular.
Holotype length = 3.7 μ m.

'Marginally spoked and centrally hubbed' central area. Fenestrate part of the grill results from 50–90% of the proximal shield elements attenuating into the central opening. The tube cycle is a wall of distally expanding laths with parallel sutures.

Remarks: Edwards (1973) examined thousands of light and scanning electron micrographs of forms which he assigned to *R. hampdenensis*, being unable to find natural criteria to split them. He demonstrated how *R. hampdenensis* could be distinguished from *R. dictyoda* and *R. placomorpha*, but the holotype is thought to be a specimen of *R. coenura*, making *R. hampdenensis* a junior synonym of *R. coenura*. Detailed examination of topotype material revealed an assemblage containing *R. coenura*, *R. foveolata*, *R. callida* and *R. daviesii* where Edwards (1973) had only listed *R. hampdenensis*.

---

(17) *R. haqii*
Backman 1978
Vera Basin, sample 11.

Distal and proximal shields equal in size and built up of 40–50 elements each. The narrow crystal elements are not discernible in the light microscope.

Early Miocene to Late
Pliocene (NN1–NN18).
Small, elliptical.
Length = 3.0–5.0 μ m,
central opening = 1.0–1.5 μ m.

Tube cycle may distally consist of coarse crystallites, but is thin in many specimens. Small central opening, in which a grill may be seen on the proximal side.

Remarks: resembles *R. pseudoumbilicus* in possessing a central opening surrounded by a tube cycle and in having rather similar crossed-nicols images, but it is distinguished by its consistently smaller size. Backman (1980) separated *R. haqii* from *R. minuta* and *R. minutula* on the basis of very small differences in the overall diameter and central opening diameter. Pujos (1987) recognised *R. haqii* and *R. minutula* as ecotypes of the same species. As their spatial and temporal distributions are closely similar there is no reason to separate *R. haqii* from *R. minutula*; thus the former is considered a junior synonym of the latter.

---

(18) *R. hesslandii*
(Haq 1966) Roth 1970
Tabqa, NW Syria.

39 elements in proximal shield, 40–45 elements in the distal shield. Roth (1970) quoted 60 elements in the proximal shield. Sutures of distal shield are inclined sharply clockwise adjacent to the tube cycle.

Middle Eocene to Oligocene (NP16–NP25).
Very variable in size, shape and number of shield elements.
Usually circular to subcircular.
Diameter = 2.8–9.1 μm,
central opening = 0.5 μm.

Coarse central grill. Central opening is approximately 20% of the total size of the coccolith. The grill is composed of irregular, twisting bars. The tube cycle consists of 6–8 large, irregular elements.

Remarks: distinguished from *C. floridanus* by the presence of a coarse grill in the central area. It is obvious that the taxonomy of this species is in considerable doubt as Backman (1987) considered it to be synonymous with *R. scrippsae*, whilst Pujos (1987) separated them on the basis of overall size; *R. scrippsae* 7.6–10.0 μm, *R. hesslandii* 2.8–6.3 μm, and suggested that *R. hesslandii* evolved from *R. scrippsae*.

---

(19) *R. hillae*
Bukry & Percival 1971
Shubuta member,
Yazoo Clay, Chickasaway R.,
Mississippi, USA.

Wide shields very similar to those of *R. umbilicus*. The distal shield elements are slightly imbricated, whilst those in the slightly smaller proximal shield are more or less radial (same number in each shield).

Middle Eocene to Early Oligocene (NP16–NP22).
Large elliptical outline, wide elliptical tube cycle and small central area.

Thick, wide, elliptical tube cycle. Small central opening, approximately 33% of the total length of the coccolith. Fine central area grill visible in the electron microscope.

Remarks: *R. hillae* is differentiated from *Cyclicargolithus abisectus* by its elliptical shape and larger central opening. The distinction from *R. umbilicus*, however, based on a wider tube cycle and correspondingly smaller central opening is less certain. Backman and Hermelin (1986) conducted a thorough morphometric study on these two species and found a complete gradation in terms of overall size and central opening size. They concluded that *R. hillae* may be an ecophenotype of *R. umbilicus*, but for biostratigraphic purposes a diameter of < 14 μm should be used to recognise *R. hillae* and > 14 μm for *R. umbilicus*.

---

(20) *R. inclinata* Roth 1970 JOIDES 5, Blake Plateau, 484ft below top.	Chevron-shaped elements in distal shield; 34–40 tabular and dextrally imbricate. Suture lines inclined counterclockwise in outer half of shield, then turn sharply in inner half of shield. Proximal shield is slightly smaller than the distal shield, but has the same number of elements.
Oligocene (NP21–NP25). Narrow elliptical outline. Length = 3.8 μm.	Central grill has nearly straight bars with long intermediate slits. Central area occupies approximately 66% of the total length of the coccolith. No apparent tube cycle.

Remarks: the identity of this species is in some doubt because of the poor preservation of the holotype (tube cycle absent). The lack of published references since its original description is testimony to its somewhat ambiguous position.

---

(21) *R. insignita* Roth & Hay in Hay *et al*, 1967 JOIDES 5, Blake Plateau, 410ft below top.	Large number of fine elements in the distal shield. Proximal shield is constructed of approximately 50 elements separated by sutures with slight counterclockwise inclination in proximal view.
Oligocene (NP21–NP25). Narrowly elliptical. Length = 3.7 μm.	Slits around the margin of the central area. 36 round perforations in the central grill which takes up approximately 50% of the total area of the coccolith. The tube cycle consists of narrow rectangular elements.

Remarks: *R. insignita* is distinguished from *R. gartneri* by having a more narrowly elliptical outline, relatively larger central area, and by having a ring of slits around the margin of the central area. It is difficult to imagine how these two species can be confused when Pujos (1987) compares *R. insignita* to *R. minuta*, and *R. gartneri* to *R. pseudoumbilicus* — a potential size difference of 10.0 μm. *R. insignita* closely resembles *R. foveolata* and may prove to be a junior synonym of it. Pujos (1987) also discussed the possibility that *R. insignita* may have been ancestral to *R. minuta*.

---

(22) *R. laevis* Roth & Hay in Hay *et al*. 1967 JOIDES 5, Blake Plateau, 337ft 11in below top.	Distal shield subcircular and smooth, composed of approximately 140 poorly defined elements. Proximal shield is 85% of the size of the distal shield and also has a large number of elements. Margin finely serrate and sutures inclined counterclockwise.
Oligocene (NP21–NP25). Subcircular outline. Holotype length = 6.4 μm.	Small round central opening spanned by a grill of approximately 30–40 small round perforations. Tube cycle elements show strong imbrication.

Remarks: the distal shield has an outer cycle of 60–80 elements, not 140 as stated in the type description. It is also described as the most nearly circular species of *Reticulofenestra*, but this does not take account of *R. reticulata* which was described in the same year as *R. laevis* and which is almost perfectly circular. It is possible that the close similarity in morphologies of these two species represents an evolutionary link. However, there is a gap between their published ranges; thus this link remains tentative.

---

(23) *R. lockeri*
Müller 1970
Farve I, 149.0–150.0 m

Distal shield has 50–60 fine, radially arranged elements. The proximal shield has a similar number of elements as the distal shield, but is slightly smaller.

Oligocene (NP23/24). Mildly elliptical outline. Diameter = 8.0–12.0 μm.

Large elliptical central area grill which has a number of more or less regular pores. In the tube cycle there are 20–22 almost rectangular elements.

Remarks: Waghorn (1981, p. 53) considered this form to be a junior synonym of *R. coenura*. The central area of the holotype as illustrated by Müller (1970) is preservationally altered, and thus difficult to classify. Overall size, shape and shield structure appear consistent with placement in *R. coenura*.

---

(24) *R. martinii*
(Hay & Towe 1962)
nov. comb.
Tuilerie de Donzacq,
Landes, France.

Distal and proximal shields with the same number of elements, some of which contribute to the central area grill. Proximal shield slightly smaller than the distal shield with approximately the same number of elements.

Eocene (NP11–NP15).
Shields narrow, elliptical outline.
Holotype length = 2.2 μm.
Length = 1.9–2.9 μm,
width = 1.3–2.4 μm.

Tube cycle has 25 wedgeshaped, dextrally imbricate elements, the sutures of which have clockwise inclination. Each element of the proximal shield contributes to the central area grill.

Remarks: a little-known species characterised by a lack of reticulation in the central area grill, which consists of a relatively small number of coarse rods meeting along the long axis of the coccolith.

---

(25) *R. minuta*
Roth 1970
10ft above base of Red Bluff
Fmn, Lone Star Cement Qy,
Alabama, USA.

16–22 elements in distal shield which are dextrally imbricate and wedgeshaped. The sutures are inclined in a clockwise direction. The proximal shield is almost the same size as the distal and consists of 16–26 wedgeshaped elements with counterclockwise inclined sutures.

Early Eocene to Pliocene
(NP13–NN18).
Very small, elliptical.
Length = 2.3–4.0 μm.

Ring of elements surrounds the central opening (= tube cycle). 10–15 pores in central grill separated by coarse bars. Central opening occupies approximately 33% of the total area of the coccolith.

Remarks: *R. foveolata* has a larger central area and more numerous pores than *R. minuta* which appears to lack a central area grill under crossed-nicols. Backman (1980, p. 44–45) separated *R. minuta* and *R. minutula* on the basis of the former being < 3.0 μm in size and the latter being > 3.0 μm in size. Pujos (1987) refers Palaeogene examples of *R. minuta* to *R. insignita* (see (21)).

---

(26) *R. minutula*
(Gartner 1967) Haq &
Berggren 1978
200 cm depth, core 64–A–9–5E,
Sigsbee knolls, Gulf of Mexico.

Shields have approximately 45 imbricate elements, with subradial sutures. Each element of the proximal shield is curved in the proximal direction.

Early Miocene to Pliocene
(NN1–NN18).
Small, elliptical.
Length = 3.5 μm.

Tube cycle protrudes distally and is continuous with the proximal shield. Central area is open distally, but contains a simple grill proximally.

Remarks: differs from *E. huxleyi* in not having a gap between adjacent elements of the shield. Gartner (1967) stated that its small size separated it from associated species in the light microscope, but this fails to take into account *R. minuta* and *R. haqii*. Backman (1980) and Pujos (1987) have discussed the taxonomy of these three species in detail and with conflicting conclusions. It is probably reasonable to include *R. haqii* as a junior synonym of *R. minutula* and to separate *R. minuta* on the criteria suggested by Backman (1980) (see (25)).

---

(27) *R. oamaruensis*
(Deflandre & Fert 1954)
Stradner in Stradner &
Edwards 1968
Diatomite, William's Bluff,
Oamaru, New Zealand.

Distally convex, 80–100 wedgeshaped elements. Proximal shield very slightly smaller than the distal shield, with a similar number of elements.

Late Eocene to Early
Oligocene (NP18–NP21).
Elliptical outline.
Length = up to 18.0 μm;
width, up to 15.0μm.

A wide reticulate 'membrane' with 9–13 concentric rows of oblique pores covers the central opening. The diameter of the 'membrane' is greater than 50% of the total diameter of the coccolith. The tube cycle is unique in that the narrow, imbricate elements project high above the level of the distal shield.

Remarks: *R. oamaruensis* is differentiated from *R. umbilicus* by the different proportions of the diameters of central area grill and shield, and by the number, shape and size of pores. *R. oamaruensis* has no marginal fenestration, only reticulation in a strict geometrical pattern. Investigation of topotype material has revealed that this species has a distally projected tube cycle unlike that in any other species of *Reticulofenestra*.

---

(28) *R. onusta*
(Perch-Nielsen 1971)
Wise 1983
Orby, Denmark.

Distal shield has 60–70 almost radial elements, as does the smaller proximal shield.

Eocene (NP14–NP16).
Large, elliptical.
Length = 8.0–11.0 μ m.

Relatively wide and relatively flat central area of two layers of net-forming elements. Central area grill slightly raised; a suture runs along the long axis of the grill. A groove separates the central area from the rim elements. The tube is sometimes absent, but usually consists of a low ring of imbricate, rectangular laths.

Remarks: differs from *R. callida* and *R. daviesii* in the structure of the central area. Lacks a 'lath covering in line with the central area' (a tube cycle) according to Wise (1983). However, micrographs shown by Wise (1983) and Perch-Nielsen (1971b) clearly illustrate the presence of a tube cycle. It is the author's experience that the tube cycle is often absent in this species, but it certainly exists. The stratigraphic position and complex structure of this species might be indications of it being an ancestor of *R. oamaruensis*.

---

(29) *R. ornata*
Müller 1970
Bohrung Fussung I, 521.8m,
Bandermergel.

Proximal shield has 60–70 fine elements, almost radially disposed. The characteristics of the distal shield were not described and are not known from other sources.

Oligocene (NP23).
Slightly elliptical. Length =
6.0–7.0 μ m (may be up to
10.0 μ m).

Grill on proximal side of central area and a fine 'net' on the distal side with numerous pores. The 'net' rods are arranged like the hands of a clock. The central area grill covers approximately 50% of the total coccolith area.

Remarks: Waghorn (1981, p. 53) considered this species to be a junior synonym of *R. coenura*; however, it shows more affinity to *R. scissura*. The rods of calcite in the central area probably result from preservational effects.

---

(30) *R. pectinata*
Roth 1970
13ft above base of Red Bluff
Fmn, Lone Star Cement Qy,
Alabama, USA.

Distal shield constructed of 50–60 wedgeshaped, dextrally imbricate elements with clockwise inclined sutures. The margin is serrate. The proximal shield appears to be smaller than the distal, with a similar arrangement of elements.

Middle Eocene–Late
Oligocene (NP14–NP23).
Small.

Tube cycle has 40–50 wedgeshaped sinistrally imbricate elements. Wide, coarse grill with long slits between bars which are continuous with every second element. Central opening covers 60% of the total coccolith area.

Remarks: Roth (1970, p. 851) distinguished this species from *R. alabamensis* on the basis of its having a relatively larger central opening and more numerous elements in the shield. Examination of the micrographs of both holotypes reveals that *R. alabamensis* has a central opening which occupies approximately 60% of the total length of the coccolith, as does *R. pectinata,* and both species have approximately 45 elements in the shield. Hence, these two forms are synonymous and both are junior synonyms of *R. foveolata*.

(31) *R. pelycomorpha*
(Reinhardt 1966) Perch-
Nielsen 1985
Bohrung Dolgen 1/63.

Distal shield has 80–90 slightly imbricate elements. The proximal shield is 25% smaller, with a similar number of radially arranged elements.

Middle Eocene–Early
Oligocene (NP16–22).
Elliptical. Length-to-width
ratio = 10:8.
Length = 11.0–14.0 μ m

Central area is 33–50% of the total length of the coccolith and contains a fine grill. The tube cycle consists of a ring of imbricate elements around the central area grill.

Remarks: this form is considered separate from *R. placomorpha* by Reinhardt (1966) on the basis that the percentage of the total length taken up by the central area is different. It is clear, however, that these two species are synonymous. The combination by Perch-Nielsen (1985a) is invalid (ICBN Art. 33.2).

---

(32) *R. perplexa*
(Burns 1975) Wise 1983
DSDP Site 125-16-6,
depth 25–26 cm.

Medium-sized, finely striate distal shield, prominent in cross-polarised light. Composed of many thin inclined elements. The proximal shield is smaller than the distal and the two are not closely adpressed.

Early to Late Miocene (NN1–
NN10). Medium sized,
elliptical. Length = 5.0–6.0
μ m.

Medium-sized central area, bright between crossed-nicols. No central perforation, but a 'tortuous' line can be seen in the centre of the central area. The central area on the distal surface is composed of many small elements arranged in a counterclockwise direction emanating from the distal part of the tube cycle.

Remarks: distinguished by its prominent centre, lack of central perforation and particularly by the tortuous line in the central area. Backman (in van Heck 1981, p. 40) pointed out that *D. antarcticus* Haq is a junior synonym of this species. This escaped most authors' attention according to Wise (1983) because of scaling errors in Burns' (1975) original description. All of the magnifications are overstated by a factor of approximately 2.5, and hence the holotype of '*D. perplexa*' should measure 5.0–6.0 μ m in length, not 18.0 –20.0 μ m as originally quoted.

---

(33) *R. placomorpha*
(Kamptner 1948) Stradner in
Stradner & Edwards 1968
Inner Alps, Wiener Basin.

Distal shield has 60–100 imbricate, wedgeshaped elements. Proximal shield is smaller, but has the same number of elements.

Middle Eocene to Early
Oligocene (NP16–NP22).
Elliptical. Length, up to 15.0
μ m; width, up to 12.0 μ m.

Large central opening spanned by a grill with elongate fenestration marginally and reticulate fenestration centrally. Imbricate elements of the tube cycle may extend towards the centre and close the central area distally.

Remarks: when first introduced by Kamptner (1948) *Tremalithus placomorphus* was invalid as it had only been conditionally erected. Deflandre (1952) is quoted as having validated the name, but there may be some dispute in this under Art. 32.2 of the ICBN. However, the specific name was validated by Kamptner (1956) as *Coccolithus placomorphus* by reference to previously published descriptions and figures. This species is often put in synonymy with *R. umbilicus,* and it is the author's opinion that detailed morphometric analysis is necessary to establish whether these are the same species or not.

---

(34) *R. productella*
(Bukry 1975) nov. comb.
Pacific Ocean.

The larger distal shield is composed of approximately 28–40 wedge-shaped, slightly imbricate elements. The sutures between elements are slightly inclined. The periphery of the shield is serrate. The proximal shield which also has a serrate edge is concave and contains the same number of elements as the distal shield.

Late Miocene to Recent
(NN10–NN21).
Small, elliptical.
Length = 3.5–4.5 μ m.

Central area closed distally by a covering of irregular calcite laths emanating from the tube cycle (details of which are obscured). A slit divides the central area along its long axis. The extinction figure in the central area has a distinctive swastika shape.

Remarks: this species is distinguished from *R. perplexa* on the basis of overall size. The name *R. producta* has been widely used for this species in the past, but is not valid. In 1963 Kamptner erected the taxon *Ellipsoplacolithus productus* which is invalid because it was conditionally erected (Art. 34.1). Sachs and Skinner (1973) combined this species name into the genus *Coccolithus,* claiming that it would be validated by doing so (Art. 34). However, this is not the case, and it was not until Bukry (1975) erected the name *Crenalithus productellus* with reference to his figure of *Gephyrocapsa producta* in Bukry (1971b) that the form became legitimately named.

---

(35) *R. pseudoumbilicus*
(Gartner 1967) Gartner 1969
250 cm depth, core 64-A-9-5E,
Sigsbee knolls, Gulf of Mexico.

About 70 radial elements in both proximal and distal shields. Very similar light microscope image to that of *R. umbilicus* and *R. hillae.*

Middle Miocene to Pliocene
(NN4–NN18).
Relatively large elliptical
outline. Holotype length =
8.0 μ m.

Central area size varies and contains a reticulate grill. Tube cycle may have large coarse elements. The central opening generally takes up less than 50% of the total coccolith area.

Remarks: resembles *R. umbilicus* and *R. dictyoda* somewhat, from which it has a large temporal separation. More relevant is its similarity to *R. gartneri* from which it may well have evolved. The first occurrence datum of *R. pseudoumbilicus* corresponds to the last occurrence datum of *R. gartneri.* Varol (pers. commun.) used the name *R. tenuistriata* for forms between 5.0–7.0 μ m in length, and *R. pseudoumbilicus* for forms > 7.0 μ m.

---

(36) *R. reticulata*
(Gartner & Smith 1967)
Roth & Thierstein 1972

50–70 imbricate elements in the distal shield with radial sutures which incline and curve slightly clockwise. Sutures of proximal shield are irregular near the centre but become straight near the periphery, and

Yazoo Fmn, Louisiana, USA.

may imbricate slightly dextrally and incline slightly counter-clockwise.

Middle to Late Eocene (NP15–NP18). Circular. Maximum diameter = 6.0–9.0 μ m.

Central opening spanned by a grill and surrounded by a tube cycle of strongly imbricate elements as seen in distal view. Grill is slightly arched. Unique extinction cross in centre under cross-polarised light.

Remarks: Gartner and Smith (1967) differentiated *R. reticulata* from *R. dictyoda* on the basis of its greater number of elements per shield. However, the distinction is much clearer than that, in so far as the circular outline of *R. reticulata* and its prominent central area grill with obvious extinction cross under cross-polarised light are specific characteristics which make it one of the most readily recognisable species of *Reticulofenestra*.

---

(37) *R. rotaria* Theodoridis 1984 DSDP Site 219, Indian Ocean.

Birefringent distal and proximal shields. No electron micrographs of this species have been published, and thus no comment on the number and arrangement of elements in the shields can be made.

Late Miocene, *R. rotaria* Zone, (NN11). Small, circular. Maximum diameter = 5.0–7.0 μ m.

Relatively large, circular central opening. No central area grill is present and no tube cycle has been described. If a tube cycle can be proven it is likely that this species belongs to the genus *Cyclicargolithus*.

Remarks: *R. rotaria* cannot be a species of *Reticulofenestra* as no central area grill has been illustrated or described. It is probably better allocated to *Cyclicargolithus*.

---

(38) *R. samodurovii* (Hay, Mohler & Wade 1966) Roth, 1970 Nal'chik, Nal 11.

Distal shield has 54–55 imbricate elements. Sutures have slight clockwise inclination in distal view. Proximal shield not as wide as distal shield, but has a similar number of elements.

Middle Eocene–Early Oligocene (NP16–NP22). Broadly elliptical. Length = 5.0–12.0 μ m, width = 4.0–11.0 μ m.

Central opening is 33% of total length of the coccolith. Tube cycle has 54–55 tall lath-like elements having sinistral imbrication in distal view. Finely constructed central area grill.

Remarks: differentiated from *R. umbilicus* by being smaller and having a relatively smaller central opening. Considered by many authors to be synonymous with *R. umbilicus*, though in view of Backman and Hermelin's (1986) morphometric study it may better be considered a junior synonym of *R. hillae*.

---

(39) *R. scissura*
Hay, Mohler & Wade 1966
Nal'chik, Nal 11.

40–66 wedgeshaped elements with strong sinistral imbrication in proximal view. Proximal shield is smaller than the distal, but has the same number of elements and straight sutures.

Middle Eocene to Late Oligocene (NP16–NP25). Elliptical outline. Holotype length = 8.0 μ m, holotype width = 7.4 μ m, length, 6.4–10.0 μ m, width = 5.4–9.0 μ m.

Central opening occupies approximately 25% of the total coccolith diameter and is spanned by 8–11 anastomosing laths which produce elongate fenestration. The tube consists of a ring of flat, imbricate crystals which may project over the central opening to obscure the proximally positioned grill.

Remarks: *R. scissura* and *R. bisecta* are considered synonymous despite the assertion of Hay *et al.* (1966) that *R. scissura* is slightly smaller and has a wider central area grill. Many of the specimens figured by Stradner and Edwards (1968) as *R. dictyoda* are herein considered to be *R. scissura*.

---

(40) *R. scrippsae*
(Bukry & Percival 1971)
Roth 1973
Red Bluff Clay, Chickasaway
R., Shubuta, Mississippi, USA.

Similar construction to *R. scissura*. The distal and proximal shields have numerous wedgeshaped elements. The proximal shield is smaller, but has the same number of elements with straight sutures.

Middle Eocene to Late Oligocene (NP16–NP25). Small, elliptical. Diameter = 6.0–12.0 μ m.

Solid central area composed of radial calcite laths (= covering over central area grill), extending from the imbricate, rectangular elements of the tube cycle.

Remarks: in the type description of *R. scrippsae* it is differentiated from *R. scissura* on the basis of its smaller size, more elliptical outline and continuous, sharply bent extinction lines. Backman (1987) considered this species to be a junior synonym of *R. hesslandii*, but a consideration of overall size (see (18)) determines whether these forms are split or not. It appears that *R. hesslandii* is smaller than *R. scrippsae* which is smaller than *R. scissura*, though no formal parameters have yet been established. Such a relationship of size ranges may represent an evolutionary scheme or just taxonomic oversplitting.

---

(41) *R. tokodensis*
Baldi-Beke 1982
Borehole Many 181, 338 m.

80–100 elements in the distal shield. Straight sutures and straight extinction cross. Proximal shield 15–20% smaller than distal shield, with the same number of elements.

Middle Eocene (NP14–NP18). Large, broadly elliptical. Length = 8.0–13.0 μ m.

Large, empty central area under cross-polarised light. Thick tube cycle. Central opening 30–60% of the total coccolith area.

Remarks: similar to *R. ornata* according to Baldi-Beke (1982, p. 300), but has a straight extinction cross, wider central area and a thicker tube cycle. The micrographs of the holotype and isotypes figured by Baldi-Beke (1982, 1984) are of very poor quality and illustrate specimens obviously greatly altered by preservational effects. The true assignation of this form (similar to *R. hillae, R. umbilicus*) is in doubt until better preserved specimens can be adequately figured.

---

(42) *R. umbilicus* (Levin 1965) Martini & Ritzkowski 1968 Sample 27A, 42.6–109.5 WUMC 000023, Yazoo Fmn, Mississippi, USA.	Large proximal and distal shields, very thin sutures visible. The peripheral margin is slightly serrate. Each shield contains many elements, radially arranged.
Middle Eocene to Early Oligocene (NP16–NP22). Large circular to slightly elliptical. Length $> 14.0 \mu$m (Backman and Hermelin 1986).	Robust tube cycle of many imbricate laths surrounds a large central opening (up to 70% of total coccolith area). The central area grill is a very fine meshwork of thin laths. Each proximal shield element contributes to the construction of the central area grill.

Remarks: an extensive synonymy list for this species is given by Waghorn (1981, pp. 71–73). It is usually a very distinctive species, though morphometric studies such as that by Backman and Hermelin (1986) have proved necessary to distinguish it from similar forms.

3.7 SUMMARY LIST OF *RETICULO-FENESTRA* SPECIES

Species are listed alphabetically. Senior synonyms are followed by a list of junior synonyms and possible junior synonyms; whilst junior synonyms are followed by the reference number of their senior synonym. * indicates a species not quoted in the compendium owing to infrequency of use, but with connections to a more well-known species.

(1) *R. alabamensis* Roth 1970 = see no.13.
(2) *R. ampliumbilicus* Theodoridis 1984
(3) *R. bisecta* (Hay *et al.* 1966) Roth 1970 = see no.34.
(4) *R. callida* (Perch-Nielsen 1971) Bybell 1975.
(5) **R. caucasica* Hay *et al.* 1966 = see no.45.
(6) *R. clatrata* Müller 1970 = see no.13.
(7) *R. coenura* (Reinhardt 1966) Roth 1970 = *R. hampdenensis, R. lockeri?*
(8) *R. danica* (Black 1967) Roth 1970 = see no.41.
(9) *R. daviesii* (Haq 1968) Haq 1971.
(10) *R. dictyoda* (Deflandre & Fert 1954) Stradner in Stradner and Edwards 1968.

(11) *R. falcata* (Gartner & Smith 1967) Roth 1970 = see no.41.
(12) *R. floridana* (Roth and Hay in Hay *et al.* 1967) Theodoridis 1984 = *Cyclicargolithus floridanus* (Roth & Hay 1967) Bukry 1971.
(13) *R. foveolata* (Reinhardt 1966) Roth 1970 = *R. alabamensis, R. clatrata?, R. insignita?, R. pectinata.*
(14) *R. gabrielae* Roth 1970.
(15) *R. gartneri* Roth & Hay in Hay *et al.* 1967 = *R. tenuistriata?*
(16) *R. gelida* (Geitzenauer 1971) Backman 1978 = see no. 37.
(17) *R. hampdenensis* Edwards 1973 = see no. 7.
(18) *R. haqii* Backman 1979 = see no. 27.
(19) *R. hesslandii* (Haq 1966) Roth 1970 = see no. 42.
(20) *R. hillae* Bukry & Percival 1971 = *R. samodurovii?, R. tokodensis?*
(21) *R. inclinata* Roth 1970; identity in doubt, but no obvious synonym.
(22) *R. insignita* Roth & Hay in Hay *et al.* 1967 = see no. 13.
(23) *R. laevis* Roth & Hay in Hay *et al.* 1967.
(24) *R. lockeri* Müller 1970 = see no. 7.

(25) *R. martinii* (Hay & Towe 1961) nov. comb.

(26) *R. minuta* Roth 1970.

(27) *R. minutula* (Gartner 1971) Haq & Berggren 1978 = *R. haqii*.

(28) *R. oamaruensis* (Deflandre & Fert 1954) Stradner in Stradner & Edwards 1968.

(29) *R. onusta* (Perch-Nielsen 1971) Wise 1983.

(30) *R. ornata* Müller 1970 = see no. 41.

(31) *R. pectinata* Roth 1970 = see no. 13.

(32) *R. pelycomorpha* (Reinhardt 1966) Perch-Nielsen 1985 (invalid) = see no. 34.

(33) *R. perplexa* (Burns 1975) Wise 1983 = *Dictyococcites antarcticus* Haq 1976.

(34) *R. placomorpha* (Kamptner 1956) Stradner in Stradner & Edwards 1968 = *R. pelycomorpha*.

(35) **R. prebisecta* Aubry 1983 = see no. 42.

(36) *R. productella* (Bukry 1975) nov. comb.

(37) *R. pseudoumbilicus* (Gartner 1967) Gartner 1969 = *R. gelida, R. tenuistriata*.

(38) *R. reticulata* (Gartner & Smith 1967) Roth & Thierstein 1972.

(39) *R. rotaria* Theodoridis 1984

(40) *R. samodurovii* (Hay et al. 1966) Roth 1970 = see no. 20.

(41) *R. scissura* Hay *et al.* 1966 = *R. bisecta, R. danica, R. falcata, R. ornata?*

(42) *R. scrippsae* (Bukry & Percival 1971) Roth 1973 = *R. hesslandii, R. prebisecta.*

(43) **R. tenuistriata* (Kamptner 1963) Martini 1979 = see nos. 15 and 37.

(44) *R. tokodensis* Baldi-Beke 1982 = see nos. 20 and 45.

(45) *R. umbilicus* (Levin 1965) Martini & Ritzkowski 1968 = *R. caucasica, R. samodurovii?, R. tokodensis?*

3.8 ACKNOWLEDGEMENTS

This work was carried out during the tenure of a Natural Environmental Research Council (NERC) research studentship (CASE award with Shell UK Exploration and Production). The support of NERC and the co-operation of Shell are gratefully acknowledged. The author would also like to express thanks to Dr. P. Bown, Dr. A. R. Lord and colleagues in the Postgraduate Unit of Micropalaeontology at University College London for helpful discussion and constructive criticism of the manuscript.

PLATE 3.1
Scanning electron micrographs. All scale bars represent 1.0 μ m.

Plate 3.1, Fig. 1. *R. umbilicus* broken specimen, oblique proximal view showing the continuity of structure between the distal shield, proximal shield and tube cycle. The central area grill septa are seen to extend from the margin of the proximal shield and tube cycle. UCL–2553–31. Whitecliff Bay, Bracklesham Group, Selsey Sand Formation, Fisher Bed XVII, 0.5 m above base. Middle Eocene.

Plate 3.1, Fig. 2. *R. scissura* detail of central area structure, proximal view showing the continuous nature of the rim elements into the central area grill. UCL–2216–07. North Sea Well No. 29/10–1, depth 7122 ft. Late Oligocene.

Plate 3.1, Fig. 3. *R. umbilicus* oblique proximal view illustrating the delicate structure of the central area grill and the continuous nature of elements from the distal shield to the proximal shield. UCL–2423–06. Whitecliff Bay Bracklesham Group, Selsey Sand Formation, Fisher Bed XVII, 0.5 m above base. Middle Eocene.

Plate 3.1, Fig. 4. *R. umbilicus* corroded specimen, oblique distal view showing the direction of imbrication of the tube cycle elements. UCL–2423–19. Whitecliff Bay Bracklesham Group, Selsey Sand Formation, Fisher Bed XVII, 0.5 m above base. Middle Eocene.

Plate 3.1, Fig. 5. *R. dictyoda* distal view of an 'open' reticulofenestrid. UCL–2553–32. Whitecliff Bay Bracklesham Group, Selsey Sand Formation, Fisher Bed XVII, 0.5 m above base. Middle Eocene.

Plate 3.1, Fig. 6. *R. scissura* distal view of a 'closed' reticulofenestrid. UCL–2216–05. North Sea Well No. 29/10–1, depth 7122 ft. Late Oligocene.

Plate 3.1, Fig. 7. *R. minuta* proximal view of two specimens of a 'small' reticulofenestrid. UCL–2553–13. Whitecliff Bay Bracklesham Group, Selsey Sand Formation, Fisher Bed XVII, 0.5 m above base. Middle Eocene.

Plate 3.1, Fig. 8. *R. foveolata* oblique distal view of a 'coarse' reticulofenestrid. UCL–2349–33. S136/898. William's Bluff, Oamaru, New Zealand. Upper Eocene.

Pl. 3.1]   *Reticulofenestra*: A critical view of taxonomy   69

PLATE 3.2
Scanning electron micrographs. All scale bars represent 1.0 μ m.

Plate 3.2, Fig. 1. *Toweius* sp. side view of a highly corroded specimen showing the position of the outer tube cycle (inclined) relative to the proximal shield elements (straight) and the inner tube cycle. UCL–2583–27. AG16, Thanet Formation, Kent. Upper Palaeocene.

Plate 3.2, Fig. 2. *Toweius* sp. same specimen as Fig. 1, but an oblique proximal view. Some of the inner tube cycle elements can be seen behind the inclined outer tube cycle elements and the proximal shield elements are seen to continue into the central area as grill septa. UCL–2583–28.

Plate 3.2, Fig. 3. *Toweius* cf. *T. pertusus* corroded specimen (distal shield completely missing), oblique distal view. The inner tube cycle elements can be seen to be continuous with the proximal shield elements. UCL–2583–05. AG16, Thanet Formation, Kent. Upper Palaeocene.

Plate 3.2, Fig. 4. *Toweius* cf. *T. pertusus* same specimen as Fig. 3. but in side view. A few inclined outer cycle elements can be seen abutting against the inner tube cycle elements and proximal shield. UCL–2583–07.

Plate 3.2, Fig. 5. *Prinsius dimorphosus* coccosphere. Very simple arrangement of shield and tube cycle elements. UCL–2462–35. North Sea Well No. 30/19–2, depth 9096 ft. Upper Palaeocene.

Plate 3.2, Fig. 6. *Reticulofenestra pseudoumbilicus* coccosphere. In contrast to Fig. 5 the elements are highly organised and closely packed. UCL–2578–19. Sample No. 9099, Dtrymou, Cyprus. Middle Miocene.

Plate 3.2, Fig. 7. *Reticulofenestra oamaruensis* distal view. Very well preserved specimen with a coarsely reticulate central area grill. UCL–2391–19. S136/898, William's Bluff, Oamaru, New Zealand. Upper Eocene.

Plate 3.2, Fig. 8. *Reticulofenestra oamaruensis* same specimen as Fig. 7, but in side view. The unusually high tube cycle 'wall' is well illustrated. UCL–2391–21.

Pl. 3.2]   *Reticulofenestra*: A critical view of taxonomy   71

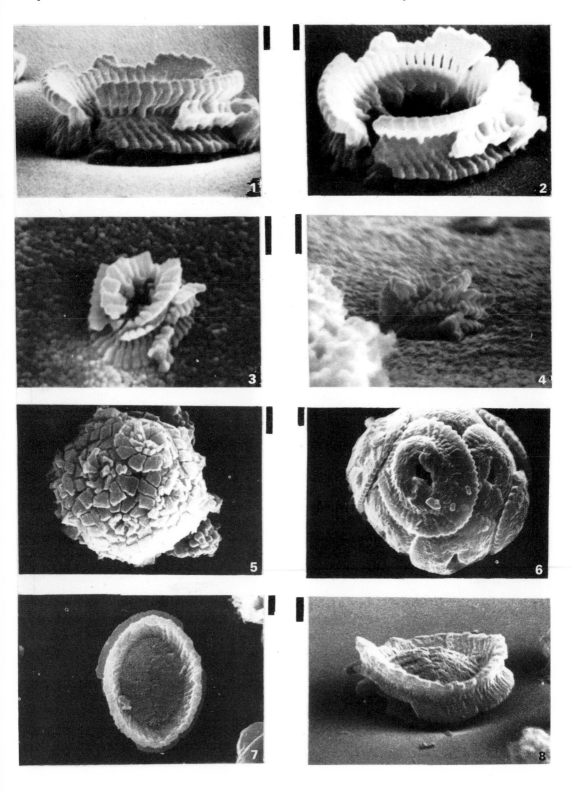

PLATE 3.3

Scale bars represent 1.0 μ m. Figs. 1–4 are scanning electron micrographs; Figs. 5–12 are light micrographs.

Plate 3.3, Fig. 1. *R. scissura* proximal view of a well-preserved specimen. See Plate 3.1, Fig. 2, for central area detail. UCL–2216–06. North Sea Well No. 29/10–1, depth 7122 ft. Upper Oligocene.

Plate 3.3, Fig. 2. *R. umbilicus* proximal view with a well-preserved central area grill. UCL–2423–05. Whitecliff Bay, Bracklesham Group, Selsey Sand Formation, Fisher Bed XVII, 0.5 m above base. Middle Eocene.

Plate 3.3, Fig. 3. *R. callida* distal view, well-defined central area plug. UCL–2601–32. North Sea Well No. 49/9–1, depth 2026 ft. Middle Eocene.

Plate 3.3, Fig. 4. *R. foveolata* distal view, etched tube cycle. UCL–2335–18. North Sea Well No. 21/11–1, depth 3050 ft. Upper Eocene.

Plate 3.3, Figs. 5, 9. *R. umbilicus*. Phase contrast (Fig. 5) and crossed-nicols (Fig. 9) views. UCL–2295–31 (Fig. 5) and UCL–2295–32 (Fig. 9). North Sea Well No. 21/11–1, depth 3410 ft. Upper Eocene.

Plate 3.3, Figs. 6, 10. *R. callida*. Phase contrast (Fig. 6) and crossed-nicols (Fig. 10) views. UCL–2254–04 (Fig. 6) and UCL–2254–03 (Fig. 10). North Sea Well No. 21/11–1, depth 3050 ft. Upper Eocene.

Plate 3.3, Figs. 7, 11. *R. reticulata*. Phase contrast (Fig. 7) and crossed-nicols (Fig. 11) views. UCL–2479–04 (Fig. 7) and UCL–2479–03 (Fig. 11). North Sea Well No. 49/9–1, depth 1288 ft. Upper Eocene.

Plate 3.3, Figs. 8 12. *R. scissura*. Phase contrast (Fig. 8) and crossed-nicols (Fig. 12) views. UCL–2187–04 (Fig. 8) and UCL–2187–03 (Fig. 12). North Sea Well No. 29/10–1, depth 7122 ft. Upper Oligocene.

Pl. 3.3] *Reticulofenestra*: A critical view of taxonomy 73

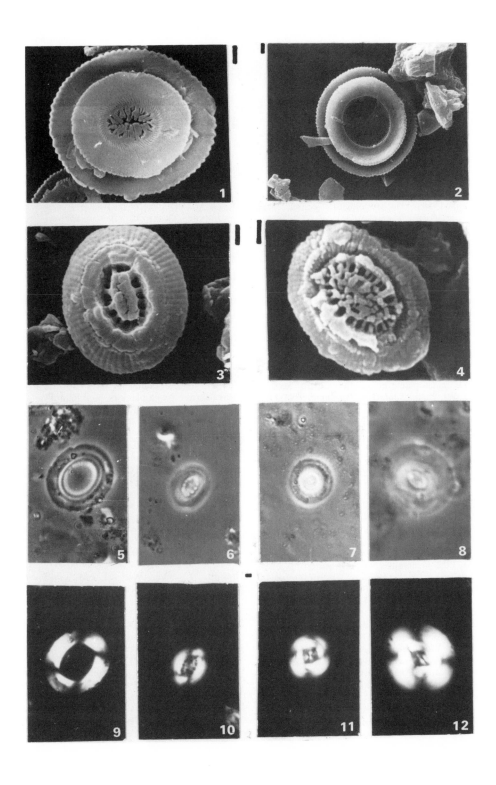

3.9 REFERENCES

Aubry, M.-P. 1983. Biostratigraphie du Paléogène épicontinental de l'Europe du nord-ouest. Etude fondee sur les nannofossiles calcaires. *Doc. Lab. Géol. Lyon.*, **89**, 1–317.

Backman, J. 1978. Late Miocene–Early Pliocene nannofossil biochronology and biogeography in the Vera Basin, SE Spain. *Stockholm Contrib. Geol.*, **32**, 93–114.

Backman, J. 1980. Miocene–Pliocene nannofossils and sedimentation rates in the Hatton–Rockall Basin, NE Atlantic Ocean. *Stockholm Contrib. Geol.*, **36**, 1–91.

Backman, J. 1987. Quantitative calcareous nannofossil biochronology of Middle Eocene through Oligocene sediment from DSDP sites 522 and 523. *Abh. Geol. B.-A.*, **39**, 21–32.

Backman, J. and Hermelin, J. O. R. 1986. Morphometry of the Eocene nannofossil *Reticulofenestra umbilicus* lineage and its biochronological consequences. *Palaeogeogr., Palaeoclimatol., Palaeoecol.*, **57**, 103–116.

Baldi-Beke, M. 1982. A new nannoplankton species from sediments overlying the Eocene coal seams in Transdanubia. *M. All. Foldt. Int. Evi Jel. 1980-Rol.*, 297–308.

Baldi-Beke, M. 1984. The nannoplankton of the Transdanubia Paleogene formations. *Geol. Hungarica Ser. Pal.*, **43**, 153–223.

Black, M. 1967. New names for some coccolith taxa. *Proc. Geol. Soc. Lond.*, **1640**, 139–145.

Bown, P. 1987. The biostratigraphy, evolution and distribution of Early Mesozoic Nannofossils. *Thesis*, University College London (unpublished).

Bramlette, M. N. and Wilcoxon, J. A. 1967. Middle Tertiary calcareous nannoplankton of the Cipero Section, Trinidad, WI. *Tulane Stud. Geol. Paleontol.*, **5**, 93–131.

Bukry, D. 1971a. Cenozoic calcareous nannofossils from the Pacific Ocean. *San Diego Soc. Nat. Hist., Trans.*, **16**, 303–328.

Bukry, D. 1971b. Coccolith stratigraphy. *Init. Rep. DSDP.*, **6**, 965–1004.

Bukry, D. 1975. Coccolith and silicoflagellate stratigraphy, Northwestern Pacific Ocean. *Init. Rep. DSDP*, **32**, 677–692.

Bukry, D. and Percival, S. F. 1971. New Tertiary calcareous nannofossils. *Tulane Stud. Geol. Paleontol.*, **8**, 123–146.

Bybell, L. M. 1975. Middle Eocene calcareous nannofossils at Little Stave Creek, Alabama. *Tulane Stud. Geol. Paleontol.*, **11**, 177–247.

Deflandre, G. 1952. Generalités sur les Protistes; Flagellés; Thecamoebiens. In J. Piveteau (Ed.), *Traité de Paleontologie*, **1**, Masson, Saint-Germain, 89–132.

Deflandre, G. and Fert, C. 1954. Observations sur les Coccolithophoridés actuels et fossiles en microscopie ordinaire et électronique. *Ann. Paléontol.*, **40**, 115–176.

Edwards, A. R. 1973. Key species of New Zealand calcareous nannofossils. *N.Z. J. Geol. Geophys.*, **16**, 68–89.

Ellis, C. H., Lohmann, W. H. and Wray, J. L. 1972. Upper Cenozoic calcareous nannofossils from the Gulf of Mexico (Deep Sea Drilling Project, Leg 1, Site 3). *Quarterly J. Colorado School of Mines*, **67**, 1–103.

Gallagher, L.T. in prep. Calcareous nannofossils and their biostratigraphic application in the Tertiary of the Central and Southern North Sea Basin.

Gartner, S. 1967. Calcareous nannofossil from Neogene of Trinidad, Jamaica and Gulf of Mexico. *Univ. Kansas Paleontol. Contrib.*, **29**, 1–7.

Gartner, S. 1969. Correlation of Neogene planktonic foraminifera and calcareous nannofossil zones. *Trans. Gulf Coast Assoc. Geol. Soc.*, **19**, 585–599.

Gartner, S. 1971. Calcareous nannofossils from the JOIDES Blake Plateau cores, and revision of the Paleogene Nannofossil Zonation. *Tulane Stud. Geol. Paleontol.*, **8**, 101–121.

Gartner, S. and Smith, L. A. 1967. Coccoliths and related calcareous nannofossils from the Yazoo Formation (Jackson, Late Eocene) of Louisiana. *Univ. Kansas Paleontol. Contrib.*, **20**, 1–7.

Geitzenauer, K. R. 1972. The Pleistocene calcareous nannoplankton of the sub-antarctic Pacific Ocean. *Deep Sea Res.*, **19**, 45–61.

Haq, B. U. 1966. Electron microscope studies on some Upper Eocene calcareous nannoplankton from Syria. *Stockholm Contrib. Geol.*, **15**, 23–37.

Haq, B. U. 1968. Studies on Upper Eocene calcareous nannoplankton from NW Germany. *Stockholm Contrib. Geol.*, **18**, 13–74.

Haq, B. U. 1971. Paleogene calcareous nannoflora Parts I–IV. *Stockholm Contrib. Geol.*, **25**, 1–158.

Haq, B. U. 1976. Coccoliths in cores from the Bellinghausen abyssal plain and Antarctic continental rise. *Init. Rep. DSDP*, **35**, 557–567.

Haq, B. U. and Berggren, W. A. 1978. Late Neogene calcareous plankton biochronology of the Rio Grande Rise (South Atlantic Ocean). *J. Paleontol.*, **52**, 1167–1194.

Haq, B. U. and Lohmann, G. P. 1976. Early Cenozoic calcareous nannoplankton biogeography of the Atlantic Ocean. *Mar. Micropaleont.*, **1**, 119–194.

Hay, W. W., Mohler, H. P., Roth, P. H., Schmidt, R. R. and Boudreaux, J. E. 1967. Calcareous nannoplankton zonation of the Cenozoic of the Gulf Coast and Caribbean–Antillean area, and transoceanic correlation. *Trans. Gulf Coast Assoc. Geol. Soc.*, **17**, 428–480.

Hay, W. W., Mohler, H. P. and Wade, M. E. 1966. Calcareous nannofossils from Nal'chik (northwest Caucasus). *Eclogae. Geol. Helv.*, **59**, 379–399.

Hay, W. W., and Towe, K. M. 1962. Electron microscope examination of some chalkliths from Donzacq (France). *Eclogae. Geol. Helv.*, **55**, 497–517.

Heck, S. E. van. 1981. Bibliography and taxa of calcareous nannoplankton. *INA Newsletter*, **3**, 40.

Hodson, F. and West, I. M. 1975. Calcareous nannoplankton from an Upper Bracklesham horizon at Fawley, Hampshire. *Rev. Micropaleontology*, **13**, 165–187.

Jerkovič, L. 1970. *Noelaerhabdus* nov. gen. type d'un nouvelle familie de coccolithophoridés fossiles: Noelaerhabdaceae du Miocene supérieur de Yougoslavie. *C.R. Acad. Sci. Paris*, **270D**, 468–470.

Kamptner, E. 1963. Coccolithineen-Skelettreste aus Tiefseeablagerungen des Pazifischen Ozeans. *Ann. Naturhist. Mus. Wien.*, **66**, 139–204.

Levin, H. L. 1965. Coccolithophoridae and related microfossils from the Yazoo Formation (Eocene) of Mississippi. *J. Paleontol.*, **39**, 265–272.

Martini, E. and Ritzkowski, S. 1968. Die Grenze Eozän/ Oligozän in der Typus-Region des Unteroligozäns (Helmstedt-Egeln-Latdorf). *Mém. Bur. Rech. Geol. Minières,* **69**, 233–237.

Müller, C. 1970. Nannoplankton-Zonen der unteren Meeresmolasse Bayerns. *Geol. Bavar.,* **63**, 107–118.

Perch-Nielsen, K. 1971a. Dursicht Tertiärer Coccolithen. In A. Farinacci (Ed.), *Proc. II Plankt. Conf., Roma, 1970,* **2**, Tecnoscienza, Rome, 939–980.

Perch-Nielsen, K. 1971b. Elektronenmikroskopische Untersuchungen an Coccolithen und verwandten Formen aus dem Eozän von Dänemark. *Biol. Skr. Danske Vidensk. Selsk.,* **18**, 1–76.

Perch-Nielsen, K. 1972. Remarks on Late Cretaceous to Pleistocene coccoliths from the North Atlantic. *Init. Rep. DSDP,* **12**, 1003–1069.

Perch-Nielsen, K. 1985a. Cenozoic calcareous nannofossils. In H. M. Bolli, J. B. Saunders, and K. Perch-Nielsen (Eds.), *Plankton Stratigraphy,* Cambridge University Press, Cambridge, 428–554.

Perch-Nielsen, K. 1985b. Notes on Prinsiaceae, a Cenozoic calcareous nannoplankton Family. *INA Newsletter,* **7**, 85–88.

Pujos, A. 1987. Late Eocene to Pleistocene medium-sized and small-sized "Reticulofenestrids". *Abh. Geol. B.-A.,* **39**, 239–278.

Reinhardt, P. 1966. Fossile Vertreter coronoider und styloider Coccolithen (Familie Coccolithaceae Poche 1913). *Monatsber. Deutsch. Akad. Wiss. Berlin,* **8**, 513–524.

Romein, A. J. T. 1979. Lineages in early Paleogene calcareous nannoplankton. *Utrecht Micropaleont. Bull.,* **22**, 1–231.

Roth, P. H. 1970. Oligocene calcareous nannoplankton biostratigraphy. *Eclogae Geol. Helv.,* **63**, 799–881.

Roth, P. H. 1973. Calcareous nannofossils—Leg 17, Deep Sea Drilling Project. *Init. Rep. DSDP.,* **17**, 695–795.

Roth, P. H. and Thierstein, H. R. 1972. Calcareous nannoplankton: Leg 14 of the Deep Sea Drilling Project. *Init. Rep. DSDP,* **14**, 421–485.

Sachs, J. B., and Skinner, C. H. 1973. Late Pliocene–Early Pleistocene nannofossil stratigraphy in the north Central Gulf Coast area. *Tulane Stud. Geol. Paleontol.,* **10**, 113–162.

Sherwood, R. W. 1974. Calcareous nannofossil systematics, paleoecology, and biostratigraphy of the Middle Eocene Weches Formation of Texas. *Tulane Stud. Geol. Paleontol.,* **11**, 1–79.

Steinmetz, J. C. and Stradner, H. 1984. Cenozoic calcareous nannofossils from Deep Sea Drilling Project Leg 75, southeast Atlantic Ocean. *Init. Rep. DSDP,* **75**, 671–753.

Stradner, H. and Allramm, F. 1981. The nannofossil assemblages of the Deep Sea Drilling Project Leg 66— Middle America Trench. *Init. Rep. DSDP,* **66**, 589–639.

Stradner, H. and Edwards, A. R. 1968. Electron microscopic studies on Upper Eocene coccoliths from Oamaru Diatomite, New Zealand. *Jahrb. Geol. B.-A.,* **13**, 1–66.

Theodoridis, S. A. 1984. Calcareous nannofossil biozonation of the Miocene and revision of the Helicoliths and Discoasters. *Utrecht Micropaleont. Bull.,* **32**, 1–271.

Waghorn, D. B. 1981. New Zealand and Southwest Pacific Late Eocene and Oligocene calcareous nannofossils. Part 2. Taxonomy, Appendices, References. *Thesis,* University of Wellington, New Zealand (unpublished).

Wise, S. W. 1983. Mesozoic and Cenozoic Nannofossils. *Init. Rep. DSDP,* **71**, 481–550.

Young, J. R. 1987. Neogene nannofossils from the Makran Region of Pakistan and the Indian Ocean. *Thesis,* Imperial College London (unpublished).

4

The optical properties and microcrystallography of Arkhangelskiellaceae and some other calcareous nannofossils in the Late Cretaceous

Shimon Moshkovitz and Kenneth Osmond

A short review of the optical properties and phenomena as seen in some Cretaceous calcareous nannofossils (Arkhangelskiellaceae, *Quadrum, Micula, Lithastrinus, Tegulalithus, Nannoconus, Microrhabdulus* and *Bukryaster*) in the Atlantic region and in the eastern Tethys, Israel, is presented by means of the scanning electron microscope and the light microscope (with the aid of the gypsum Red I plate). The microcrystallographic characteristics of the above-mentioned taxa, including the famous 'windmill effect' in Arkhangelskiellaceae and its evolution during the Late Cretaceous, are discussed.

A new species, *Arkhangelskiella paucipunctata* from the latest Maastrichtian *Micula prinsii* Zone, presumably the end member of the family before its extinction at the end of the Cretaceous Period, is described.

4.1 INTRODUCTION

First discoveries of calcareous nannobodies and debates over their origin of formation date back to the early and middle parts of the 19th century (Ehrenberg 1936, Huxley 1868, Wallich 1861). Sorby (1861) is to be credited in this respect for his suggestion that they are to be regarded as part of the organic world. However, it was not before the pioneering works, by means of the optical microscope, of Arkhangelsky (1912), Kamptner (1927, 1928) and especially the painstaking studies of Deflandre (Deflandre 1952, 1959, Deflandre and Deflandre 1967), that a systematic and taxonomic knowledge of the fossil members of this group was developed. Eventually their great potential as a new biostratigraphical tool was discovered (Bramlette and Riedel 1954, Stradner 1961, 1963, Bramlette and Sullivan 1961, Bramlette and Martini 1964, Stover 1966, Čepek and Hay 1969). With the advent of the electron microscopes in the 1950s (the transmission electron microscope and especially the scanning electron microscope) which revolutionised our ideas and knowledge about the microstructure of the nannofossils (Kamptner 1950, Deflandre and Fert 1952, 1954, Braarud et al. 1952), the interest in this group of fossils has increased to a great extent for the benefit of various fields of geology and the oil industry.

Because of their tiny size, so close to the light resolution limit (≈ 0.25 μm), study of their microstructure with the light microscope alone is not very satisfactory and, ideally, the best results are achieved through a combined study of the same object in both the light microscope and the scanning electron microscope (Thierstein and Roth 1972, Moshkovitz 1974, Hansen et al. 1975, Smith 1975). In practice, however, and for rapid stratigraphical considerations, especially in the oil industry, most work is done by means of the light microscope, which probably will long remain the chief practical investigational instrument. In this respect it was shown that some idea regarding the microstructure and arrangement of the subunits of the nannofossils could be gained by using polarised light and the accessory plates, e.g. gypsum Red I (Kamptner 1952, Deflandre and Fert 1954, Bramlette and Martini 1964, Stover 1966, Prins 1969, Reinhardt 1972, and especially Romein 1979). Such information is based on some basic principles concerning optics and the interaction of light with the mineral calcite (of which the calcareous nannofossils are known to be made).

4.2 REVIEW OF SOME OPTICAL PROPERTIES

According to Maxwell's electromagnetic wave theory, visible light (as part of the electromagnetic spectrum) travels through space in continuous waves along straight lines with transverse wave motion. The surface to which light has spread in a given time interval is called the ray velocity surface. Substances are referred to as isotropic if the light travels with equal velocities in all directions (e.g. liquids, glasses, crystals of the isometric—cubic or regular—system). In these substances the ray velocity surface derived from a point source is a sphere, the radii of which are the light rays. In other substances, including crystals that do not belong to the isometric system, light travels with different velocities in different directions. Such substances are described as anisotropic. Non-isometric crystals include uniaxial and biaxial types. Uniaxial crystals (tetragonal and hexagonal systems) are divided into two optical groups: positive and negative (Fig. 4.1). These crystals are positive when the ordinary ray 'o' possesses greater velocity and are negative when the extraordinary ray 'e' has the greater velocity.

An important feature is the optical indicatrix, a three-dimensional geometrical figure that serves to illustrate the optical characteristics of a crystal. This helps to visualise the relation of the refractive indices (n) and their vibration directions (which are perpendicular to the direction of a propagation of light through the crystal). Birefringence is defined as the difference between the maximum and the minimum refractive indices. Fig. 4.2(a) shows a positive uniaxial indicatrix (prolate spheroid of rotation since n_E is greater than n_O), of which quartz and rutile are examples, and Fig. 4.2(b) shows a negative uniaxial indicatrix (oblate spheroid of rotation with n_E smaller than n_O), calcite and apatite being examples (in calcite $n_E = 1.486$, $n_O = 1.658$; hence its birefringence is 0.172).

When monochromatic light strikes an air–calcite interface in any direction other than parallel to the c axis, the ray entering the transparent mineral is polarised and broken into two components that will travel with different velocities in defined paths according to their ray velocity surfaces (Fig. 4.3). One, the ordinary (slow) ray, vibrates in a basal plane, perpendicular to the c axis and parallel to the long diagonal of the rhombic face of the cleavage rhombohedron. The other, the extraordinary (fast) ray, vibrates at right angles to it, i.e. in a plane which also includes the c axis and is parallel to the short diagonal. This type of plane of which there is an infinite number is called the principal section (Fig. 4.4). Such vibrations in an Iceland spar nicol prism are shown in Fig. 4.5.

(a) Uniaxial crystals between crossed-nicols

When a wave of ordinary light is forced to vibrate in a single plane, light is said to be polarised. We use this phenomenon in the polarising microscope by applying the polariser (inserted below the stage of the microscope and transmitting plane-polarised light vibrating in a N–S direction) and the analyser (which is inserted in the tube above the stage, with light vibrating in an E–W direction only). With such an orientation (known as cross-polarised light or crossed-nicols) there is extinction and no light should reach the observer. This is true when

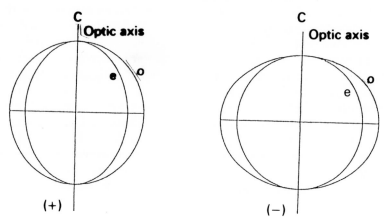

Fig. 4.1 — Ray velocity surface of uniaxial crystals: (+), positive; (−), negative; e, extraordinary ray; o, ordinary ray.

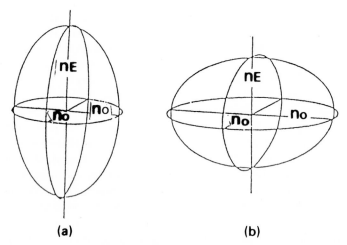

Fig. 4.2 — Optical indicatrix in uniaxial crystals: (a) positive; (b) negative; n_O, refractive index of ordinary ray; n_E, refractive index of extraordinary ray.

viewing an isotropic object between crossed-nicols, since light velocities are the same in all directions and there is no double refraction.

In anisotropic crystals, however, extinction occurs when (a) light moves parallel to the optical axis so that light from the polariser passes through the crystal as through isotropic matter and is cut off by the analyser, or (b) the vibration direction of light coming from the polariser coincides with the vibration direction of the crystal. In that case, the light (O ray or E ray) leaving the crystal will be eliminated by the analyser (which is superimposed with its plane of vibration at right angles to that of the polariser). In such cases, as the crystal is rotated on the revolving stage of the microscope, this extinction position occurs four times (once every 90°) during a complete revolution of 360°.

Since light travels in anisotropic crystals with different velocities (except for the case when parallel to the c axis), a phase difference is created causing the two rays (O and E) that are permitted to pass the analyser (i.e. those vibrating in an E–W plane) to interfere. Depending on the path

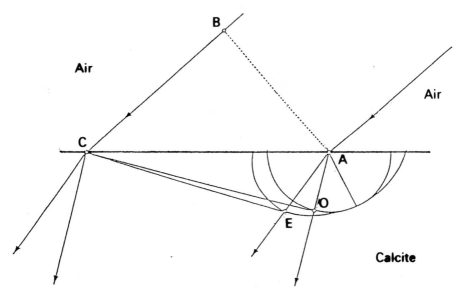

Fig. 4.3 — Double refraction in calcite.

A–B, wave front at the moment the ray reaches A.

A–O, travelling direction and distance of ordinary ray at the moment the ray reaches C (slow ray). This is also the radius of the ray velocity surface.

A–E, travelling direction and distance of extraordinary ray at the moment the ray reaches C (fast ray). This is also the ray velocity surface.

C–O, C–E, new wave fronts of ordinary and extraordinary rays respectively.

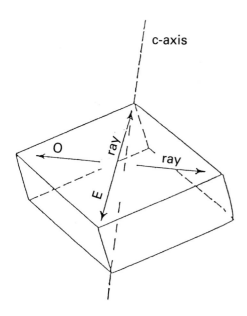

Fig. 4.4 — Calcite rhombohedron. Short (c axis) and long diagonals.

difference between these rays, constructive or destructive interference results in light or darkness and, consequently, interference colours are produced—*vide* the Michel–Levy colour chart. The nature and intensity of these colours depend on the orientation of the studied object, its thickness and the birefringence or retardation (the product of the crystal thickness and its birefringence).

Accessory plates (e.g. quartz wedge, mica and gypsum plates) used with the polarising microscope assist in determining some important optical and crystallographic characteristics of studied objects (e.g. slow or fast ray vibration direction, exact extinction direction, relative thickness of mineral). The gypsum plate (sometimes called the sensitive tint or first-order red) is found to be very useful in determining the optical sign of very thin crystals (less than $10\,\mu$ m) when low-order interference colours (gray, white) occur. The gypsum plate is cut to a thickness that gives in white light a first-order red

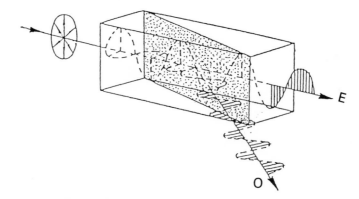

Fig. 4.5 — Ray paths and their vibrations in the nicol prism. o, ordinary ray, e, extraordinary ray.

(∼violet) interference colour, with retardation of 550nm. It is inserted into the microscope below the analyser (in most microscopes usually in an NW–SE direction), so that the vibration direction makes an angle of 45° with the polarisers. Usually, the trace of the vibration plane of the fast component is parallel to the long axis of the accessory plate whereas the slow component is directed at right angles. If the fast component (smaller refractive index) in the studied crystal coincides with the fast component in the gypsum plate, the two add up and the original red (∼violet) colour of the gypsum plate goes up in the spectrum into a blue. When the slow component (higher refractive index) coincides with the fast ray of the gypsum, the components are subtracted and the colours decrease into a yellow. In other words, in the case of addition, the red of the first order plus the grey will produce blue of the second order, whereas, in the case of subtraction, the red minus grey will give a yellow colour (Fig. 4.6). (It should be noted that in some old microscopes, the resulting image (and hence also the colour distribution) may be reversed with regard to the actual position of the inspected object. Therefore it is strongly suggested that the optical system of the microscope be checked first by inserting an easily identifiable object in low magnification. Another simple method is to check the slow and fast rays of a small calcitic crystal present in any of the slides and to inspect the colours along the short

and long diagonals of the rhombohedron (Plate 4.1, Figs. 1, 7, 15, 16).)

Translating these principles for the study of calcareous nannofossils (calcite, uniaxial negative—with three *a* axes at 120° to each other and in a plane at 90° to the unique *c* axis; (Fig. 4.7), we summarise as follows.

Between crossed-polarisers, the object (or the sector investigated) is

(A) dark, when (1) the *c* axis is nearly vertical (to the microscope stage) or

(2) the *c* axis and *a* plane are N and S, in either order;

(b) bright, when the *c* axis is roughly horizontal (but not E–W or N–S).

With a gypsum plate the object is

(A) dark for the same conditions as above;

(B) (1) bright yellow (first order) when the *c* axis is parallel to the gypsum slow ray (usually directed NE–SW) or

(2) bright blue (second order) when the *c* axis is parallel to the gypsum fast ray (usually directed NW–SE)

(in the case of overlapping crystals, the brighter layers show up, masking the darker layers).

By using the above principles, it is generally possible

(a) to deduce the orientation of the axial elements of the calcite crystallites in the various parts of the nannofossil,

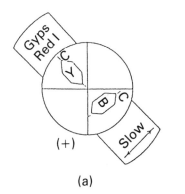 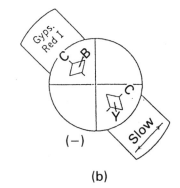

(a) (b)

Fig. 4.6 — Gypsum Red I accessory plate in positive (+) and negative (−) uniaxial crystals: Y, yellow; B, blue.

(b) to interpret the pattern of the rhombohedral arrangement, whether in rows, cycles or superimposed layers, and the relative thickness, and

(c) to determine the viewing aspect, i.e. proximal or distal, of the coccolith.

4.3 OPTICAL PROPERTIES OF SOME CRETACEOUS NANNOFOSSILS— RESULTS

Investigators of Cretaceous nannofossils (e.g. Bramlette and Martini 1964, Stover 1966, Reinhardt 1972) have already noted that various groups react very distinctively to the use of optical accessories. These include various representatives of Arkhangelskiellaceae, *Quadrum, Micula, Lithastrinus, Nannoconus* and many others.

In the following pages and illustrations we will point out some of the phenomena in the groups examined, consider the optical and crystallographic properties and discuss the various implications as well as the variations and distribution through time.

(a) Arkhangelskiellaceae Bukry 1969

Representatives of the Arkhangelskiellaceae include elliptical coccoliths with complex rims built of two to four tiers composed of numerous elements. The large central area, usually pierced by small pores, is in most genera characterised by

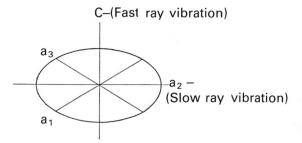

Fig. 4.7 — The hexagonal calcite axes are three *a*-axes at 90° to the unique *c*-axis.

the presence of axial or subaxial sutures or an elevated cross; a stem (or central process) may be present in the early representatives of the group. The 'windmill effect' (discernible only in polarised light and exceptionally well developed in the more advanced representatives of *Arkhangelskiella* and *Gartnerago* and less so in *Thiersteinia* and *Aspidolithus*) is a well-known characteristic of the group. It was described by Bramlette and Martini (1964, p. 297) as follows: 'The central area in *Arkhangelskiella cymbiformis* Vekshina, 1959, which is the type species of the genus, is of very distinctive character, and is divided along the longitudinal and shorter axis into four segments. Each of the four segments shows a less conspicuous radial subdivision with the larger wedge-shaped part (dextral part in proximal view) of each quarter showing strong birefringence and appearing like the blades of a windmill.'

Present studies of various representatives of

Arkhangelskiellaceae (same specimens both in light microscopy and in scanning electron microscopy) show that in distal view the bright sectors or wedges are always located dextrally and adjacent to the major and minor axes of the elliptical central areas (e.g. Plate 4.1, Figs. 2, 3). In proximal views, the bright areas are sinistral to the axes (Plate 4.1, Figs. 4, 5).

Even though these rules were correctly interpreted by some authors for *Arkhangelskiella* and *Gartnerago* (Bramlette and Martini 1964, Plate 1, Figs. 3, 8), Thierstein 1974, Plate 6, Figs. 8, 10, Plate 10, Figs. 1, 4, Smith 1981, p. 29), confusion still exists (Gartner 1968, Plate 11, Fig. 4, Plate 15, Fig. 1, Plate 21, Fig. 7) and information as to the exact orientation (proximal, distal) is often incomplete. Examples are numerous and usually most authors indicate 'plan view' or just 'crossed polarised light' (e.g. Wise and Wind 1977, Plate 50, Fig. 4, Crux 1982, Plate 5.9, Fig. 21, Hattner *et al.* 1980, Plate 4, Fig. 7, Wise 1983, Plate 16, Fig. 2, Stradner and Steinmetz 1984, Plate 18, Figs. 2, 3).

In cases of poor preservation, e.g. diagenetic overgrowth (Plate 4.3, Fig. 18), it is sometimes difficult to recognise the 'windmill effect' solely with crossed-nicols. However, even then the orientation of the coccoliths can be easily determined by study of the interference colours. This is done by inserting the gypsum Red I plate. When the long axis of the elliptical coccolith is directed N–S, the colour distribution is then as follows.

(a) In distal view, the extreme ends of the elliptical margins are yellow and central parts are blue (Plate 1, Figs. 8, 9, 13, 14).

(b) In proximal view, the extreme ends increase to blue whereas the central parts decrease to yellow (Plate 4.1, Figs. 11, 12).

Microcrystallographically, the optical phenomenon of the 'windmill effect' seems to be related to the orientation of the *c* axes in the different parts of the central area. Fig. 4.8(a) is a general interpretation of this phenomenon as seen in distal view. The elements in the outer marginal ring bear crystals with their *c* axes nearly horizontal and oriented generally radially

to the coccolith, but inclined at about 70° clockwise when viewed distally. The same 70° inclination is exhibited by the crystallites in the segments on the right of each major and minor radius (hatched areas). However, the crystallites in the alternating left-hand segments do not exhibit strong birefringence (stippled areas), presumably because they are oriented more nearly vertically (Fig. 4.8(b)).

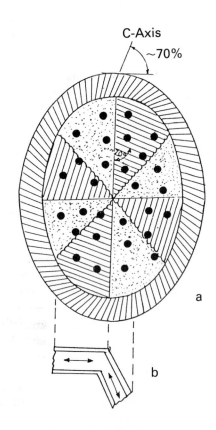

Fig. 4.8 — An interpretation of the 'windmill effect' of *Arkhangelskiella* — distal side — under crossed-nicols with gypsum Red I plate (long axis of specimen in N–S direction): (a) arrows and lines indicate direction of approximate *c* axes; dotted areas show inclined, nearly vertical *c* axes (dark sectors); (b) cross-section of areas aligned to long axis showing asymmetric disposition of *c* axes on both sides of the cross.

The pattern of retardation–addition (blue) with the gypsum accessory plate is easily anticipated by inspection of Fig. 8. The *c* axes are parallel to the fast ray of the accessory in the NW and SE quadrants of the outer ring, and in the inner segments dextral to the N–S (long) axis. Retardation–subtraction (yellow) will occur, conversely, where the c axes are aligned parallel to the slow ray of the accessory, in the NE and SW quadrants of the outer ring, and in the segments dextral to the E–W (short) axis. The positions of the yellow and blue sectors would be reversed in proximal views of the same coccolith pictured in Fig. 4.8. Distally, the minor axis of the cross is inclined a few degrees (from E–W) in a counter clockwise direction whereas, in proximal view, the inclination is in the opposite direction.

Sporadic studies of coccoliths related to various representatives of Arkhangelskiellaceae (including a new species, i.e. *Arkhangelskiella paucipunctata* from the latest Maastrichtian, see below), in different stratigraphical horizons have shown that the direction and position of their c axes vary and accordingly also the bright and dark areas of the optical 'windmill'.

In order to determine the general configuration and the tendency of this phenomenon, five different taxa were chosen in the following stratigraphical levels.

	Species		Stratigraphical level (and sample)
(1)	*Gartnerago costatum*	(1)	Turonian (DSDP 71–511–47–6,100 cm)
(2)	*Thiersteinia ecclesiastica*	(1)	Turonian (DSDP 71–511–47–6,100 cm)
(3)	*Aspidolithus parcus expansus*	(2)	Upper Santonian (DSDP 71–511–41–1,144 cm)
(4)	*Aspidolithus parcus expansus*	(3)	Upper Santonian (DSDP 71–511–40–6,16 cm)
(5)	*Aspidolithus parcus expansus*	(4)	Upper Campanian (DSDP 71–511–35–1,20 cm)
(6)	*Arkhangelskiella cymbiformis*	(5)	Uppermost Campanian–Lower Maastrichtian–*Q. trifidum* Zone, Arad area, Israel, N.7656
(7)	*Arkhangelskiella paucipunctata* sp.nov.	(6)	Uppermost Maastrichtian (DSDP 93–605–66–2, 108–110 cm)

About 25 well-preserved specimens of each of the above representatives were examined (from the distal side only, in the above levels), for their 'windmill effect', angles were measured in relation to the N–S and E–W polarisers. The mean dispositions of the dark and bright areas are shown in Fig. 4.9 and are as follows.

Species no.	Stratigraphical level no.	Bright	Dark	Bright	Dark	Bright
7	6	0–22	22–43	43–108	108–122	122–180+
6	5	0–23	23–43	43–102	102–122	122–180+
5	4	0–10	10–42	42–103	103–129	129–180+
4	3	0–7	7–40	40–102	102–127	127–180+
3	2	0–5	5–35	35–94	94–113	113–180+
2	1		0–27	27–88	88–107	107–178
1	1		0–17	17–87.5	87.5–98	98–178

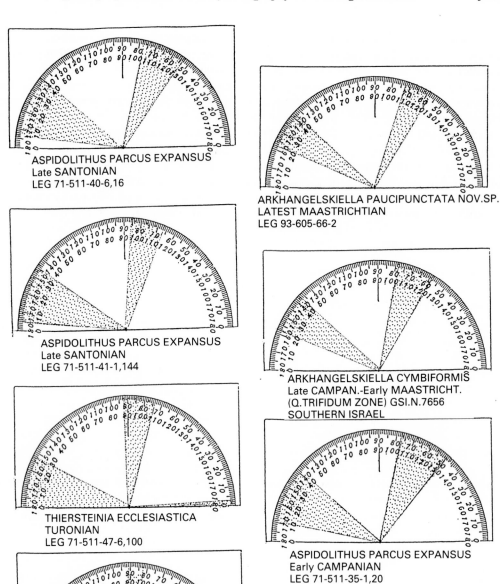

ASPIDOLITHUS PARCUS EXPANSUS
Late SANTONIAN
LEG 71-511-40-6,16

ASPIDOLITHUS PARCUS EXPANSUS
Late SANTONIAN
LEG 71-511-41-1,144

THIERSTEINIA ECCLESIASTICA
TURONIAN
LEG 71-511-47-6,100

GARTNERAGO COSTATUM
TURONIAN
LEG 71-511-47-6,100

ARKHANGELSKIELLA PAUCIPUNCTATA NOV.SP.
LATEST MAASTRICHTIAN
LEG 93-605-66-2

ARKHANGELSKIELLA CYMBIFORMIS
Late CAMPAN.-Early MAASTRICHT.
(Q.TRIFIDUM ZONE) GSI.N.7656
SOUTHERN ISRAEL

ASPIDOLITHUS PARCUS EXPANSUS
Early CAMPANIAN
LEG 71-511-35-1,20

Fig. 4.9 — Measurements of 'windmill effect' in Arkhangelskiellaceae. Bright and dark (dotted) areas, and their mean disposition in the examined representatives of Arkhangelskiellaceae (90° = N), are shown.

The results indicate that (a) the bright and dark areas are arranged asymmetrically in relation to the axes of the cross, and (b) the group as a whole shows a tendency, with time, for the bright and dark areas to move in a clockwise direction, probably as a result of change in general direction of the c axes in the central area of the coccoliths.

Arkhangelskiella paucipunctata nov.sp.
(Plate 4.1, Figs. 8, 9; Plate 4.2, Figs. 13–15; Plate 4.3, Figs. 1–3)

Derivation of name: Latin—*paucipunctata* = few pores.

Holotype: Plate 4.2, Fig. 13.

Type level: Latest Maastrichtian (*Micula prinsii* Zone) DSDP 93–605–66–2, 108–110 cm.

Depository: the collection of the Paleontological Department, Geological Survey of Israel, Jerusalem, stub f/16/4.

Diagnosis: medium to large size, low number of pores and very curved coccolith with four loosely stacked tiers as seen in proximal view.

Description: subelliptical forms measuring 8–14 μm. Distally, the wide, flat marginal area is smooth and the distal radial elements extend to the outermost rim. The central area is distinguished by a rather small number of relatively wide pores (two to four and rarely five per quadrant). Proximally and in side view, the margin is seen to be composed of four very loosely stacked and curved tiers, thus forming rather thick coccoliths, with a thickness of about 2 μm.

Remarks: *A. paucipunctata* differs from the closely related *A. cymbiformis* by its very curved form and the four loosely stacked tiers when seen in side view (compare Plate 4.2, Figs. 9–12, with Plate 4.2, Figs. 13–15), by the radial marginal elements which cover the entire margin on the distal side (Plate 4.3, Figs. 1,2) and by the relatively larger forms (increase in size of *Arkhangelskiella* towards the end of the Maastrichtian was also noted by Girgis (1987) and by Prins and Roersma (1987); the aspect was discussed in the 1987 INA meeting in London). *A. paucipunctata* seems to be one of the end members of the group before its extinction at the end of the Cretaceous Period.

Distribution: in the uppermost Maastrichtian sediments, *M. prinsii* Zone, DSDP 93–605–66–2,108–110 cm.

Other occurrences: N. Orvim, Northern Israel, uppermost Maastrichtian, *M. prinsii* Zone, GSI.N.7669.

Quadrum Prins & Perch-Nielsen in Manivit *et al.* 1977,
Micula Vekshina 1959,
Lithastrinus Stradner 1962,
Tegulalithus Crux 1986

Recently the systematic relationship, evolutionary lineages and origin of *Quadrum, Micula* and *Lithastrinus* have been discussed in a series of papers (Thierstein 1974, Perch-Nielsen 1979, 1985, Roth and Bowdler 1979, Crux 1982, Farhan 1987). In this connection the problem of the preservational state and its impact on the taxonomy of calcareous nannofossils was raised by Thierstein (1974, p. 622) and Crux (1982, p. 99).

The present notes concern some observations on the optical properties and microstructure of the above genera.

Representatives of *Quadrum (Q. gothicum, Q. trifidum, Q. sissinghii)* as well as *Micula murus* and *M. prinsii* show that the c axes are directed perpendicularly to the length of the arms (yellow colour in NW–SE sectors; blue in SW–NE sectors, Plate 4.3, Figs. 5, 6, 9, 11). Bramlette and Martini (1964, p. 320) were aware of that when describing the optical properties of *Tetralithus murus* (Bramlette and Martini 1964, Plate 6, Figs. 20, 21, which in this case resembles *M. prinsii* of the uppermost Maastrichtian). *Micula decussata* and related cube-shaped nannoliths in contrast show that the general direction of the c axes of the numerous small rhombohedra is parallel to the elongation of the arms (blue colour in NW–SE direction, parallel to the fast ray of the gypsum plate—Plate 4.3, Figs. 15, 16—see also right-hand side in Plate 4.3, Fig. 9). The same colour distribution was noted in numerous badly preserved, recrystallised specimens of *Micula*. Some Middle and Late Maastrichtian specimens however, show the presence of a small central cross on top of the main cube with a reverse colour distribution, similar to that found in the representatives of *Quadrum*. Reverse colour distribution is also found at the tips of *Micula concava*, especially in Upper Maastrichtian sediments. At this stage it is premature to speculate on that phenomenon and its bearing on the relationship of *Micula* with other groups.

Lithastrinus floralis is a widespread form in Turonian sediments. Between crossed-nicols and gypsum plate, the blue colour is distributed in the SW–NE sectors, whereas the yellow colour is located in the NW–SE sectors (Plate 4.4, Fig. 3), reversed from that found in *Cylindralithus* (Plate 4.4, Fig. 6). The microcrystallographic structure of *L. floralis* is rather difficult to reconstruct, but the closely related and recently described *Tegulalithus septentrionalis* shows that this Early Cretaceous form is characterised by its circular to subcircular form, with proximal and distal plates composed of numerous small rhombohedral crystals which overlap one another spirally like roof tiles and surround a central hole. Crux's original description (1986, Plate 1, Fig. 1, 2) shows remarkably well the position of the crystallites and Fig. 4.10 is a reconstruction to

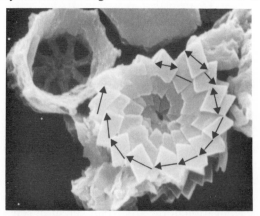

Fig. 4.10 — *Tegulalithus septentrionalis*. Reconstruction of Crux's original description (Crux 1987, Plate 1, Figs. 1,2) to show *c*-axes directions.

show the *c* axes. With the insertion of the gypsum plate, the fast rays of the crystals coincide with those of the gypsum plate in the SW and NE sectors of the nannolith, where a blue colour is expected. The closely related *L. floralis* shows the same colour distribution.

The superposition of axial elements on the crystalline morphology in Fig. 4.10 suggests that the crystal elongation, and the presumed direction of growth in these nannoliths, is normal to the 0111 face, with the result that, at an aspect angle of about 70°, the *c* axis direction and crystal elongation appear to be parallel.

(b) Nannoconaceae Deflandre 1959

The Nannoconaceae include nannoliths with thick walls composed of closely packed calcitic plates arranged spirally around a central axial cavity (Wise and Wind 1977, pl. 72). This group is widespread especially in Tethyan sediments of Early Cretaceous age where it attains its highest development.

Nannoconus Kamptner 1932

The genus *Nannoconus*, which has its FO in the Tithonian, is the main representative of the Nannoconaceae. This genus was one of the first to be used in stratigraphy and Brönnimann (1955) distinguished three different assemblages based on the shape, size and ratio of diameter of internal cavity to external diameter of the body. For the study of these parameters light microscopy is preferable to the scanning electron microscope. With the use of the gypsum plate, the colour distribution of *Nannoconus* in top view shows that the *c* axes of the calcitic plates, which are spirally arranged (as in *Lithastrinus*), run parallel to the direction of the fast ray of the gypsum plate in SW and NE sectors (blue colour) whereas, in the NW and SE areas, the *c* axes are parallel to the slow ray of the gypsum plate (yellow colour, Plate 4.4, Fig. 9). This may be helpful when trying to distinguish between nannoconids (as well as lithastrinids) and coccoliths (e.g. small *Watznaueria, Cyclagelosphaera, Ellipsagelosphaera,* with radially arranged elements, and therefore with reversed colour distribution; Plate 4.4, Fig. 19) especially when dealing with badly preserved samples of uppermost Jurassic and Lower Cretaceous sediments.

(c) Microrhabdulaceae Deflandre 1963

Microrhabdulus Deflandre 1959

The genus *Microrhabdulus* includes rod-like forms somewhat circular in cross-section. Its origin is unknown.

Microrhabdulus decoratus Deflandre 1959

M. decoratus is one of the commonest species of the genus *Microrhabdulus*. Its FO is at the base of the CC 10 Zone in the Cenomanian (Sissingh 1977, p. 436). Examined specimens of the Campanian and Maastrichtian sediments agree with that described by Deflandre (1959). The form is very conspicuous between crossed-nicols and especially with the gypsum plate (Plate 4.4, Figs.12–14,16). The crystals are oriented with their *c* axes almost radial but not perpendicular ($\approx 70°$) to the axis of the nannolith. *c* axes of neighbouring segments alternately lean towards and away from each other (Fig. 4.11). This is manifested by the blue and yellow colours of the alternate segments when oriented parallel to either nicol and has already been described by Deflandre (1959). Bramlette and Martini (1964, p. 314) state correctly that 'without the gypsum plate, the specimens in this position show all the rectangular areas bright, and alternate areas of

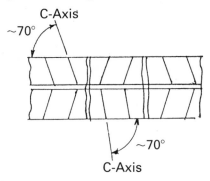

Fig. 4.11 — *Microrhabdulus decoratus*: An interpretation of the microcrystallography.

bright and dark are shown when the length of the specimens is turned dextrally about 30° from the plane of the nicol'. These properties are displayed in Plate 4.4, Figs. 12–14.

4.4 CONCLUSION

During the present study, hundreds of specimens related to the above species were checked for their optical properties and although diagenetic processes, especially overgrowth (e.g. Plate 4.3, Fig. 18), were noted in such degrees that they sometimes affected the general form of the nannolith, the original disposition of the microcrystallographic elements (e.g. orientation of *c* axes) tends to remain in place or to grow in crystallographic continuity. Therefore study of the optical properties of the calcareous nannofossils may be of great help when dealing with taxonomy, evolution, diagenesis and other problems related to this group of fossils.

4.5 ACKNOWLEDGEMENTS

The authors wish to thank the following persons: S. W. Wise, Jr. (Florida State University, Tallahassee) for providing the various samples and for the use of the scanning electron microscope through Project DPP 8414268 granted to him by the National Science Foundation; K. Riddle and S. Silvers (Florida State University, Tallahassee) and M. Dvorachek (Geological Survey of Israel, Jerusalem) for the high quality of the scanning electron microscopy (JEOL 840) pictures taken by them; G. Martinotti (Geological Survey of Israel, Jerusalem) for his assistance during preparation of the manuscript; the DSDP authorities for enabling the use of the DSDP material from the Atlantic.

S. W. Wise, Jr. (Florida State University, Tallahassee) and A. R. Lord (University College London) and the geological staff of these institutions are also to be thanked for the kind hospitality and for the various facilities extended to S. Moshkovitz during his 1986–1987 sabbatical stay at Florida State University and University College London.

PLATE 4.1

Plate 4.1, Fig. 1. Calcite rhombohedron—*c* axis in NW–SE direction, parallel to fast ray of the gypsum plate (blue colour in these sectors), × 3000.

Plate 4.1, Figs. 2,3. *Gartnerago obliquum* Early Maastrichtian. Fig. 2, scanning electron microscope, distal side. Central Israel, Hartuv-2 well, 12–15 m, GSI.N.7604, × 3000. Fig. 3, light microscope, crossed-nicols. Same specimen and position as in Fig. 2. Bright areas of 'windmill blades' are located dextrally to central cross, × 3500.

Plate 4.1, Figs. 4–6. *Gartnerago obliquum* Early Maastrichtian. Same specimen and position. Fig. 4, scanning electron microscope, proximal side. Central Israel, Hartuv-2 well, 12–15 m, GSI.N.7604, × 3200. Fig. 5, light microscope, crossed-nicols. Bright areas of 'windmill blades' are located sinistrally to the central cross, × 3500. Fig. 6, differential interference, Nomarski.

Plate 4.1, Fig. 7. Calcite rhombohedron (same as in Fig.1) with *c* axis rotated in NE–SW direction, parallel to slow ray of gypsum plate (yellow colour dominating), × 3000.

Plate 4.1, Figs. 8, 9. *Arkhangelskiella paucipunctata* Maastrichtian. Distal side (same specimen), DSDP 93–605–75–3. Fig. 8, SEM × 3500. Fig. 9, light microscope, crossed-nicols and gypsum Red I, × 3400.

Plate 4.1, Fig. 10. *Gartnerago* sp. Maastrichtian. Distal side. South Israel, north Negev, Mishor Rotem, oil-shale tunnel, GSI.N.6378, light microscope, crossed-nicols (bright areas—dextral) × 3500.

Plate 4.1, Figs. 11, 12. *Arkhangelskiella cymbiformis* Late Campanian–Early Maastrichtian.
Proximal side. South Israel, north Negev, Arad area, GSI.N.7655, *Q. trifidum* Zone. Fig. 11, light microscope, crossed-nicols (long axis of coccolith N–S) with gypsum Red I, × 3500. Fig. 12, same specimen and position as in Fig. 11, light microscope, crossed-nicols.

Plate 4.1, Figs. 13, 14. *Arkhangelskiella cymbiformis*. Same specimen as in Figs. 11, 12, turned on its distal side. Fig. 13, light microscope, crossed-nicols, with gypsum Red I plate (long axis of coccolith N–S). Fig. 14, light microscope, crossed-nicols, same position as in Fig. 13.

Plate 4.1, Figs. 15, 16. *Arkhangelskiella cymbiformis* Late Campanian–Early Maastrichtian. Proximal side. South Israel, north Negev. Arad area, GSI.N.7655, *Q. trifidum* Zone (same specimen), × 3500. Fig. 15, light microscope, crossed-nicols (long axis of coccolith 47° NW—'windmill blades' extinguished). Fig. 16, light microscopy, crossed-nicols, with gypsum Red I plate (note blue colour of small crystal at top left hand-side; its *c* axis parallels the fast ray of gypsum plate).

Plate 4.1, Figs. 17, 18. *Arkhangelskiella cymbiformis*. Same specimen as in Figs. 15, 16 turned on its distal side. Fig. 17, light microscope, crossed-nicols (long axis of coccolith 40° NE—'windmill blades' extinguished). Fig. 18, same specimen and position as in Fig.17, light microscope, crossed-nicols, with gypsum Red I plate.

Plate 4.1, Fig. 19. *Gartnerago costatum* Turonian. Distal side. DSDP 71–511–48–1, 40 cm. Light microscope, crossed-nicols, with gypsum Red I plate (long axis of coccolith N–S—'windmill blades' extinguished), × 3400.

Plate 4.1, Fig. 20. Same specimen as in Fig. 19. Light microscope, crossed-nicols (long axis of coccolith 45° NW—'windmill blades' bright dextrally to central cross—compare with *Arkhangelskiella*, Fig.15).

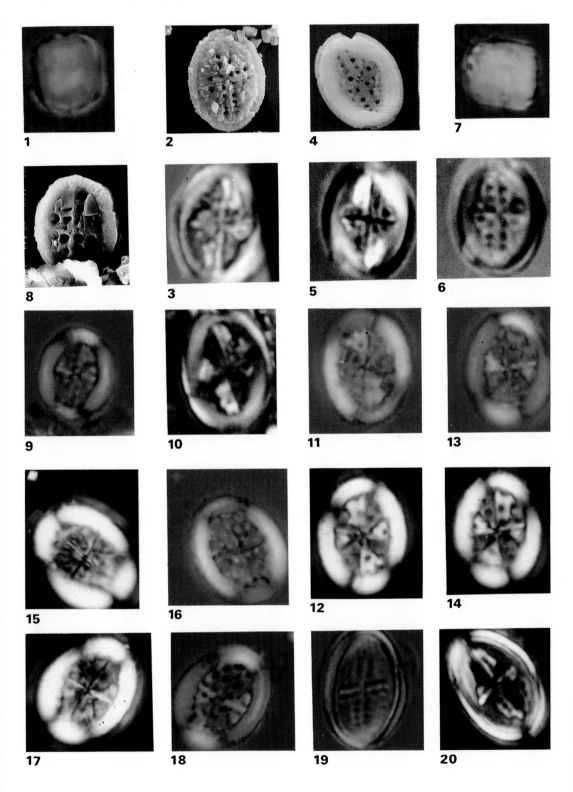

PLATE 4.2

Plate 4.2, Figs. 1, 2. *Gartnerago costatum* Turonian. DSDP 71–511–47–6, 10 cm, scanning electron microscope. Fig. 1, distal side, × 2600. Fig. 2, another specimen, proximal side, × 3500.

Plate 4.2, Figs. 3, 4. *Thiersteinia ecclesiastica* Turonian. DSDP 71–511–47–6, 16 cm scanning electron microscope (same specimen), Fig. 3, distal side, × 3500. Fig. 4, 70° tilt, side view, × 3700.

Plate 4.2, Figs. 5, 6. *Aspidolithus parcus spp. expansus* Late Santonian. Scanning electron microscope (same specimen). DSDP 71–511–40–6, 16 cm. Fig. 5, distal side, × 3900. Fig. 6, 80° tilt, side view, × 4800.

Plate 4.2, Figs. 7, 8. *Arkhangelskiella specillata* Early Maastrichtian. DSDP 36–327A–10–2, 61 cm, scanning electron microscope (same specimen). Fig. 7, distal side, × 3500. Fig. 8, 80° tilt, side view, × 3800.

Plate 4.2, Figs. 9, 10. *Arkhangelskiella cymbiformis* Early Maastrichtian. DSDP 36–327A–10–2, 61 cm, scanning electron microscope (same specimen). Fig. 9, distal side, × 3500. Fig. 10, 80° tilt, side view, × 3800.

Plate 4.2, Figs. 11, 12. *Arkhangelskiella cymbiformis* Early Campanian. Central Israel, Judean Mts., Mahaseya, GSI.N.7978. Scanning electron microscope (same specimen). Fig. 11, proximal side, × 3500. Fig. 12, 80° tilt, side view, × 3800.

Plate 4.2, Figs. 13–15. *Arkhangelskiella paucipunctata* Latest Maastrichtian. DSDP 93–615–66–2, scanning electron microscope (different specimens). Fig. 13, holotype, distal side, × 2600. Fig. 14, proximal side, showing four tiers, × 2800. Fig. 15, proximal–side view showing four loose, very curved tiers, × 3500.

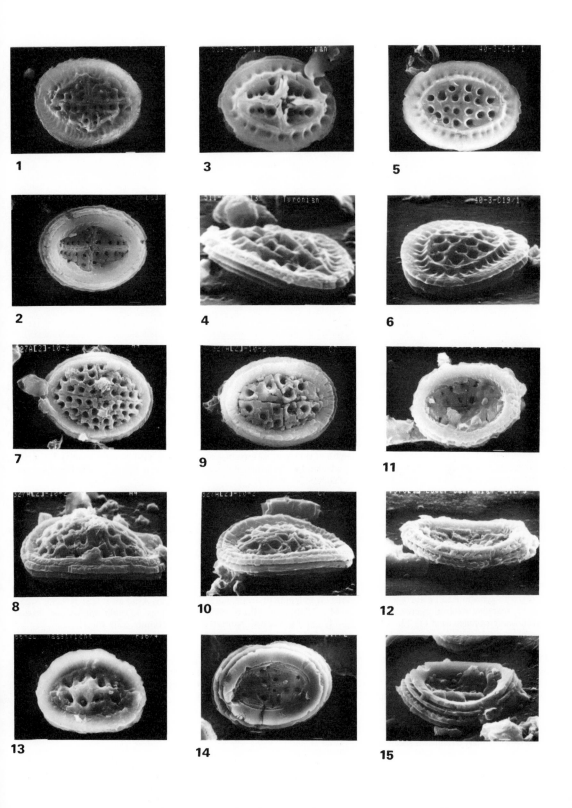

PLATE 4.3

Plate 4.3, Figs. 1–3. *Arkhangelskiella paucipunctata* Latest Maastrichtian. DSDP 93–605–66–2, scanning electron microscope. Fig. 1, two specimens showing proximal (left) and distal (right) sides, × 2400. Fig. 2, 80° tilt of same specimens as in Fig.1, side views, × 2500. Fig. 3, proximal side view showing very curved four loose tiers, × 3200.

Plate 4.3, Figs. 4–7. *Quadrum sissinghii* Late Campanian–Early Maastrichtian. *Q. trifidum* Zone, Israel. Fig. 4, light microscope, crossed-nicols, southern Negev, G.Rehavam, GSI.N.6778, × 3500. Fig. 5, light microscope, crossed-nicols with gypsum Red I plate (*c* axes perpendicular to length of arms), north-east Negev, Arad area, GSI.N.7655, × 3500. Fig. 6, light microscope, crossed-nicols with gypsum Red I plate, another specimen of the same sample as in Fig.5, × 3500. Fig. 7, scanning electron microscope, Judean Mts., Tarqumie, GSI.N.6624, × 3800.

Plate 4.3, Figs. 8–10. *Quadrum trifidum* Late Campanian–Early Maastrichtian, Israel. Fig. 8, light microscope, crossed-nicols. Judean Desert, north Gorfan, GSI.N.8048, × 3500. Fig. 9, light microscope, crossed-nicols, with gypsum Red I plate (*c* axes perpendicular to length of arms; note different colour distribution of small *Micula decussata* to the right), north-east Negev, Arad area, GSI.N.7655, × 3500. Fig. 10, scanning electron microscope, Judean Mts., Tarqumie, GSI.N.6624, × 3500.

Plate 4.3, Figs. 11–14. *Quadrum gothicum* Late Campanian–Early Maastrichtian, Israel. Fig. 11, light microscope, crossed-nicols, with gypsum Red I plate (*c* axes perpendicular to length of arms), north-east Negev, Arad area, GSI.N.7655, × 2500. Fig. 12, scanning electron microscope, Judean Mts., Tarqumie, GSI.N.6605, × 3000. Figs. 13, light microscope, crossed-nicols, same specimen as in Fig.12, × 3600. Fig. 14, light microscope, Nomarski, same specimen as in Fig.12.

Plate 4.3, Figs. 15–17. *Micula decussata* Campanian and Maastrichtian, Israel. Fig. 15, light microscope, crossed-nicols with gypsum Red I plate (general direction of *c* axes parallel to length of arms), north-east Negev, Arad area, GSI.N.7655, × 2500. Fig. 16, scanning electron microscope, Hartuv-2 well, 12–14 m depth, GSI.N.7604, Early Maastrichtian, × 3200. Fig. 17, light microscope, crossed-nicols, north Negev, Mishor Rotem, oil-shale tunnel, GSI.N.6377, Maastrichtian, × 3500.

Plate 4.3, Fig. 18. *Gartnerago* sp.cf. *G. costatum* Early Maastrichtian. DSDP 36–327A–10–2, scanning electron microscope, distal side (strongly overgrown specimen).

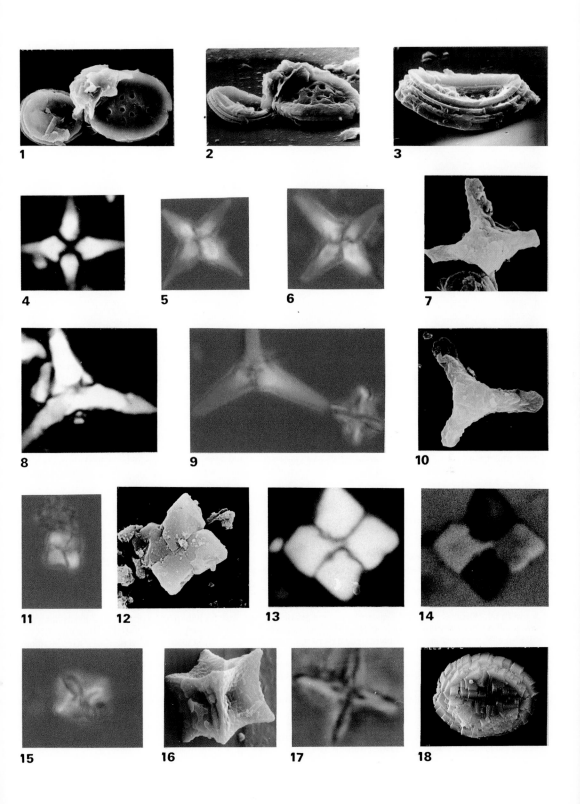

PLATE 4.4

Plate 4.4, Figs. 1–3. *Lithastrinus floralis* Turonian, DSDP 71–511–47–6, 100 cm. Fig. 1, scanning electron microscope side view, × 3800. Fig. 2, light microscope, crossed-nicols, × 3000. Fig. 3, light microscope, crossed-nicols, with gypsum Red I plate, same specimen as in Fig. 2. (*c* axes arranged concentrically, similar to that in *Tegulalithus*—cf. Fig. 4.10), × 3000.

Plate 4.4, Figs. 4–7. *Cylindralithus biarcus* Santonian and Campanian, Israel. Fig. 4, scanning electron microscope, Judean Mts., Zanoah, GSI.N.8161, Late Santonian, × 4000. Fig. 5, scanning electron microscope, same specimen as in Fig.4, tilted, × 4000. Fig. 6, light microscope, crossed-nicols, with gypsum Red I plate (*c* axes arranged radially), same sample as in Fig. 4, × 2500. Fig. 7, light microscope, crossed-nicols, Negev, north Zin, GSI.N.8213, Early Campanian, × 3000.

Plate 4.4, Figs. 8, 9. *Nannoconus colomii* Valanginian. Fig. 8, light microscope, crossed-nicols, with gypsum Red I plate, side view, central Israel, Gevar'am 1 well, Core 1, GSI.N.3310, × 3500. Fig. 9, light microscope, crossed-nicols, with gypsum Red I plate, plan view (*c* axes arranged concentrically), same sample as in Fig.8, × 3500.

Plate 4.4, Fig. 10. *Watznaueria barnesae* Santonian. Light microscope, crossed-nicols, with gypsum Red I plate (*c* axes arranged radially). Central Israel, Judean Mts., Zanoah, GSI.N.8161, × 3500.

Plate 4.4, Figs. 11–16. *Microrhabdulus decoratus* Campanian and Maastrichtian. Fig. 11, scanning electron microscope. Central Israel, Hartuv-2 well, 12–15 m depth, Maastrichtian, × 12000. Fig. 12, light microscope, crossed-nicols (long axis of nannofossil N–S). Central Israel, Judean Mts., Mahaseya, GSI.N.7978, Early Campanian, × 3000. Fig. 13, light microscope, crossed-nicols, with gypsum Red I plate (long axis of nannofossil N–S, showing alternate directions of *c* axes in adjoining units), DSDP 93–605–66–2, Latest Maastrichtian, × 3000. Fig. 14, light microscope, crossed-nicols (long axis of nannofossil − 25° NW), same sample as in Fig. 12, × 3000. Fig. 15, scanning electron microscope, same sample as in Fig. 12, × 4500. Fig. 16, light microscope, crossed nicols, with gypsum Red I plate (long axis of nannofossil 45° NW; in this direction most *c* axes are parallel to slow ray of gypsum), same sample as in Fig. 13, × 3000.

Plate 4.4, Figs. 17, 18. *Bukryaster hayi* Early Campanian. Fig. 17, light microscope, crossed-nicols, with gypsum Red I plate (*c* axes perpendicular to length of arms), distal side, Mississippi, USA, Mooreville Fm., × 3200. Fig. 18, scanning electron microscope, proximal side, Central Israel, Judean Mts., Mahaseya, GSI.N.7978, × 4000.

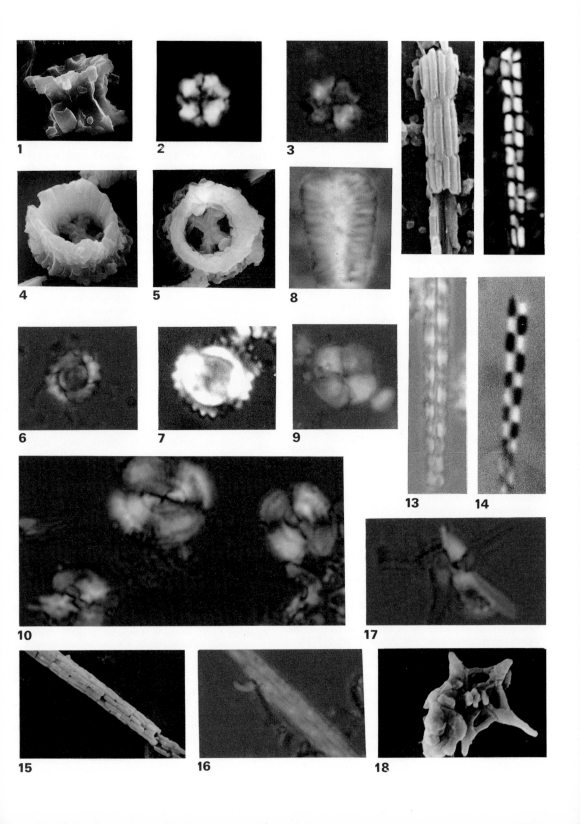

4.6 REFERENCES

Arkhangelsky, A. D. 1912. Upper Cretaceous deposits of east European Russia. *Mater. Geol. Russ.*, **25**, 1–631.

Braarud, T., Gaarder, K. R., Markali, J. and Nordli, E. 1952. Coccolithophorids studied in the electron microscope. Observations on *Coccolithus huxleyi* and *Syracosphaera carterae*. *Nytt. Mag. Botanik*, **1**, 129–134.

Bramlette, M. N. and Martini, E. 1964. The great change in calcareous nannoplankton fossils between the Maastrichtian and Danian. *Micropaleontology*, **10**, 291–322.

Bramlette, M. N. and Riedel, W. R. 1954. Stratigraphic value of discoasters and some other microfossils related to recent coccolithophores. *J. Paleontol.*, **28**, 385–403.

Bramlette, M. N. and Sullivan, F. R. 1961. Coccolithophorids and related nannoplankton of the early Tertiary in California. *Micropaleontology*, **7**, 129–188.

Brönnimann, P. 1955. Microfossils *incertae sedis* from the Upper Jurassic and Lower Cretaceous of Cuba. *Micropaleontology*, **1**, 28–51.

Bukry, D. 1969. Upper Cretaceous coccoliths from Texas and Europe. *Univ. Kansas Paleontol. Contrib.*, **51**, 1–79.

Čepek, P. and Hay, W. W. 1969. Calcareous nannoplankton and biostratigraphical subdivision of the Upper Cretaceous. *Trans. Gulf Coast Assoc. Geol. Soc.*, **19**, 323–336.

Crux, J. A. 1982. Upper Cretaceous (Cenomanian to Campanian) Calcareous nannofossils. In A. R. Lord (Ed.), *A Stratigraphical Index of Calcareous Nannofossils*, Ellis Horwood, Chichester, 81–135.

Crux, J. A. 1986. *Tegulalithus*. A new genus of Early Cretaceous calcareous nannofossils. *INA Newsletter*, **8**, 88–92.

Deflandre, G. 1952. Classe des coccolithophoridés. In P. P. Grassé (Ed.), *Traité de Zoologie*, **1**, Masson, Paris, 439–470.

Deflandre, G., 1959. Sur les nannofossiles calcaires et leur systématique. *Rev. Micropaléontol.*, **2**, 127–152.

Deflandre, G. 1963. Sur les microrhabdulidés, famille nouvelle de nannofossiles calcaires. *C.R. Acad. Sci. Paris*, **256**, 3484–3486.

Deflandre, G. and Deflandre, M. 1967. *Fichier Micropaléontologique Général. Ser.17, Nannofossiles Calcaires I*, Centre National de la Recherche Scientifique, 3423–3830.

Deflandre, G. and Fert, C. 1952. Sur la structure fine de quelque coccolithes fossiles observés au microscope électronique. *Acad. Sci. C.A., Paris*, **234**, 2100–2102.

Deflandre, G. and Fert, C. 1954. Observations sur les Coccolithophoridés actuels et fossiles en microscopie ordinaire et électronique. *Ann. Paleontol.*, **40**, 115–176.

Ehrenberg, C. G. 1836. Bemerkungen ueber feste mikroskopische anorganische Formen in den erdigen und derben Mineralien. *Kön. Akad. Wiss. Berlin: Ber.*, 84–85.

Farhan, A. 1987. Evolutionary trend of the genus *Lithastrinus* to the genus *Uniplanarius*. *Abh. Geol. B.-A.*, **39**, 57–65.

Gartner, S. 1968. Coccoliths and related calcareous nannofossils from Upper Cretaceous deposits of Texas and Arkansas. *Univ. Kansas Paleontol. Contrib.*, **1**, 1–56.

Girgis, M. H. 1987. A morphometric analysis of the *Arkhangelskiella cymbiformis* group and its stratigraphic significance. *INA Newsletter*, **9**, 55 (Abstract).

Hansen, H. J., Schmidt, R. R. and Middelsen, N. 1975. Convertible techniques for the study of the same nannoplankton specimen. *Proc. Kon. Ned. Akad. Wet. Ser. B*, **78**, 226–230.

Hattner, J. G., Wind, F. H. and Wise, S. W. 1980. The Santonian–Campanian boundary: Comparison of nearshore–offshore calcareous nannofossil assemblages. *Cah. Micropaléontol.*, **3**, 9–26.

Huxley, T. H. 1868. On some organisms living at great depths in the North Atlantic Ocean. *Royal Microsc. Soc. Quart. J. Ser. 2*, **15**, 203–212.

Kamptner, E. 1927. Beitrag zur Kenntnis adriatischer Coccolithophoriden. *Arch. Protistenk.*, **58**, 173–184.

Kamptner, E. 1928. Über das System und Phylogenie der Kalkflagellaten. *Arch. Protistenk.*, **64**, 19–43.

Kamptner, E. 1950. Über den submikroskopische Aufbau der Coccolithen *Anz. Osterr. Akad. Wiss. Math. Naturw. Kl.*, **87**, 152–158.

Kamptner, E. 1952. Das mikroskopische Studium des Skelettes der Coccolithineen (Kalkflagellaten). Uebersicht der Methoden und Ergebnisse. I. Die Gestalt des Gehauses und seiner Bauelemente. *Mikroskopie*, **7**(1), 232–244, **7**(2), 375–386.

Manivit, H., Perch-Nielsen, K., Prins, B. and Verbeek, J. W. 1977. Mid Cretaceous calcareous nannofossil biostratigraphy. *Proc. Kon. Ned. Akad. Wet., Ser.B.*, **80**, 169–181.

Moshkovitz, S. 1974. A new method for observing the same nannofossil specimen both by light microscope and scanning electron microscope and preservation of types. *Isr. J. Earth-Sci.*, **23**, 145–147.

Perch-Nielsen, K. 1979. Calcareous nannofossils from the Cretaceous between the North Sea and the Mediterranean. In *Aspekte der Kreide Europas. IUGS Series A*, **6**, 223–272.

Perch-Nielsen, K. 1985. Mesozoic calcareous nannofossils. In H. M. Bolli, J. B. Saunders and K. Perch-Nielsen (Eds.), *Plankton Stratigraphy*, Cambridge University Press, Cambridge, 329–426.

Prins, B. 1969. Evolution and stratigraphy of coccolithinids from the Lower and Middle Lias. In P. Brönnimann and P. Renz, (Eds.), *Proc. I Int. Conf. Plankt. Microfossils, Genève, 1967*, **2**, Brill, Leiden, 547–558.

Prins, B. and Roersma, H. J. 1987. Developments in the family Arkhangelskiellaceae and the species concept in calcareous nannoplankton. *INA Newsletter*, **9**, 60.

Reinhardt, P. 1972. Coccolithen. Kalkiges Nannoplankton seit Jahrmillionen. *Neue Brehm Bucherei*, **453**, 1–99.

Romein, A. J. T. 1979. Lineages in Early Paleogene calcareous nannoplankton. *Utrecht Micropaleont. Bull.*, **22**, 1–231.

Roth, P. H. and Bowdler, J. L. 1979. Evolution of the calcareous nannofossil genus *Micula* in the Late Cretaceous. *Micropaleontology*, **25**, 272–280.

Sissingh, W. 1977. Biostratigraphy of Cretaceous calcareous nannoplankton. *Geol. Mijnbouw.*, **56**, 37–65.

Smith, C. C. 1975. A new method for studying calcareous nannofossils using scanning electron and transmitted light optics. *Rev. Española Micropaleont.*, **7**, 43–48.

Smith, C. C. 1981. Calcareous nannoplankton and stratigraphy of Late Turonian, Coniacian, and Early Santonian Age of the Eagle Ford and Austin Groups of Texas. *U.S. Geol. Survey, Prof. Paper*, **1075**, 1–198.

Sorby, H. C. 1861. On the organic origin of the so called 'crystalloids' of the chalk. *Annals Mag. Nat. Hist., Ser. 3*, **8**, 193–200.

Stover, L.E. 1966. Cretaceous coccoliths and associated nannofossils from France and the Netherlands. *Micropaleontology*, **12**, 133–167.

Stradner, H. 1961. Vorkommen von Nannofossilien im Mesozoikum und Alttertiär. *Erdöl. Z.*, **77**, 77–88.

Stradner, H. 1962. Über neue und wenig bekannte Nannofossilien aus Kreide und Alttertiär. *Verh. Geol. B.-A.*, **2**, 363–377.

Stradner, H. 1963. New contributions to Mesozoic stratigraphy by means of nannofossils. *Proc. 6th World Petrol. Congr. Sec. 1*, paper 4 (preprint), 1–16.

Stradner, H. and Steinmetz, J. 1984. Cretaceous calcareous nannofossils from the Angola Basin, DSDP, Site 530. *Init. Rep. DSDP*, **75**, 565–649.

Thierstein, H. R. 1974. Calcareous nannoplankton. Leg 26. DSDP *Init. Rep. DSDP*, **26**, 619–667.

Thierstein, H. R. and Roth, P. H. 1972. Scanning electron and light microscopy of the same small object. *Micropaleontology*, **17**, 501–502.

Vekshina, V. N. 1959. Kokkolithoforidy maastrikhtskikh otlozheniy zapadno-Sibirskoy nizmennosti (Cocco-lithophoridae of Maastrichtian deposits of the western Siberian lowland). *Sibir. Nauchno-Issled. Inst. Geologii, Geofisiki Mineralnovo Syrya Trudy*, **2**, 56–77.

Wallich, G. C. 1861. Remarks on some novel phases of organic life, and on the boring powers of minute annelids, at great depths in the sea. *Annals Mag. Nat. Hist., Ser. 3*, **6**, 457–458.

Wise, S. W. 1983. Mesozoic and Cenozoic calcareous nannofossils recovered by Deep Sea Drilling Project Leg 71 in the Falkland Plateau Region, Southwest Atlantic Ocean. *Init. Rep. DSDP*, **71**, 481–550.

Wise, S. W. and Wind, F. H. 1977. Mesozoic and Cenozoic calcareous nannofossils recovered by DSDP Leg 36 Drilling on the Falkland Plateau, Southwest Atlantic Sector of the Southern Ocean. *Init. Rep. DSDP*, **36**, 269–492.

5

Conical calcareous nannofossils in the Mesozoic

Paul R. Bown and M. Kevin E. Cooper

The calcareous nannofossils *Eoconusphaera zlambachensis, Mitrolithus jansae* and *Conusphaera mexicana* all possess remarkably similar morphologies consisting of an outer casing of thin, vertical elements and an inner core of radially arranged lamellae. All three are abundant in Tethyan nannofossil assemblages but are rarely found outside the Tethyan Realm. At present these conical nannofossils are thought to have non-concurrent ranges in the Rhaetian, Lower Jurassic and Upper Jurassic–Lower Cretaceous respectively. The classification of these forms will be discussed in the light of their morphology, range, distribution and possible evolutionary links. These three examples highlight the problems of constructing a biological classification based on morphology in a group where homeomorphy often occurs and data are often stratigraphically and geographically limited. A new calcareous nannofossil, *Pseudoconus enigma*, and one subspecies of *Conusphaera mexicana*, *C. mexicana* spp. *minor*, are described.

5.1 INTRODUCTION

This paper is primarily concerned with three Mesozoic calcareous nannofossils, *Eoconusphaera zlambachensis, Mitrolithus jansae* and *Conusphaera mexicana*, which possess strikingly similar cone-like morphologies, occur abundantly (often dominating) in their respective Tethyan (low-latitude) assemblages

and yet at present are recorded with non-concurrent ranges in the Late Triassic–Early Cretaceous time interval. In attempting a classification for nannofossils which is not purely morphological, the presence of stratigraphical gaps separating forms with similar morphologies raises the problem of homeomorphy. This problem is made more acute by data which are often geographically and stratigraphically limited. The three nannofossils in question will be described and compared and the problems of their classification, i.e. phylogeny or homeomorphy, will be discussed. The discussion is aided by the consideration of a number of additional nannofossils, *Conusphaera rothii*, *Pseudoconus enigma* gen. et sp. nov. and nannoconids, which possess analogous morphologies and occur within the same time interval.

(a) *Eoconusphaera zlambachensis*
E. zlambachensis has been recorded from the alpine Triassic by Moshkovitz (1982), Jafar (1983, = *Eoconusphaera tollmanniae*), Posch and Stradner (1987) and Bown (1985, 1987). It has a Norian–Rhaetian range and frequently occurs very abundantly (i.e. 10–20 specimens/field of view at × 1000, 50% of the assemblage); it is not yet reported away from the alpine Austrian–south-west German area (i.e. Tethyan).

  E. zlambachensis possesses a truncated conical form (tapered cylinder) with an outer casing (cf. a rim) of 10–15 vertically arranged plates or elements and a longitudinally continuous inner core of around 40 radial lamellae (Plate 5.1, Figs. 1–8). The core lamellae have a clockwise twist and protrude at the broader, distal end as a domed surface. The core possesses a fine median line when observed in the light microscope, which represents the point from which the lamellae radiate. Size, particularly height, may vary considerably within the assemblage.

(b) *Mitrolithus jansae*
M. jansae was first recorded by Prins (1969, = *M. irregularis*—nomen nudum) from the Upper Pliensbachian–Lower Toarcian of north-west Europe (location(s) not given) but not recorded again until 1984 by Wiegand (= *Calcivascularis jansae*) from DSDP Leg 79 Site 547 (north-west Morocco, continental edge). Bown (1987) records *M. jansae* as the dominant component of Mediterranean–Tethys assemblages from Lower Sinemurian to Lower Toarcian, often occurring very abundantly (10–20 specimens/field of view at × 1000, > 50% of the assemblage) and characterising a Lower Jurassic Mediterranean–Tethys nannofloral realm. It is only very rarely and sporadically found in high latitudes (i.e. north-west Europe)—the only consistent occurrence being recorded around the Pliensbachian–Toarcian boundary in the Mochras Borehole, West Wales (Bown 1987). It has also been found reworked in Oxfordian and Kimmeridgian rocks of the Dorset coast. *M. jansae* has not yet been found in the Paris or German Basins.

   M. jansae has a truncated cone morphology with an outer rim of 10–16 vertical plates and an inner core, best seen in the light microscope, which divides longitudinally into two units (Plate 5.1, Figs. 9–16). The upper part of the core is a rounded 'boss' formed from a number of superimposed cycles of radiating laths–lamellae which protrude distally as a domed surface. The lower core unit appears to be formed from radial lamellae surrounding a central axial canal. The dimensions of *M. jansae*, particularly height, are very variable within assemblages (i.e. height 2.3–5.6 μ m).

(c) *Conusphaera mexicana*
Described by Trejo (1969) from the Upper Jurassic of Mexico, *C. mexicana* is a well-known nannofossil which occurs abundantly, often dominantly, in low-latitude (Tethyan) Tithonian assemblages and ranges into the Hauterivian.

   C. mexicana has a truncated cone morphology with 12–18 vertical outer rim plates and an inner core formed from two longitudinally continuous, concentric, near-radial cycles of lamellae—the innermost cycle twisting anticlockwise, the outer twisting clockwise (Plate 5.2, Figs. 1–12).

The similarities and differences between these three nannofossils are summarised in the following lists. All three forms have
(1) truncated cone morphologies,
(2) an outer rim of 10–18 thin, vertically arranged plates,
(3) an inner core formed from radial lamellae which protrude distally,
(4) comparable size dimensions, but showing considerable variation particularly in height,
(5) (dominantly) Tethyan distributions,
(6) great abundance, often dominating their respective assemblages.
The three forms differ in their
(1) ranges,
(2) detailed inner core structure.

5.2 CLASSIFICATION

At a purely morphological level the classification of these three forms is superficially quite straightforward, all three possessing identical outer rim structures and inner cores which share similar patterns of construction but differ in detail. This coherent morphological grouping is seemingly further reinforced by the palaeobiogeographical information which shows each of the nannofossils occurring abundantly in a restricted Tethyan area. The classification of the three forms into one genus (*Conusphaera*), with successive inner core developments delineating three separate species, appears quite logical. The first problem with this reasoning arises when the stratigraphical ranges of these nannofossils are considered—at present they are recorded with mutually exclusive stratigraphic distributions (Fig. 5.1). Admittedly the interval between the last occurrence of *E. zlambachensis* (Rhaetian) and the first occurrence of *M. jansae* (Sinemurian) includes a period, the Tethyan Hettangian, from which no information is yet available (although research is continuing). The relationship between *E. zlambachensis* and *M. jansae* is further complicated by the possibility that *M. jansae* is related to *Mitrolithus elegans*. The upper core unit–boss structure of *M. jansae* was considered by Prins (1969) and Bown and Young in Young *et al.* (1986) to be analogous to the spine structure seen in *M. elegans,* i.e. both are complex, massive structures formed from

superimposed cycles of radial elements. This generic assignment for the species *jansae* is now thought to be questionable, although it is thought best to await further information before attempting a final classification of this species. Similarly, an evolutionary relationship between *E. zlambachensis* and *M. jansae* is still thought to be feasible.

The discontinuity between the last occurrence of *M. jansae* (Lower Toarcian) and the first occurrence of *C. mexicana* (Lower Tithonian) is not so easily explained away as due to restricted data. The Middle Jurassic and more so the Upper Jurassic are relatively well-studied intervals from a wide geographical area. It is still conceivable that intermediate forms continued during this

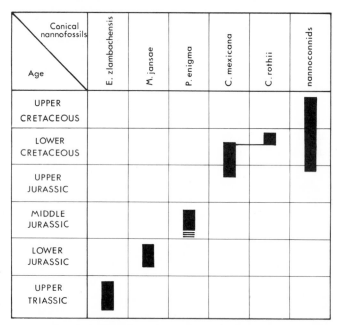

Fig. 5.1 — Stratigraphical ranges of conically shaped nannofossils in the Mesozoic.

period of time in an isolated population which has not yet been or will never be located, or that representatives of the lineage became decalcified and thus disappeared from the fossil record. Such hypotheses, while feasible and convenient explanations for discontinuities within supposed lineages, are virtually impossible to prove or disprove. However, the presence of homeomorphy throughout the history of nannofossils is

incontrovertible and probably the more likely reason for the reappearance of similar nannofossil morphologies. It is also interesting to note that the first *C. mexicana* specimens observed in the Lower Tithonian are very small forms (2 μm) and a gradual increase in size occurs through the Tithonian until more typical larger forms (8–12 μm) become established in the Upper Tithonian–Lower Berriasian (observed at DSDP Site 534A—Cooper 1987). It is now thought that *C. mexicana* gave rise to another species, *C. rothii*, in the Hauterivian. The latter species retains the outer rim but is slightly shorter and wider and possesses an inner core displaying a greater degree of twisting of the constituent lamellae (observed in the light microscope). *C. rothii* additionally displays a more cosmopolitan distribution, being found in high-latitude sites, e.g. the North Sea (Jakubowski 1986).

Thus, while a relationship between *E. zlambachensis* and *M. jansae* is yet to be conclusively proven or otherwise, it is thought that *C. mexicana* represents a later homeomorph. While it provides a striking example of homeomorphy it is by no means the only Mesozoic form to exploit the conical form. The decline of *C. mexicana* itself, after its initial abundance in the Lower Tithonian, coincided with the appearance and rapid expansion of another broadly conical group of nannofossils, the nannoconids. In the same way as the previous conical forms, the nannoconids were initially restricted to the Tethyan area (becoming more cosmopolitan in the Cretaceous) and similarly dominated nannofossil assemblages (from the Berriasian to the Valanginian). A new cone-like form, *Pseudoconus enigma,* has also been found in the Middle Jurassic (?Aalenian–Bathonian) and while its light microscope appearance is comparable with those of the three other Jurassic conical forms its structure, as revealed in the scanning electron microscope, is quite different, having a rather flattened shape, no outer plates– rim and a granular ultrastructure. Other comparable conical morphologies include the two species of *Calcicalathina, C. alta* and *C. oblongata* (which also appear to have non-concurrent ranges in the Cretaceous).

It therefore appears that the earliest nannofossils and those that followed in the Mesozoic found the conical morphology to be a highly successful evolutionary development. This is confirmed by its re-occurrence in a number of apparently unrelated lineages which were all dominant in the Tethyan area during their respective ranges (Fig. 5.1). It is possible that successive forms occupied the same or similar environmental niches. There appear to be no direct analogues in the Tertiary record although genera such as *Sphenolithus* and *Fasciculithus* may display roughly similar shapes. The reason for the success of the conical form and its restriction to low latitudes is unclear. However, the way in which these nannofossils formed around the organism is known for *C. mexicana* at least with Trejo (1969) clearly illustrating a spheroidal structure made up of contiguous *C. mexicana* cones (Trejo 1969, Plate 2, Figs. 7, 8; Plate 3, Figs. 1, 5). Such an arrangement must have added considerably to the weight of the cell and created a relatively thick exathecal layer.

5.3 CONCLUSION

The possibility of finding an evolutionary relationship between *E. zlambachensis* and *M. jansae* is at present impeded by the lack of information from the Tethyan Hettangian. Any further discussion of their classification will thus be deferred until more data are forthcoming. The long interval between the last occurrence of *M. jansae* and the first occurrence of *C. mexicana*, however, appears to preclude any close relationship between these two forms. It is interesting to note that the Tethyan Mesozoic nannofossil record is, for quite considerable lengths of time, dominated by the presence of conical forms which appear to represent a succession of highly successful nannofossil homeomorphs.

5.4 SYSTEMATIC PALAEONTOLOGY
Incertae Sedis

Genus: *Conusphaera* Trejo, 1969

Type species: *Conusphaera mexicana* Trejo, 1969.

Conusphaera mexicana Trejo, 1969 ssp. *mexicana*
Plate 5.2, Figs. 1–9

1965   Particule calcaire; Noël, Plate 28; Figs. 4, 7, 9.

1969	*Conusphaera mexicana* Trejo; pp. 6–8; Figs. 1–4; Plate 1, Figs. 1–9; Plate 2, Figs. 1–8; Plate 3, Figs. 1–7; Plate 4, Figs. 1–6.
1986	*Conusphaera mexicana* Trejo; Jakubowski, Plate 1, Figs. 28, 29.
Non 1971	*Cretaturbella rothii* Thierstein, p. 483, Plate 3, Figs. 1–5.
Non 1981	*Conusphaera mexicana* Trejo; Köthe, p. 23, Plate 4, Fig. 7.
Non 1983	*Conusphaera mexicana* Trejo; Roth, p. 611, Plate 2, Figs. 11, 12.

Diagnosis: a subspecies of *C. mexicana* Trejo which has a height always greater than 4 μm.

Remarks: *C. mexicana* spp. *mexicana* represents the more typical form of *C. mexicana* which was first described by Trejo (1969).

Range: Tithonian–Hauterivian.

Conusphaera mexicana Trejo, 1969 spp. *minor*
spp. nov.
Plate 5.2, Figs. 10–12

1983	*Conusphaera mexicana* Trejo; Roth, p. 611, Plate 2, Figs. 11, 12.

Diagnosis: a subspecies of *C. mexicana* with a height less than 4 μm.

Remarks: *C. mexicana* spp. *minor* gave rise to *C. mexicana* spp. *mexicana* in the Tithonian and thus both subspecies are coherent groups in terms of both stratigraphical distribution and morphological characteristics. As well as a difference in size, *C. mexicana* spp. *minor* possesses a morphology which is far less tapered than that of *C. mexicana* spp. *mexicana*, with parallel to subparallel sides and has a broad, squat appearance.

Holotype: UCL–1730–13 (Plate 5.2, Fig. 10).

Isotype: UCL–1730–9 (Plate 5.2, Fig. 12).

Type locality: Blake–Bahama Basin, DSDP Site 534A.

Type level: DSDP Site 534A–99–1, Lower Tithonian.

Range: Tithonian.

Conusphaera rothii (Thierstein 1971) Jakubowski
1986

1971	*Cretaturbella rothii* Thierstein, p. 483, Plate 3, Figs. 1–5.
1981	*Conusphaera mexicana* Trejo; Köthe, p. 23, Plate 4, Fig. 7.
1986	*Conusphaera rothii* (Thierstein); Jakubowski, p. 41, Plate 11, Figs. 5–8.

Remarks: the diagnostic characteristic of this species given by Jakubowski (1986) is the twisting of its inner core elements. The inner core elements of *C. mexicana* also show a certain degree of twisting and it should be emphasised that the twisting observed in *C. rothii* (light microscope) is far more extreme. *C. rothii* is also shorter and wider than *C. mexicana* spp. *mexicana* and has a more prominent distal protrusion of its inner core.

Range: Hauterivian–Aptian.

PLATE 5.1

Abbreviations: SEM, scanning electron microscope; LM, light microscope; p-c, phase contrast; c-p, crossed-polars. Scale bars represent 1 μm.

Plate 5.1, Figs. 1–8. *Eoconusphaera zlambachensis*. Weissloferbach, *marshi* Zone, Rhaetian (sample 13b). Fig. 1, SEM, distal oblique view, UCL–2036–36, × 9920. Fig. 2, SEM, as Fig. 1, proximal oblique view, UCL–2036–35, × 9810. Fig. 3, SEM, as Fig. 1, distal view, UCL–2036–37, × 11660. Fig. 4, SEM. broken specimen showing inner core lamellae, UCL–2041–4, × 10360. Fig. 5, LM, c-p, UCL–2274–27, × 3580. Fig. 6, LM, as Fig. 5, p-c, UCL–2274–28. Fig. 7, LM, c-p, field of view with seven specimens of various dimensions, UCL–2274–18, × 1160. Fig. 8, LM, as Fig. 7, p-c, UCL–2274–19.

Plate 5.1, Figs. 9–16. *Mitrolithus jansae*. Fig. 9, SEM, distal oblique view, UCL–2190–1, DSDP Site 547–20–1, Sinemurian, × 9450. Fig. 10, SEM, as Fig. 9, distal view, UCL–2190–2, × 9530. Fig. 11, SEM, side view, UCL–2046–36, DSDP Site 547–22–1, Sinemurian, × 10430. Fig. 12, SEM, broken specimen revealing inner core lamellae, UCL–2190–5, DSDP Site 547–20–1, Sinemurian, × 8150. Fig. 13, LM, c-p, UCL–2030–31, DSDP Site 547–15–1, Lower Pliensbachian, × 3680. Fig. 14, LM, as Fig. 13, p-c, UCL–2030–32. Fig. 15, LM, c-p, UCL–2093–13, Trunch borehole (England), *jamesoni* Zone (Lower Pliensbachian) (sample TR20), × 4690. Fig. 16, LM, c-p, UCL–2179–31, Brenha (Portugal), ibex Zone (Lower Pliensbachian) (sample 3531), × 4200.

Pl. 5.1]   Conical calcareous nannofossils in the Mesozoic   103

Genus: *Eoconusphaera* Jafar, 1983

Type species: *Conusphaera zlambachensis* Moshkovitz, 1982 (= *Eoconusphaera tollmanniae* Jafar, 1983).

Eoconusphaera zlambachensis (Moshkovitz 1982) comb. nov.
Plate 5.1, Figs. 1–8

Basionym: *Conusphaera zlambachensis* Moshkovitz 1982, pp. 612–613, Plate 1, Figs. 1–10.
Remarks: named *E. tollmanniae* by Jafar (1983) who was presumably unaware of the previous year's publication of Moshkovitz. It is thought appropriate to place the species *zlambachensis* in the genus *Eoconusphaera*, having concluded that the Upper Jurassic *Conusphaera mexicana* is probably an unrelated homeomorph.

Genus: *Pseudoconus* gen. nov.

Type species: *Pseudoconus enigma* gen. nov. et sp. nov.
Diagnosis: nannofossils with a truncated cone-like outline in side view which are constructed from numerous roughly equidimensional crystallites displaying varying degrees of organisation.
Remarks: the name refers to the shape of the nannofossil which resembles a truncated cone in side view. The nature of the ultrastructure suggests this form may be a holococcolith. The structure is quite unlike that seen in *Eoconusphaera* Jafar (1983), *Conusphaera* Trejo (1969) and *Mitrolithus* Deflandre (1954). It does resemble the Lower Cretaceous *Zebrashapka* Covington and Wise (1987) but is slightly less organised in structure and differs considerably in cross-section and light microscope appearance.

Pseudoconus enigma sp. nov.
(Plate 5.2, Figs. 13–20).

Diagnosis: truncated cone-like nannofossil with a subrectangular cross-section constructed from numerous, small, roughly equidimensional crystallites which may be cubic to prismatic or slightly irregular. The crystallites show some degree of organisation forming strips parallel to the top and bottom of the nannofossil. In the light microscope the structure divides into three longitudinal parts, implying the presence of a wall surrounding a central space. An additional proximal structure is seen as two bright nodes.

PLATE 5.2
Abbreviations: SEM, scanning electron microscope; LM, light microscope; p-c, phase contrast; c-p, crossed-polars. Scale bars represent 1 μm.

Plate 5.2, Figs. 1–9. *Conusphaera mexicana* spp. *mexicana*. DSDP Site 534A. Fig. 1, SEM, side view, UCL–1652–2, 534–96–1, Tithonian, × 5110. Fig. 2, SEM, distal oblique view, UCL–1676–20, 534–96–1, Tithonian, × 7030. Fig. 3, SEM, proximal oblique view, UCL–1676–19, 534–96–1, Tithonian, × 7030. Fig. 4, SEM, as Fig. 1, distal oblique view, UCL–1652–3, × 5920. Fig. 5, LM, p-c, UCL–1730–29, 534–95–5, Tithonian, × 1600. Fig. 6, LM, p-c, UCL–1730–21, 534–96–3, Tithonian, × 1600. Fig. 7, LM, p-c, UCL–1730–17, 534–96–5, Tithonian, × 1600. Fig. 8, LM, c-p, UCL–1730–16, 534–97–1, Tithonian, × 1600. Fig. 9, LM, as Fig. 8, p-c, UCL–1730–15.

Plate 5.2, Figs. 10–12. *Conusphaera mexicana* ssp. *minor*. DSDP Site 534A. Fig. 10, holotype, LM, c-p, UCL–1730–13, 534–99–1, Tithonian, × 1600. Fig. 11, LM, as Fig. 10, p-c, UCL–1730–14. Fig. 12, isotype, LM, p-c, UCL–1730–9, 534–100–1, Tithonian, × 1600.

Plate 5.2, Figs. 13–20. *Pseudoconus enigma* gen. et sp. nov. Fig. 13, holotype, SEM, side view, UCL–2512–12, Watton Cliff, *aspidoides* Zone, Upper Bathonian (sample D24), × 7090. Fig. 14, isotype, SEM, side view, UCL–2512–3, Watton Cliff (sample D24), × 7380. Fig. 15, isotype, SEM, side view, UCL–2512–11, Watton Cliff (sample D24), × 7450. Fig. 16, SEM, as Fig. 15, proximal oblique view, UCL–2512–9, × 8900. Fig. 17, LM, c-p, UCL–2477–11, Langton Hive Point, *aspidoides* Zone, Upper Bathonian (sample D25, Lord and Bown 1987), × 3840. Fig. 18, LM, as Fig. 17, p-c, UCL–2477–12. Fig. 19. LM, p-c, UCL–2477–7, Watton Cliff, Upper Bathonian (sample D24), × 5120. Fig. 20, LM, c-p, UCL–2477–5, Brenha (Portugal), Upper Bathonian (sample 3789), × 3650.

Pl. 5.2] **Conical calcareous nannofossils in the Mesozoic** 105

Description: each face of the nannofossil is formed from the equidimensional crystallites; no internal structure has been observed. In plan view at least one of the sides may have a shallow median furrow running down its length. Greater irregularity of the crystallites and general structure may be introduced owing to varying states of preservation.

Remarks: the name *enigma* refers to the puzzling structure of this nannofossil and the apparent lack of any similar forms in the Jurassic. *P. enigma* has been recorded from both Tethyan and Boreal areas and appears to have a restricted range within the Middle Jurassic.

Holotype: UCL–2512–2 (Plate 5.2, Fig. 13).

Isotypes: UCL–2512–1, UCL–2512–3 (Plate 5.2, Figs. 15, 14).

Type locality: Watton Cliff, Dorset, England.

Type level: Upper Fuller's Earth, *aspidoides* Zone, Upper Bathonian. Sample D24 (Lord and Bown 1987).

Range: Upper Bajocian–Upper Bathonian (Dorset, England). Lower to Upper Bathonian (Brenha, Portugal).

Dimensions: length 4.0–6.2 (6.0) μ m
distal width 4.0–4.9 (4.8) μ m
proximal width 2.0–3.5 (3.2) μ m
thickness 2.0–2.3 (2.0) μ m
(dimensions of holotype in parentheses)

5.5 ACKNOWLEDGEMENTS

We would like to thank Dr. Alan Lord (University College London) for supervision of the research projects from which much of these results have been drawn. The financial support of NERC is gratefully acknowledged as is the provision of samples by DSDP. Technical assistance was provided by Jim Davy (University College London). Kevin Cooper acknowledges the research support of Stratigraphic Services International.

5.6 REFERENCES

Bown, P. R. 1985. *Archaeozygodiscus*—A new Triassic coccolith genus. *INA Newsletter*, **7**, 32–35.

Bown, P. R. 1987. Taxonomy, evolution, and biostratigraphy of Late Triassic–Early Jurassic calcareous nannofossils. *Spec. Papers Palaeontology*, **38**, 116 pp.

Cooper, M. K. E. 1987. Calcareous nannofossils across the Jurassic/Cretaceous boundary. *Thesis*, University College London (unpublished).

Covington, J. M. and Wise S. W., Jr. 1987. Calcareous nannofossil biostratigraphy of a Lower Cretaceous deep-sea fan complex: DSDP Leg 93 Site 603, lower continental rise off Cape Hatteras. *Init. Rep. DSDP*, **93**, 617–660.

Deflandre, G. and Fert, C. 1954. Observations sur les coccolithophoridés actuels et fossiles en microscopie ordinaire et électronique. *Ann. Paleontol.*, **40**, 115–176.

Jafar, S. A. 1983. Significance of the Late Triassic calcareous nannoplankton from Austria and southern Germany. *Jb. Geol. Paläont. Abh.*, **166**, 218–259.

Jakubowski, M. 1986. New calcareous nannofossil taxa from the Lower Cretaceous of the North Sea. *INA Newsletter*, **8**, 35–42.

Köthe, A. 1981. Kalkiges Nannoplankton aus dem Unter-Hauterivian bis Unter-Barremian der Tongrube Moorberg/Sarstedt (Unter-Kreide, NW-Deutschland). *Mitt. Geol. Inst. Univ. Hannover*, **21**, 1–95.

Lord, A. R. and Bown, P. R. 1987. Mesozoic and Cenozoic stratigraphical micropalaeontology of the Dorset coast and Isle of Wight, southern England. *BMS Field Guide No 1*, British Micropalaeontological Society, London.

Moshkovitz, S. 1982. On the findings of a new calcareous nannofossil (*Conusphaera zlambachensis*) and other calcareous organisms in the Upper Triassic sediments of Austria. *Eclogae Geol. Helv.*, **75**, 611–619.

Noël, D. 1965. *Sur les Coccolithes du Jurassique européen et d'Afrique du Nord. Essai de Classification des Coccolithes Fossiles*, Editions du CNRS, Paris.

Posch, F. and Stradner, H. 1987. Report on Triassic nannoliths from Austria. *Abh. Geol. B.-A.*, **39**, 231–237.

Prins, B. 1969. Evolution and stratigraphy of coccolithinids from the Lower and Middle Lias. In P. Brönnimann and P. Renz (Eds.), *Proc. I Int. Conf. Plankt. Microfossils, Genève, 1967*, **2**, Brill, Leiden, 547–558.

Roth, P. H. 1983. Jurassic and Lower Cretaceous calcareous nannofossils in the Western North Atlantic (Site 534): biostratigraphy, preservation and some observations on biogeography and palaeoceanography. *Init. Rep. DSDP*, **76**, 587–621.

Thierstein, H. R. 1971. Tentative Lower Cretaceous calcareous nannoplankton zonation. *Eclogae Geol. Helv.*, **64**, 459–488.

Trejo, M. 1969. *Conusphaera mexicana*, un nuevo coccolitoforido del Jurassico Superior de Mexico. *Revista Inst. Mexicana Petrol.*, **1**, 5–15.

Wiegand, G. 1984. Two genera of calcareous nannofossils from the Lower Jurassic. *J. Paleontol.*, **58**, 1151–1155.

Young, J. R., Teale, C. T. and Bown, P. R. 1986. Revision of the stratigraphy of the Longobucco Group (Liassic, southern Italy); based on the new data from nannofossils and ammonites. *Eclogae Geol. Helv.*, **79**, 117–135.

Part II

Palaeoenvironmental and Palaeogeographical Applications

Among all the microorganisms that leave fossil records in oceanic sediments, the Coccolithophoridae probably have the greatest potential as palaeoclimatic indicators.

McIntyre 1967

6

Variations in coccolith assemblages during the last glacial cycle in the high and mid-latitude Atlantic and Indian oceans

Gunilla Gard

The coccolith assemblages during the last glacial cycle (oxygen isotope stages 1–5) are dominated by small placoliths (*Gephyrocapsa* spp. and *Emiliania huxleyi*). The abundance and distribution of *Calcidiscus leptoporus* and *Helicosphaera carteri* are correlated with sea surface temperatures. They are rare in subpolar waters but increase in relative abundance with increasing surface water temperatures, reaching maximum relative abundances in mid latitudes.

Coccolithus pelagicus occurs in high abundances in subpolar waters but is excluded from warm transitional and tropical waters. Its present rareness in the Subantarctic and during the last interglacial in the North Atlantic suggests that factors other than surface water temperatures also control the distribution of *C. pelagicus*.

Maximum production of coccoliths occurred during interglacial oxygen isotope stages 1 (the Holocene) and 5e (the Eemian).

6.1 INTRODUCTION

The potential of coccoliths for indicating palaeo-oceanographical surface water conditions is still comparatively unexplored. Coccoliths are produced by planktonic coccolithophorid algae which live in the upper 200 m of the water column. Thus, fossil coccolith assemblages can be used for delineating past oceanographical and climatic conditions in the surface waters.

Comprehensive studies on the biogeography of living coccolithophorids have been made by Honjo (1977), McIntyre and Bé (1967), McIntyre et al. (1970), Okada and Honjo (1973) and Okada and McIntyre (1977, 1979). These studies show that the species compositions can be separated into geographical zones. Geitzenauer et al. (1977) and McIntyre and Bé (1967) studied the distribution of coccoliths in surface sediments and McIntyre (1967) used coccoliths as palaeoclimatic indicators. McIntyre et al. (1972) and Ruddiman and McIntyre (1976) used both nannofloral and planktonic foraminiferal associations to delineate the intensity and extent of polar front migrations in the Late Pleistocene North Atlantic. These studies have highlighted the potential of coccoliths for research into Pleistocene climatic and oceanographical environments.

The aim of this paper is to correlate abundance variations in coccolith assemblages from the last glacial cycle with past surface water conditions.

6.2 MATERIALS AND METHODS

The study material consists of eight cores collected from polar to tropical environments. Details of locations are shown in Fig. 6.1 and in

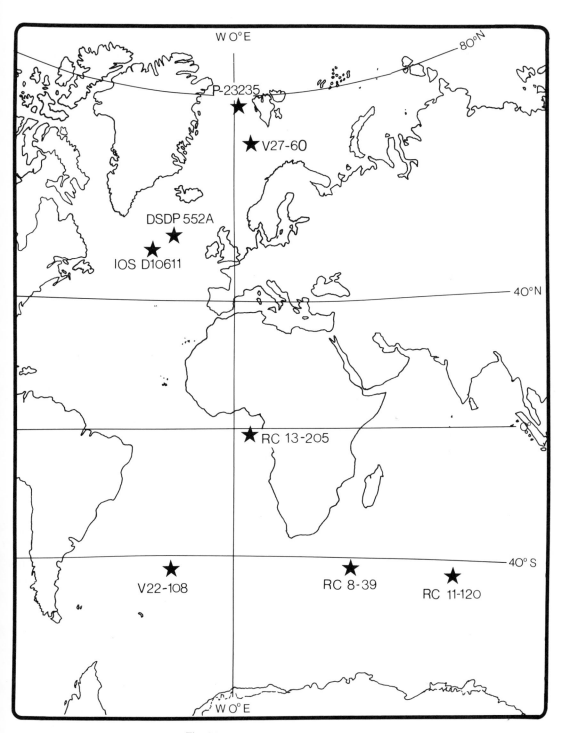

Fig. 6.1 — Location map of studied cores.

Table 6.1. These cores were chosen because they contain well-preserved sequences of the last glacial cycle and are dated by oxygen isotope stratigraphy. The cores were chosen to represent polar, subpolar, transitional and tropical regimes. The emphasis is put on studying the transitional regime, as this environment should experience strong climatic fluctuations during a glacial cycle and, if not surpassed by the polar front, also continuously support a coccolithophorid flora.

Table 6.1 — Positions and water depths for studied cores

Core	Latitude	Longitude	Water depth (m)
DSDP Hole 552A	56° 03′ N	23° 13′ W	2311
IOS D10611	52° 31′ N	31° 42′ W	3897
P-23235	78° 52′ N	01° 20′ E	2525
RC 8-3	42° 53′ S	42° 21′ E	4430
RC 11-120	43° 31′ S	79° 52′ E	3193
RC 13-205	02° 17′ S	05° 11′ E	3731
V22-108	43° 11′ S	03° 15′ W	4171
V27-60	72° 11′ N	08° 35′ E	2525

Piston core P-23235 was collected by R/V 'Polarstern' from the Fram Strait between Svalbard and Greenland (Augstein *et al.* 1984). The Fram Strait is the main passageway for water exchange between the Arctic and the rest of the world's seas. During most of the year, the site is located north of the polar front but during a couple of months in the summer it is influenced by relatively warm and saline waters from the West Spitsbergen Current, which carries North Atlantic surface waters into the Arctic Ocean. Thus, this site is for most of the year located in the arctic oceanographical regime but during the summer in a subarctic environment, close to the polar front. It is difficult to obtain a reliable chronology in sediments from the arctic environment (Gard 1987, Løvlie *et al.* 1986), but core P-23235 represents one of the best known sequences. It has been dated by calcium carbonate, Th:U, magneto- and nannofossil stratigraphies (Botz pers. commun., Bleil pers. commun., Gard in press), although there are still question marks as to the exact positions of the

oxygen isotope stage boundaries.

The location of piston core V27-60 from the Norwegian Sea is also ice covered for most of the year but the Norwegian Current pushes the polar front northwards during the subarctic summer, which is slightly longer at this site than at the location of P-23235. This core has yielded an excellent oxygen isotope stratigraphy for stages 1–6 (Labeyrie and Duplessy 1985).

From the northern North Atlantic, DSDP Hole 552A and IOS piston core D10611 were studied. Both cores are located under North Atlantic Drift water and the present-day oceanographic setting can be described as transitional between subarctic and subtropical (Ruddiman and McIntyre 1977). The oxygen isotope stratigraphy of DSDP Hole 552A was described by Shackleton and Hall (1984). The oxygen isotope stratigraphy of D10611 was provided by N. J. Shackleton (pers. commun.)

In order to assess whether the coccolith assemblages influenced by Antarctic glaciations varied similarly to the North Atlantic assemblages, three piston cores from transitional waters in the southern hemisphere were studied. Core V22-108 was collected from the South Atlantic and the oxygen isotope stratigraphy of this core was provided by N. J. Shackleton (pers. commun.). From the Indian Ocean, cores RC8-39 and RC11-120 were chosen as they have been shown to comprise excellent records of the last glacial cycle. Details of the oxygen isotope stratigraphy in RC11-120 can be found in Martinson *et al.* (1987). The oxygen isotope stratigraphy of RC8-39 was provided by N. J. Shackleton (pers. commun.). These three cores are all located under the Antarctic West Wind Drift Current.

In order to establish whether the observed differences in the coccolith assemblages are related to environmental variations rather than general evolutionary trends, a reference core from the tropical environment was also analysed. Core RC13-205 is located just south of the equator off West Africa in the Gulf of Guinea, an area which is influenced by the tropical South Eastern Central Current. The oxygen isotope stratigraphy of this core was provided by N. J. Shackleton (pers. commun.).

The palaeo-oceanographical conditions during the last glacial cycle are comparatively well known in the study area (CLIMAP Project Members 1976, Hays *et al.* 1976, Ruddiman and McIntyre 1977). Thus it should be possible to correlate coccolith abundance variations with palaeo-oceanographical parameters.

All cores have been sampled with about 10 cm intervals, except RC8-39 which had a significantly higher sedimentation rate and was sampled at about 20 cm intervals. Smear slides were prepared and studied under the light microscope at a magnification of × 800. The numbers of specimens in 25 fields of view of the following taxonomic categories were recorded: placoliths < 5 μm, *Coccolithus pelagicus, Calcidiscus leptoporus* and *Helicosphaera carteri*. The abundances are expressed as the number of specimens per square millimetre in the smear slide. Detailed descriptions of this technique can be found in Backman and Shackleton (1983). The placoliths < 5 μm consists almost exclusively of *Emiliania huxleyi* and *Gephyrocapsa* spp. These were grouped as they are difficult to differentiate using light microscope techniques. *Emiliania huxleyi* evolved from the *Gephyrocapsa* complex during oxygen isotope stage 8 (McIntyre 1970, Thierstein *et al.* 1977). These taxa are obviously closely related because coccoliths of both *E. huxleyi* and *Gephyrocapsa* spp. occasionally occur on the same coccospheres (Clocchiatti 1971, Winter *et al.* 1979). The genus *Gephyrocapsa* evolved in the Pliocene (Samtleben 1980) from a complex of similarly constructed small placoliths. *Coccolithus pelagicus* is an extremely long-ranging species as it appeared in the Early Palaeocene. *Calcidiscus leptoporus* and *H. carteri* evolved in the Miocene (Perch-Nielsen 1985).

The species recorded represent the main components of the fossil coccolith assemblages (except *Florisphaera profunda*) and are, except for rare *Syracosphaera pulchra*, the only species that are present in sediments from the northernmost study area.

6.3 EFFECTS OF DIFFERENCES IN SEDIMENTATION RATES

The studied cores show some variations in sedimentation rates, between both each other and different oxygen isotope stages. The largest difference, which appears as a common character in all cores except in RC11-120, is that oxygen isotope substage 5e shows a relatively high sedimentation rate and that substages 5a–5d show a comparatively low sedimentation rate. In RC11-120, oxygen isotope stage 2 shows the highest sedimentation rate and stage 5e the lowest. All cores, except RC8-39, have a relatively high sedimentation rate also during oxygen isotope stage 1. Abundances of coccoliths per time unit recorded from intervals with high sedimentation rates are suppressed owing to dilution by other particles. Coccoliths are concentrated in the oxygen isotope substage 5a–5d interval where the sedimentation rate has been low.

Figs. 6.2–6.6 show the abundances of coccoliths recorded per square millimetre in smear-slides from each level, multiplied by the sedimentation rate (cm/kyr) calculated for each oxygen isotope stage at that particular site. Thus the effects of coccolith abundance variations being caused by differences in sedimentation rates are avoided.

6.4 COCCOLITH ABUNDANCE VARIATIONS

The small placoliths are generally the dominant component of Quaternary coccolith assemblages, although their dominance in volume is not so pronounced as suggested by the diagrams (Figs. 6.2–6.6) because they are about half the size, or less, of the other coccoliths studied.

The abundance variations of the small placoliths through time in the studied cores are shown in Figs. 6.2–6.3. In the two northernmost cores (P-23235, V27-60), the small placoliths are only present in significant numbers during the relatively warm oxygen isotope stages 1 and 5. These cores are virtually barren of all coccoliths during oxygen isotope stages 2, 3 and 4. In the

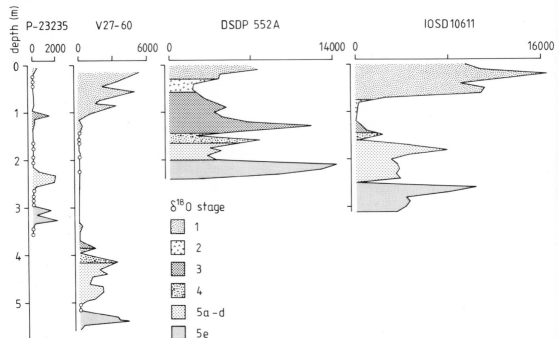

Fig. 6.2 — Abundance variations (specimens/mm² × sedimentation rate) of placoliths < 5 µm (*Emiliania huxleyi* and *Gephyrocapsa* spp.) through oxygen isotope stages 1–5 in P-23235, V27-60, DSDP Hole 552A and in IOS D10611.

transitional cores, the total abundance of the small placoliths varies strongly between different oxygen isotope stages. The small placoliths seem to have favoured conditions during oxygen isotope substage 5e, around the 5–4 boundary and during the first half of oxygen isotope stage 3. They occur with low abundances during the fully glacial stages 2 and 4.

Coccolithus pelagicus (Fig. 6.4) is together with the small placoliths (in this case *E. huxleyi*) the dominant species in the North Atlantic and Subarctic area at present. *Coccolithus pelagicus*, however, is at present absent in the low latitudes and occurs only rarely in the mid and high latitudes of the southern hemisphere (McIntyre and Bé 1967, Winter pers. commun.). Below stage 1 in the North Atlantic and Subarctic area, *C. pelagicus* shows a second, but much smaller, abundance peak around the stage 5–4 boundary. In cores from the southern hemisphere, *C. pelagicus* is most abundant during oxygen

isotope stage 3, particularly during the later half of this stage, but is relatively rare during all other time periods studied. In this area, it occurs in low abundances during the interglacial stages, 1 and 5.

Calcidiscus leptoporus (Fig. 6.5) is never abundant in the Subarctic area. Its only significant occurrences are during oxygen isotope stages 1 and 5e. *Calcidiscus leptoporus* follows this general pattern in the North Atlantic by having abundance peaks during stages 1 and 5e. Core RC13-205 shows no significant variations in the abundance of *C. leptoporus*. In the South Atlantic, *C. leptoporus* is currently, together with the small placoliths, the dominant species and has been so during most of oxygen isotope stage 1. A second abundance peak occurs in oxygen isotope substage 5e. In the Indian Ocean cores, *C. leptoporus* is moderately abundant during oxygen isotope stages 1–5a but more common during the warm substage 5e. However,

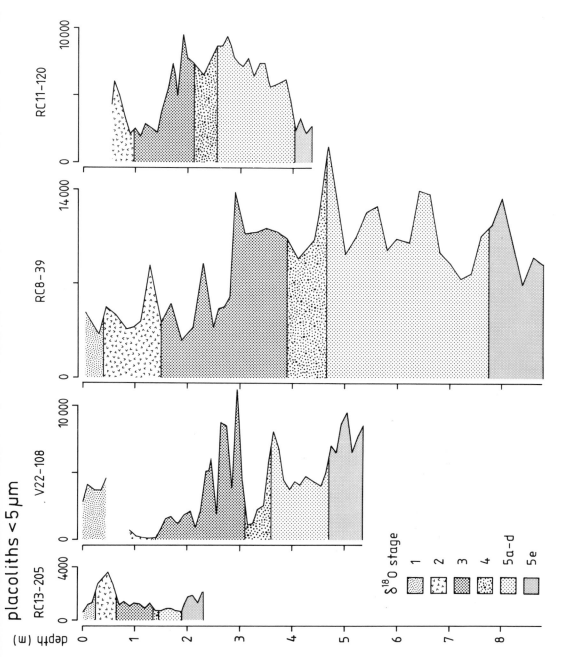

Fig. 6.3 — Abundance variations (specimens/mm^2 × sedimentation rate) of placoliths <5 μm (*Emiliania huxleyi* and *Gephyrocapsa* spp.) through oxygen isotope stages 1–5 in RC13-205, V22-108, RC8-39 and in RC11-120.

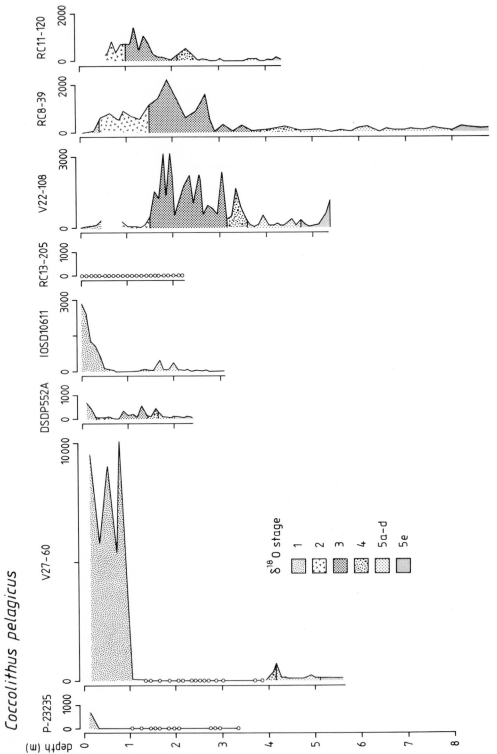

Fig. 6.4 — Abundance variations (specimens/mm^2 × sedimentation rate) of *Coccolithus pelagicus* through oxygen isotope stages 1–5.

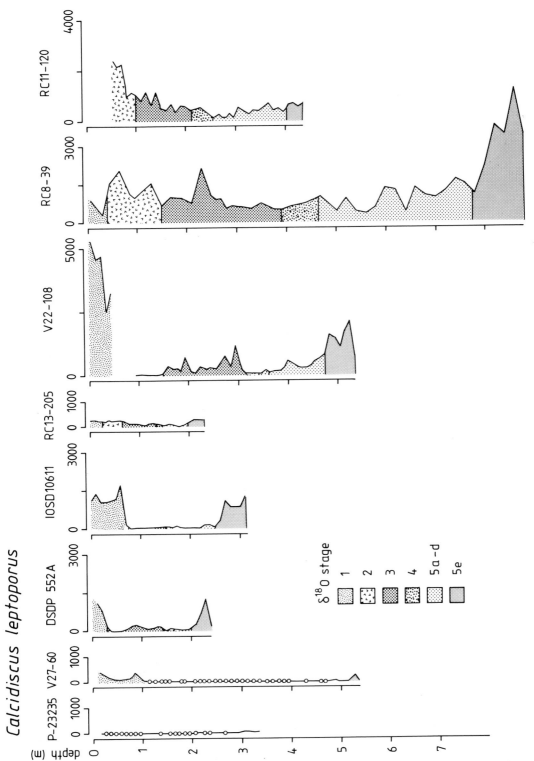

Fig. 6.5 — Abundance variations (specimens/mm^2 × sedimentation rate) of *Calcidiscus leptoporus* through oxygen isotope stages 1–5.

Helicosphaera carteri

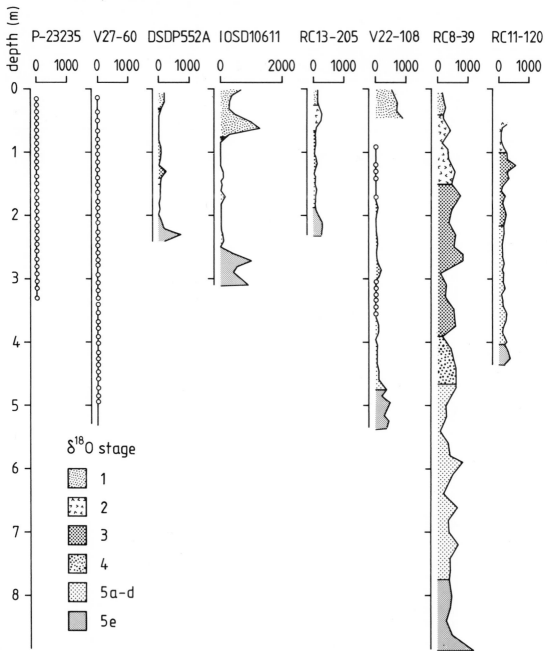

Fig. 6.6 — Abundance variations (specimens/mm^2 × sedimentation rate) of *Helicosphaera carteri* through oxygen isotope stages 1–5.

C. leptoporus shows increasing abundances in the later part of oxygen isotope stage 2 in RC11-120, which may indicate that *C. leptoporus* occurs in high abundances in this area today. Samples from the top of RC11-120 were not available.

Helicosphaera carteri (Fig. 6.6) essentially follows the abundance patterns of *C. leptoporus* but is consistently less frequent. *Helicosphaera carteri* occurs only rarely in V27-60 and is not observed in P-23235.

6.5 PALAEO-OCEANOGRAPHICAL INTERPRETATIONS

The fossil coccolith assemblages do not completely reflect the actual coccolithophorid community that lived in the surface waters at the time. The life-cycles of most coccolithophorids are not known, but may in some species include stages where coccolith formation does not occur or when the coccoliths formed are of such a nature that they will not be preserved as fossils. Studies of *C. pelagicus* have shown that two types of coccoliths are produced, the *C. pelagicus* type of coccolith which easily becomes fossilised and a *Crystallolithus hyalinus* type of coccolith which does not (Parke and Adams 1960). These two forms bloom at different times of the year (Okada and McIntyre 1979). Comparisons of plankton and fossil assemblages have demonstrated that a significant part of the coccolithophorid flora living in the surface waters is constituted by species which rarely become preserved in the sediment (Geitzenauer *et al.* 1977). Different coccolithophorid species also form varying numbers of coccoliths. Thus, abundance fluctuations of coccoliths in the sediment may not directly correspond to actual differences in numbers of living cells.

The numbers of coccoliths in the sediments per time unit are also influenced by relative dilution from other particles. Thus, comparisons of abundances of coccoliths between different sites and time periods should take into consideration changes in sedimentation rates. However, the strong abundance fluctuations that occur in the fossil coccolith assemblages, both over time and

area, undoubtedly can be used to portray surface water conditions.

The lack of variation in the coccolith assemblages recorded in RC13-205 suggests that the equatorial environment has not varied strongly over the last glacial cycle. The abundance variations observed in the mid and high latitudes are probably induced by climatic fluctuations.

The lowest total abundances of coccoliths occur in the northernmost core (P-23235) and in the equatorial core (RC13-205). Core P-23235 is located close to the northern tolerance limit of coccolithophorids and obviously the harsh environment suppresses productivity. Going southwards into the Norwegian Sea, total abundances are increased to values closer to the mid-latitude cores, implying that coccolithophorids thrive in the subpolar environment. Total abundances of coccoliths in the equatorial core are remarkably low over the entire glacial cycle. This core is collected from over 3700 m water depth and it is possible that some calcium carbonate dissolution has occurred at this site.

Coccolithophorids are not known to live in polar waters and McIntyre *et al.* (1972) interpreted coccolith barren zones in cores located above the CCD in the North Atlantic to have been caused by the polar front extending across the sites. The absence of coccoliths during the glacial stages in the Fram Strait and the Norwegian Sea simply indicates that this area experienced fully polar conditions during most of oxygen isotope stages 2–4. The two North Atlantic cores show pronounced minima during glacial stages 2 and 4, although the coccoliths never disappear completely. The polar front surpassed the location of these cores during stages 2 and 4, but Ruddiman and McIntyre (1976) explained the presence of coccoliths in stages 2 and 4 as being mixed into these short intervals from neighbouring layers. Core V22-108 from the South Atlantic contains a barren interval during oxygen isotope stage 2. Hays *et al.* (1976) concluded from studies of neighbouring cores at shallower depths that this barren interval was caused by carbonate dissolution and that the Antarctic polar front never passed the location of V22-108. Hays *et al.* (1976) also concluded that the Antarctic polar front moved a few degrees northwards during stage 2 but did not pass the locations of RC8-39 or RC11-120.

The small placoliths reach their highest total abundances in the subpolar and transitional environments. The observed abundance variations of *C. leptoporus* and *H. carteri*, with respect to both time and space, appear to be most simply explained by temperature gradients. These two species clearly increase in abundance with increasing sea surface temperatures. Geitzenauer (1969) observed in cores from the Pacific that increased abundances of *C. leptoporus* correlated to time periods of warm surface water temperatures estimated from radiolarians.

6.6 THE *COCCOLITHUS PELAGICUS* PROBLEM

The geographical and stratigraphical distribution of *C. pelagicus* is enigmatic. Currently, this cold-water-preferring species favours conditions in the northern North Atlantic and in the Subarctic area. However, it occurs only in low abundances in the transitional waters of the southern hemisphere. The core of the problem lies in the fact that *C. pelagicus* was not so abundant during oxygen isotope stage 5 as it is today in the North Atlantic and Subarctic area. Substage 5e could possibly have been too warm in the North Atlantic, but this could hardly have been the case in the Subarctic area. In the southern hemisphere, *C. pelagicus* seems to have favoured the intermediate glacial stage 3, which at the location of the studied cores at that time would have offered a subpolar–cold transitional environment (Hays *et al.* 1976). The low abundances today and during stage 5 in this area suggest that these time periods are too warm for *C. pelagicus*. So in the southern hemisphere cores, *C. pelagicus* abundances follow a logical pattern with high values during times of subpolar conditions and low values during times of warmer conditions.

The life-cycle of *C. pelagicus* includes also the motile *C. hyalinus* phase whose coccoliths are not preserved as fossils. It cannot be excluded that some intervals where this species is absent arise because the *C. pelagicus* phase was never formed. As the two phases bloom at different times of the year (McIntyre 1979), variations in for example insolation possibly could inhibit the bloom of the *C. pelagicus* phase.

McIntyre *et al.* (1970) suggest that *C. pelagicus* is rare in the southern hemisphere today because, during the post-glacial climate optimum, its niche in subpolar-transitional waters disappeared. They argued that the climatic gradient in this area was extremely steep and that polar waters met subtropical waters. This interpretation leads to the speculation that a similar situation in the northern hemisphere could explain the low abundances of *C. pelagicus* during oxygen isotope stage 5e. A rapid expansion of subtropical waters into the northern North Atlantic before the ice-cover had retreated very far in the early part of stage 5e would wipe out the niche of *C. pelagicus*. There are indications that this species started to regain its position in oxygen isotope stage 5a (Fig. 6.4) but, before reaching populations comparable with those of today, the onset of glaciation in stage 4 temporarily exterminated coccolithophorids from the northern North Atlantic. However, as *C. pelagicus* was able to repopulate the area rapidly in Holocene times, it seems unlikely that the *C. pelagicus* population was not able to recover faster in oxygen isotope substage 5a–5d.

Raffi and Rio (1981) studied the significance of *C. pelagicus* as a palaeotemperature indicator in the Late Pliocene and Pleistocene Mediterranean. They concluded that high abundances of *C. pelagicus* were indicative of cold surface waters during the Late Pliocene. The species disappeared from the Mediterranean during a warm period in the Early Pleistocene but did not repopulate the area despite low temperatures in the mid and Late Pleistocene.

From coccolith studies in the Subarctic area, it is known that *C. pelagicus* only occurs in high abundances two times during the last 500 kyrs, which includes several glacial-interglacial cycles. Those periods are during oxygen isotope stage 1 and during a short interval in oxygen isotope stage 7 or 8 (Gard in press). It can hardly be assumed that subpolar-transitional conditions have occurred only twice in the Subarctic area during the past 500 kyrs.

Excluding oxygen isotope stage 1 in the northern hemisphere cores, it is obvious that *C. pelagicus* and *C. leptoporus* have an inverse relationship, particularly pronounced in V22-108 (Figs. 6.4 and 6.5). A relationship with increased abundances of *C. pelagicus* and *E. huxleyi* together may exist. *Emiliania huxleyi* increased in abundance in mid latitudes at about oxygen isotope stage 5a–4 (Thierstein *et al.* 1977). This is the same time period when *C. pelagicus* formed an abundance peak in the northern hemisphere. From scanning electron microscopy studies it is known that the present small placolith category in the Subarctic area is almost exclusively composed of *E. huxleyi* (Gard 1986). In the Subantarctic area, however, Thierstein *et al.* (1977) failed to recognise a reversal in dominance between *Gephyrocapsa* spp. and *E. huxleyi* in core RC8-39. They interpreted this as a result of selective dissolution of *E. huxleyi* as this species has been reported to be common in subantarctic surface waters.

Looking at the total coccolith composition, it is obvious that the assemblages recorded during the later half of oxygen isotope stage 3 in the southern hemisphere cores are similar to those recorded during oxygen isotope stage 1 in the Subarctic area. This suggests, in agreement with Hays *et al.* (1976), that the southern hemisphere cores experienced subpolar conditions during this stage. The coccolith assemblage also suggest that the southern hemisphere cores during oxygen isotope stage 1 and today are slightly influenced by subtropical waters. The oceanographical gradients around the Antarctic are very steep in comparison with those in the Subarctic area. Temporary influences of subtropical waters could explain the absence of *C. pelagicus* and the comparatively high abundances of *C. leptoporus* and *H. carteri*.

In summary, it is obvious that *C. pelagicus* is a cold-water-preferring species that cannot exist in subtropical, or warmer, waters. Its rareness in areas which seem to offer reasonable living conditions for this species can have several reasons; (1) its niche in the area has recently been destroyed by influx of warm waters and the *C. pelagicus* population has not yet recovered; (2) the normally subpolar or transitional water masses are influenced by subtropical waters during particularly warm summers; (3) the *C. pelagicus* phase in the species life-cycle was never formed; (4) its distribution is also controlled by some unknown factor such as nutrient levels or competition from other plankton.

6.7 CONCLUSIONS

Fossil coccolith assemblages, both in the species composition and in relative and total abundances, reflect palaeo-oceanographical conditions in the surface waters.

High total abundances of coccoliths are found in the mid and high latitudes. However, in the immediate vicinity of the polar front, the production is suppressed. Coccoliths are absent under polar waters.

The small placoliths are the dominant taxonomic category in the entire area and time period studied. Their relative dominance is particularly pronounced in cold transitional water masses.

Calcidiscus leptoporus and *H. carteri* increase in relative abundance with increasing sea surface temperatures. The distribution of *C. pelagicus* is also related to sea surface temperatures. This species reaches maximum abundances in subpolar waters and decreases in abundance towards lower latitudes and is completely absent at the equator. However, *C. pelagicus* is rare in the Subantarctic area at present and in the Subarctic area during the last interglacial. Thus, reasons other than surface water temperatures also are responsible for the pattern of *C. pelagicus* abundances.

6.8 ACKNOWLEDGEMENTS

I thank J. Backman and P. Weaver for suggestions for improvements to this manuscript. I thank the Institute of Oceanographical Sciences, Deacon Laboratory, for samples from core IOS D10611. I thank U. Bleil for providing samples from core P-23235 and R. Botz for sharing unpublished results of this core. I am indebted to N. J. Shackleton, who provided samples from all other cores and who also provided details of unpublished oxygen isotope stratigraphies. I thank D. Q. Bowen for offering research facilities.

Financial support was given by the University of Stockholm and the Swedish Natural Science Research Council.

6.9 REFERENCES

Augstein, E., Hempel, G., Schwarz, J., Thiede, J. and Weigel, W. 1984. Die Expedition ARKTIS II des FS 'Polarstern' 1984. *Reports on Polar Research, Alfred-Wegener-Institute for Polar Research* **20**, 1–192.

Backman, J. and Shackleton, N. J. 1983. Quantitative biochronology of Pliocene and early Pleistocene calcareous nannofossils from the Atlantic, Indian and Pacific oceans. *Mar. Micropaleont.*, **8**, 141–170.

CLIMAP Project Members. 1976. The surface of the Ice-Age Earth. *Science*, **191**, 1131–1133.

Clochiatti, M. 1971. Sur l'existence de coccosphères portant des coccolithes de *Gephyrocapsa oceanica* et de *Emiliania huxleyi* (coccolithophoridés). *C.R. Acad. Sci. Paris.* **273(D)** 318–321.

Gard, G. 1986. Calcareous nannofossil biostratigraphy of late Quaternary Arctic sediments. *Boreas*, **15**, 217–229.

Gard, G. 1987. Late Quaternary calcareous nannofossil biostratigraphy and sedimentation patterns: Fram Strait, Arctica. *Palaeo-oceanography*, **2**, 519–529.

Gard G. in press. Late Quaternary calcareous nannofossil biozonation, chronology and palaeo-oceanography in areas north of the Faeroe-Iceland Ridge. *Quat. Sci. Rev.*

Geitsenauer, K. R. 1969. Coccoliths as late Quaternary palaeoclimatic indicators in the subantarctic Pacific Ocean. *Nature*, **223**, 170–172.

Geitzenauer, K. R., Roche, M. B., and McIntyre, A. 1977. Coccolith biogeography from North Atlantic and Pacific surface sediments. A comparison of species distribution and abundances. In A. T. S. Ramsay (Ed.), *Oceanic Micropalaeontology*, **2**, Academic Press, London, 973–1008.

Hays, J. D., Lozano, J. A., Shackleton, N. and Irving, G. 1976. Reconstruction of the Atlantic and western Indian Ocean sectors of the 18000 BP Antarctic Ocean. *Geol.Soc.Amer.Mem.*, **145**, 337–372.

Honjo, S. 1977. Biogeography and provincialism of living coccolithophorids in the Pacific Ocean. In A. T. S.

Ramsay (Ed.), *Oceanic Micropalaeontology*, **2**, Academic Press, London, 951–972.

Labeyrie, L. D. and Duplessy, J. C. 1985. Changes in the oceanic $^{13}C/^{12}C$ ratio during the last 140000 years: high latitude surface waters records. *Palaeogeog., Palaeoclimatol., Palaeoecol.*, **50**, 217–240.

Løvlie, R., Markussen, B., Sejrup, H.-P. and Thiede, J. 1986. Magnetostratigraphy in three Arctic Ocean sediment cores; arguments for geomagnetic excursions within oxygen isotope stages 2–3. *Phys.Earth Planet. Inter.*, **43**, 173–184.

Martinson, D. G., Pisias, N. G., Hays, J. D., Imbrie, I., Moore, T. C., Jr., and Shackleton, N. J. 1987. Age dating and the orbital theory of the ice-ages; Development of a high-resolution 0 to 300,000-year chronostratigraphy. *Quat. Res.*, **27**, 1–29.

McIntyre, A. 1967. Coccoliths as paleoclimatic indicators of Pleistocene glaciation. *Science*, **158**, 1314–1317.

McIntyre, A. 1970. *Gephyrocapsa protohuxleyi sp. n.* a possible phyletic link and index fossil for the Pleistocene. *Deep Sea Res.*, **17**, 187–190.

McIntyre, A. and Bé, A. W. H. 1967. Modern coccolithophoridae of the Atlantic Ocean I. Placoliths and cyrtoliths. *Deep Sea Res.*, **14**, 561–597.

McIntyre, A., Bé, A. W. H. and Roche, M. B. 1970. Modern Pacific coccolithophorida: a paleontological thermometer. *Trans. N.Y. Acad. Sci.*, **32**, 720–731.

McIntyre, A., Ruddiman, W. F. and Jantzen, R. 1972. Southward penetrations of the North Atlantic polar front: Faunal and floral evidence of large-scale surface water mass movements over the last 225000 years. *Deep Sea Res.*, **19**, 61–77.

Okada, H. and Honjo, S. 1973. The distribution of oceanic coccolithophorids in the Pacific. *Deep Sea Res.* **20**, 355–374.

Okada, H. and McIntyre, A. 1977. Modern coccolithophores of the Pacific and North Atlantic oceans. *Micropaleontology*, **23**, 1–55.

Okada, H. and McIntyre, A. 1979. Seasonal distribution of modern coccolithophorids in the western North Atlantic Ocean, *Mar. Biol.*, **54**, 319–328.

Parke, M. and Adams, I. 1960. The motile (*Crystallolithus hyalinus* Gaarder and Markali) and non-motile phase in the life-history of *Coccolithus pelagicus* (Wallich) Schiller. *J. Mar. Biol. Ass. UK.*, **39**, 263–274.

Perch-Nielsen, K. 1985. Cenozoic calcareous nannofossils. In H. M. Bolli, J. B. Saunders and K. Perch-Nielsen (Eds.), *Plankton Stratigraphy*, Cambridge University Press, 427–554.

Raffi, I. and Rio, D. 1981. *Coccolithus pelagicus* (Wallich): a paleotemperature indicator in the late Pliocene Mediterranean deep sea record. In F. C. Wezel (Ed.), *Sedimentary Basins of Mediterranean Margins.* CNR Italian Project of Oceanography, Tecnoprint, Bologna, 187–190.

Ruddiman, W. F. and McIntyre, A. 1976. Northeast Atlantic paleoclimatic changes over the past 600000 years. *Geol. Soc. Amer. Mem.*, **145**, 111–146.

Ruddiman, W. F. and McIntyre, A. 1977. Late Quaternary surface ocean kinematics and climatic change in the high-latitude North Atlantic. *J. Geophys. Res.*, **82**, 3877–3887.

Samtleben, C. 1980. Die Evolution der Coccolithophoriden-Gattung *Gephyrocapsa* nach Befunden im Atlantik. *Paläontol. Z,* **54**, 91–127.

Shackleton, N. J. and Hall, M. A. 1984. Oxygen and carbon isotope stratigraphy of Deep Sea Drilling Project Hole 552A: Plio-Pleistocene glacial history. *Init. Rep. DSDP.* **81**, 599–609.

Thierstein, H. R., Geitzenauer, K. R., Molfino, B. and Shackleton, N. J. 1977. Global synchroneity of late Quaternary coccolith datum levels: Validation by oxygen isotopes. *Geology,* **5**, 400–404.

Winter, A., Reiss, Z. and Luz, B. 1979. Distribution of living coccolithophore assemblages in the Gulf of Elat (Aqaba). *Mar. Micropaleontol.,* **4**, 197–223.

7

Temperature-controlled migration of calcareous nannofloras in the north-west European Aptian

Jörg Mutterlose

A rich and diverse calcareous nannoflora occurs in the 'middle' Aptian (Gargasian) of north-west Europe, while the lower- and uppermost Aptian (Bedoulian and Clansayesian) contain rather poor nannofossil assemblages. These rich and diverse nannofloras coincide with the occurrence of the '*Nannoconus truittii* assemblage', which is restricted to the marly Gargasian. This Boreal '*N. truittii* assemblage' shows close affinities to Tethyan ones, which have been described from throughout the Aptian. In north-west Germany these beds are furthermore dominated by *Rhagodiscus asper*, which outnumbers *Watznaueria barnesae*, normally the most abundant species in the Lower Cretaceous. The planktonic foraminifera (*Hedbergella infracretacea*) show similar close affinities with Tethyan faunas. A marly horizon at DSDP Site 511 (South Atlantic, Falkland Plateau) of presumably 'middle' Aptian age yielded a comparable '*N. truittii* assemblage', again associated with planktonic foraminifera of warm–subtropical affinities (*Ticinella roberti*).

This hypothesis of an influx of Tethyan derived floras and faunas into the Boreal Realm is supported by occurrences of the belemnite *Duvalia grasiana* and of the ammonites *Phylloceras* and *Zürcherella zürcheri*, all of which were immigrants from Tethys and normally uncommon in the Boreal Realm. This distribution pattern suggests an influx of warm water floras and faunas from Tethys into the North Sea and over the Falkland Plateau. For north-west Europe these warm water floras and faunas came presumably from the west via the English Proto-Channel and influenced both the English and north-west German 'middle' Aptian. The rich nannofloras and foraminifera of the marly layer in the Aptian part of Site 511 are explained in a similar way. It seems possible that this 'middle' Aptian influx of warm water floras and faunas occurred simultaneously in both areas. The abundant occurrence of *Nannoconus* spp is suggested to be an indicator of warm surface water.

7.1 INTRODUCTION

A threefold subdivision can be used in order to describe the lithology of the north-west European Aptian. Dark clays in the lowermost part (Bedoulian) of Early Aptian age and pale to varicoloured marls in the middle part (Gargasian, Hedbergellen-Mergel, *ewaldi-*, *clava-* and *inflexus*-Mergel of north-west Germany and Speeton, Sutterby Marl of Lincolnshire) of 'middle' Aptian age, are overlain by black clays of latest Aptian (Clansayesian; *jacobi-*, *nolani-* clays) and Early Albian age (*tardefurcata*-clay). In this context the Gargasian, which spans the uppermost Lower Aptian and the lowermost Upper Aptian, is referred to as 'middle' Aptian. The change in lithology is accompanied by a drastic change of the faunal and floral assemblages. The black clays either are rather poor in macrofossils or contain rich boreal ammonite faunas, while the intercalated marls yield a rich and diverse fauna in part of Tethyan origin. This distribution pattern suggests a major change in the palaeo-geographical and palaeo-oceanographical conditions in the north-west European Aptian.

A similar separation, based on lithological and palaeontological criteria, was used in France to distinguish the Bedoulian (lowermost Aptian), the Gargasian ('middle' Aptian) and the Clansayesian (Upper Aptian – Lower Albian), indicating that the changes in lithology, fauna and flora are not only restricted to north-west Europe, but are more widespread.

The distribution of the calcareous nanno-fossils was examined throughout the marl and parts of the dark clay facies, in order to detect changes in the assemblages. As a result of the occurrence of *Nannoconus* spp. within the marl facies of north-west Germany and based on further micro- and macropalaeontological evidence, warm water conditions were postulated for the 'middle' Aptian (Kemper 1987, Mutterlose 1987a). Since then further material from north-west Europe and the South Atlantic has been examined. New micro- and macropalaeontological data strongly support this idea and suggest that the palaeobiogeo-graphical distribution of the genus *Nannoconus* in the Aptian outside the Tethys is at least partly controlled by surface water temperature and palaeogeographical factors.

7.2 SECTIONS STUDIED

The material examined includes samples from outcrops in north-west Germany (Gott section, Rethmar section), samples collected by scuba divers in the southern North Sea (Helgoland), material from the Skegness borehole in eastern England and a DSDP borehole in the South Atlantic (Site 511). In addition to this, published data on the KMS section of north-west Germany

(Čepek 1982) were compiled. For the location of the sampled sites see Fig. 7.1.

For the examination of the calcareous nannofossils under the light microscope, slides of the Gott, Rethmar and Helgoland sections were prepared according to the method described by Hay (1965) and Čepek (1981), whereas simple smear slides were made for DSDP Site 511 and the Skegness borehole.

The nannofloras were examined under a Zeiss polarising light microscope. For each sample 300 identifiable individuals or a maximum of 200 fields of view were counted at $1500 \times$. Some of the samples were investigated under a scanning electron microscope.

(a) Gott section, north-west Germany (Figs. 7.2–7.4)

Location: about 30 km south of Hanover. TK 25 Sarstedt, no. 3725, re: 35 60 400, h: 57 90 650

Stratigraphy: this pit exposes Upper Aptian sediments (Gargasian, Clansayesian) resting unconformably upon Upper Barremian black clays; the Bedoulian (Lower Aptian) and the lower part of the 'middle' Aptian (Gargasian) are missing (Figs. 7.2 and 7.3). A detailed description of the lithology, stratigraphy and palaeontology of this section is given by Mutterlose (1984, 1987a).

Figs. 7.2 and 7.4 clearly show that the general distribution of the calcareous nannofossils reflects the main changes in lithology, supporting the idea of a threefold division of this sequence.

Nannofossils: the nannofossil assemblages from the Upper Barremian (samples 197/2 to 199/2) and uppermost Aptian black clays (sample 214/3) are poor in both diversity and abundance, while the intermediate *inflexus*-Mergel (samples 200/1 to 208/1) contains a rich nannoflora (Fig. 7.4). The diversity for samples derived from the *inflexus*-Mergel varies between 12–28 species, while the under- and overlying beds yielded only a very few species (*Rhagodicus asper, Watznaueria barnesae* and species of *Cretarhabdus*). The abundance follows a similar pattern: in the *inflexus*-Mergel the ratio I/mm² (individuals per mm² on a slide) ranges between 11–111, while the

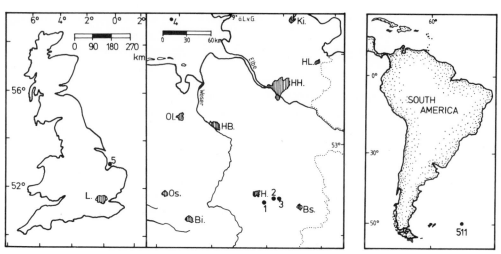

Fig. 7.1 — General map showing the geographical situation of the sections mentioned in the text. 1 = Gott, 2 = Rethmar, 3 = KMS, 4 = Helgoland, 5 = Skegness borehole, 511 = DSDP Site 511. The numerals refer to those in Figs. 7.2 and 7.3

dark clays are nearly barren. Fig. 7.4 demonstrates further that the genus *Nannoconus* only occurs in the pale and varicoloured marls, making up to 12% of the total assemblage in sample 203/1. The black clays immediately below and above the marls are barren. *Nannoconus* is represented by *N. truittii* ssp. *truittii*, *N. truittii* ssp. *frequens*, *N. boletus* and *N. bucheri*, of which *N. truittii* is strongly dominant. Another peculiarity of this sequence is the sudden dominance of *R. asper* in the *inflexus*-Mergel. This species is present throughout most of the Boreal Lower Cretaceous; however, it never outnumbers the most common species *W. barnesae* and small *Biscutum*, as was observed in this interval. In samples 203 (30%), 204 (50%) and 205 (24%) *R. asper* is by far the dominant species, playing the role *W. barnesae* usually does.

Foraminifera: the foraminifera of these beds were examined by Lutze (1968) who postulated warm water to subtropical conditions during the deposition of the *inflexus*-Mergel. Some of the results are compiled in Figs. 7.2 and 7.4; further information can be summarised as follows.

(1) The diversity and abundance of the benthic foraminifera and ostracods is highest in the *inflexus*-Mergel (samples 201–202).

(2) The benthic foraminifera consist of large-sized specimens in these beds.

(3) Samples 201–205 are characterised by a mass occurrence of planktonic foraminifera (up to 5000 specimens/g sediment).

(4) Within planktonic foraminifera right-hand coiled forms dominate (>60%), suggesting warm water conditions.

(b) Rethmar section, north-west Germany (Figs. 7.2, 7.3, and 7.5)

Location: temporary section about 20 km south-east of Hanover. TK 25 Haimar, no. 3626, re: 35 68 320, h: 59 77 580.

Stratigraphy: in 1976, excavations near Rethmar exposed Aptian clays. Samples from this section cover the upper part of the Bedoulian and the lower part of the Gargasian, i.e. the Lower–Upper Aptian boundary interval (Figs. 7.2 and 7.3)

Nannofossils: the drastic increase of diversity and abundance from sample 64238 to sample 64239 is obvious (Fig. 7.5), corresponding to the change in lithology from dark clays to pale marls. While sample 64238 (Bedoulian) yields only 18 species, the diversity in samples 64239–64245 (*ewaldi*-Mergel; lowermost Gargasian) varies between 26–30 species. The abundance increases from 32

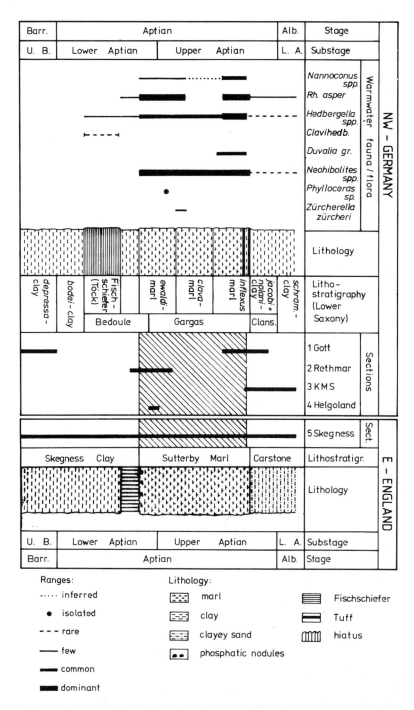

Fig.7.2 — Bio- and lithostratigraphy of the north-western European Aptian, showing the ranges of the studied sections. For lithology compare legend. The hatched areas indicate those parts of the sections which yielded *Nannoconus* spp. Barr. = Barremian, Alb. = Albian, U.B. = Upper Barremian, L.A. = Lower Albian, *Rh. asper* = *Rhagodiscus asper*, *Clavihedb.* = *Clavihedbergella*, *Duvalia gr.* = *Duvalia grasiana*, *schram.* -clay = *schrammeni* - clay.

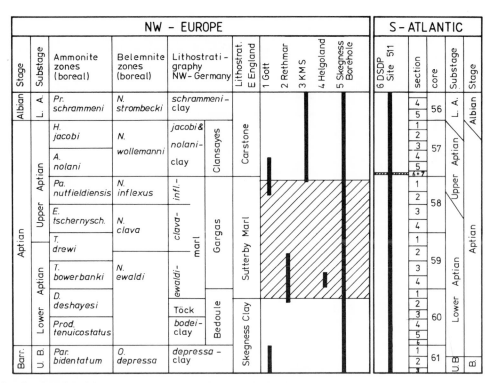

Fig. 7.3 — Stratigraphy of the Aptian of north-west Europe and the South Atlantic. The hatched areas indicate those parts of the sections which yielded *Nannoconus* spp. Barr. = Barremian, U.B. = Upper Barremian, L Alb. = Lower Albian, *Par.* = *Parancyloceras*, *Prod.* = *Prodeshayesites*, *D.* = *Deshayesites*, *T.* = *Tropaeum*, *E.* = *Epichelaniceras*, *Pa.* = *Parahoplites*, *A* = *Acanthoplites*, *H.* = *Hypacanthoplites*, *Pr.* = *Proleymeriella*, *O.* = *Oxyteuthis*, *N.* = *Neohibolites*.

I/mm² (sample 64238) to more than 300 (samples 64239–64234) decreasing to 221 and 166 I/mm² in samples 64244 and 64245.

Nannoconus (*N. elongatus*, *N. globulus*, *N. truittii*) is only present in the *ewaldi*-Mergel. The percentage of *Nannoconus* is relatively limited, recording a maximum of 3.5% (sample 64240); the most abundant *Nannoconus* is again *N. truittii*. While the lower samples are dominated by *W. barnesae*, the upper part of the section shows an increase in *R. asper*, which in one case outnumbers *W. barnesae* (sample 64243). *Watznaueria barnesae* shows a steady decline from more than 60% in the lowermost sample (64238) to less than 30% in sample 64245. However, *R. asper* is not so clearly characterised by a continuous increase, although the values for the *ewaldi*-Mergel are higher than those of the Bedoulian.

Foraminifera: in contrast to sample 64238 which yielded an impoverished foraminiferal fauna, samples 64239–64245 contain rich calcareous foraminifera including abundant *Hedbergella infracretacea*.

(c) KMS section, north-west Germany (Figs. 7.2, 7.3 and 7.6).
Location: the samples come from three different localities, K (Mittellandkanal) Vö (Vöhrum, new pit) and Alt (Altwarmbüchen, abandoned pit), all of them in between 5–30 km east and south-east of Hanover.
K: TK 25 Hämelerwald no. 3626, from re: 35 75 900, h: 57 98 180 to re: 35 78 180, h: 57 98 240.
Vö TK 25 Hämelerwald no. 3626, re. 35 78 800, h: 58 00 000.
Alt: TK 25 Großburgwedel. no. 3525, re: 35 57 860, h: 58 12 300.

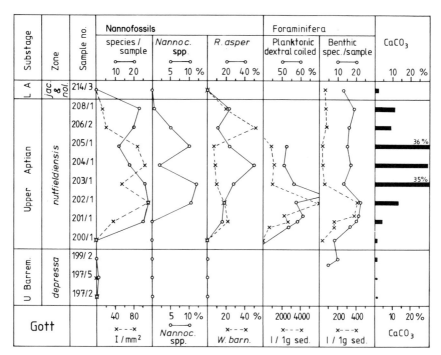

Fig. 7.4 — Chart for the Upper Barremian and Upper Aptian of the Gott section (north-west Germany), showing the distribution of calcareous nannofossils. The distribution of the planktonic and benthonic foraminifera was adopted after Lutze (1968, Fig. 4). I/mm² = individuals/mm² per slide: I/1g sed. = individuals/1 g sediment.

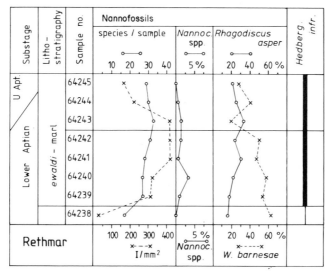

Fig. 7.5 — Chart for the Lower–Upper Aptian boundary interval of the Rethmar section (north-west Germany), showing the distribution of calcareous nannofossils. The distribution of the planktonic foraminifera was adopted from an unpublished report of the BGR (Hanover). I/mm² = individuals/mm² per slide.

Stratigraphy: the samples from this combined section cover the upper part of the marl facies (Gargasian; samples K1–K4) and the Upper Aptian black clay facies (Clansayesian; K5–Alt3) continuing into the Albian (Figs. 7.2 and 7.3). The calcareous nannofossils of this section have been studied by Čepek (1982), and a detailed analysis of the sequence is given by Kemper (1982). Fig. 7.6 was compiled after a detailed range chart given by Čepek (1982). The lower part of the KMS section (samples K1–K4) the *nutfieldiensis* Zone, can be correlated with the uppermost part of the same zone in Gott (approximately sample 208).

Nannofossils: the diversity decreases near the facies change and the remaining part of the Upper Aptian–Lower Albian is characterised by assemblages of low abundance and diversity (Fig. 7.6). The most diverse floras are to be found in the lower part of the sequence (samples K1–K12); the overall diversity does not, however, exceed 13 species. Within this part of the section a decrease of species becomes obvious. The abundance follows a similar pattern, although the concentration of nannofossils in samples K1–K5, K8, K9 and K11 is unusually high. The nannofossil content of these sediments is amongst the highest observed in the entire Lower Cretaceous of north-west Germany. A single specimen of *Nannoconus* was noticed in only one sample of the marl facies (K2). The percentage of *R. asper*, which outnumbers *W. barnesae* in the lower part (samples K1–K3), diminishes in the black clays (samples K4–Alt3), where *W. barnesae* makes up to 50–90% of the total assemblage.

(d) Helgoland, southern North Sea (Figs. 7.2 and 7.3)

Locality: a few samples were collected by scuba divers east of Helgoland from the 'Skit Gatt' on the sea bed.

Stratigraphy: the samples were attributed to the 'middle' Aptian *ewaldi*-Mergel (Gargasian). Since there is no proper section, no detailed range charts can be given.

Nannofossils: the nannofossil assemblages are diverse and show high abundances. *Nannoconus elongatus* and *N. truittii* have been observed.

(e) Skegness borehole, eastern England (Figs. 7.2 and 7.3)

Location: the Skegness borehole is situated in the north-eastern part of the Wash. Grid Ref. TF 5711 6398.

Stratigraphy: this well recovered about 4 m of Aptian sediments of a shallow water facies (Figs. 7.2 and 7.3). The Lower Aptian is represented by the Skegness Clay (1.98 m) and the 'middle' Aptian by the Sutterby Marl (1.98 m), here considered to be coeval with the marl facies of north-west Germany. This correlation is based on the presence of *N. ewaldi (sensu* Swinnerton, 1935) throughout the Sutterby Marl. Swinnerton (1935) included three species (*N. ewaldi, N. clava, N. inflexus*) in *N. ewaldi*; thus the Sutterby Marl is the equivalent of the north-west German *ewaldi-, clava-* and *inflexus*-Mergel. The Skegness Clay is the equivalent of the lower black clay sequence (Bedoulian) of north-west Germany. Lithology and stratigraphy were described by Gallois (1975), and the calcareous nannofossils were examined by Taylor (1982). Eight additional samples were studied by the author.

Nannofossils: while the Skegness Clay (lowermost Aptian, Bedoulian) is barren, except for rare *W. barnesae*, the overlying Sutterby Marl ('middle' Aptian, Gargasian) contains a rich nannoflora. This includes several species of *Nannoconus (N. globulus, N. steinmannii, N. truittii, N. elongatus, N. boletus)*, making up between 2% and 4% of the total assemblage. *Rhagodiscus asper* is abundant, but it does not outnumber *W. barnesae*.

(f) DSDP Site 511, South Atlantic (Figs. 7.3 and 7.7)

Locality: Site 511 (DSDP Leg 71) is located on the Falkland Plateau. 51° 00.28′ S, 46° 58.30′ W.

Stratigraphy: about 124 m of Lower Cretaceous sediments were recovered from this site. The Upper Barremian–Lower Albian sediments mainly consist of black shales with some thin

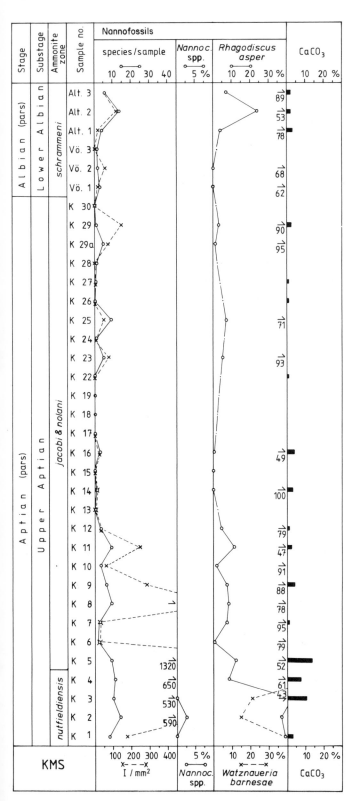

Fig. 7.6 — Chart showing the stratigraphy and distribution of calcareous nannofossils of the uppermost Aptian and lowermost Albian of the KMS section (north-west Germany), according to Čepek (1982). I/mm² = individuals/mm² per slide.

marly layers, while the 'middle' Albian is characterised by reddish to brown chalks. A rough lithological description was given in the site Report, calcareous nannofossils were described by Wise (1983), planktonic foraminifera by Krasheninnikow and Basov (1983) and macrofossils by Jeletzky (1983). As shown in Figs. 7.3 and 7.7, the positions of the Barremian–Aptian and the Aptian-Albian boundaries in this site are rather questionable. According to Krasheninnikow and Basov (1983) the Aptian–Albian boundary should be placed at the base of section 511-57-6. This includes the whole of section 511-57-6, which will be discussed later, in the lowermost Albian. This chronostratigraphic limit, based on the first occurrence of *Ticinella roberti* in section 511-57-6, was adopted for DSDP Initial Report 71 (Basov *et al.* 1983). Jeletzky (1983, p. 954), in the very same volume, suggested a slightly different position for the Aptian–Albian boundary. Because of the occurrence of the ammonite genus *Cheloniceras* (section 511-57-4) 'the upper part of this interval (sections 511, 57-3 to 57-5) is unreservedly assigned a Late Aptian age'. On the basis of these data and new investigations on the nannobiostratigraphy different boundaries are suggested in the present study (Figs. 7.3 and 7.7) The Barremian–Aptian boundary is tentatively drawn below the first occurrence of *Micrantholithus* sp. 1 (Perch-Nielsen 1979) in sample 511-61-4, 140–141 cm. It is possible that the Aptian extends further down; the Aptian–Albian boundary is based on ammonites.

Nannofossils: 37 samples from the Upper Barremian–Lower Albian interval (cores 56–62) were examined in respect to the calcareous nannofossils. Both diversity and abundance vary throughout the section; rich nannofloras characterise the intervals from 511-61-5 to 511-61-3, from 511-58-3 to 511-57-6 and from 511-57-4 to 511-56-1. There is, however, no general trend (see Fig. 7.7)

Rare specimens of *Nannoconus* were observed only in a few samples, whereas samples 511-57-6, 11–13 cm, 15–16 cm and 18–20 cm show nannofossil assemblages consisting up to 6.5%

of *Nannoconus* including *N. elongatus, N. truittii, N. multicadus, N. circularis* and *N. vocontiensis.* These occurrences correspond to an overall increase in diversity.

The calcium carbonate data, which have been determined for all the examined samples, show that this peak of *Nannoconus* in the Upper Aptian is obviously restricted to a thin, 20 cm thick marly horizon, which is intercalated in the black shales. *W. barnesae* is dominant throughout the sequence, always outnumbering *R. asper*, which is rare in most samples. The latter becomes only more abundant from 511-57-6, 18–20 cm to 511-57-6, 11–13 cm, making up between 10–20% of the total assemblage.

Foraminifera: Fig. 7.7 gives the ranges of the planktonic foraminifera in the interval discussed. According to Krasheninnikow and Basov (1983) the foraminiferal assemblages consist mainly of different species of *Hedbergella*, which range throughout the Aptian. There are, however, rare occurrences of *T. roberti* in samples 511-57-6, 11–13 cm, 511-56-1, 100–101 cm, 511–55, CC and 51-55-6, 34–36 cm (the latter two are not plotted on Fig. 7.7), whereas the rest of the Lower Cretaceous, including all the Barremian and Albian up to core 49, did not yield this species. In tropical–subtropical areas the total range of *Ticinella* occupies the entire Albian. Based on the different stratigraphical range in the tropical–subtropical areas and the Falkland Plateau, Krashenninnikow and Basov (1983, p. 801) consider *Ticinella* to be a stenotherm genus, which penetrated into the area of the Falkland Plateau during periods of warming or with incursions of warm water surface currents. Note the co-occurrence of *Nannoconus* spp. and *T. roberti* in samples 511-57-6, 11–13 cm, 15–16 cm, 18–20 cm.

7.3 NANNOBIOSTRATIGRAPHY

Several zonation schemes based on nannofossils have been suggested, in order to subdivide the sequence from the Upper Barremian to the lowermost Albian. An overview is given by Perch-Nielsen (1985, p. 341). The most recent zonation, based on material from the North Sea,

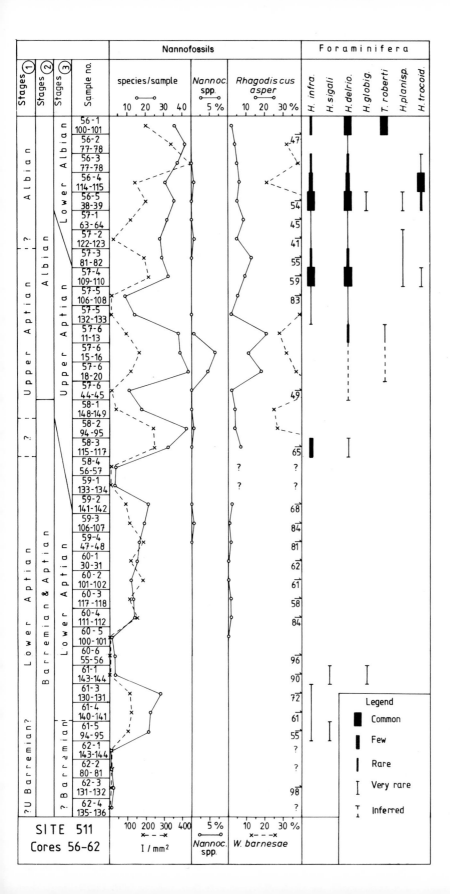

Fig. 7.7 — Stratigraphy and distribution of selected calcareous nannofossils from the Aptian of DSDP Site 511 (South Atlantic, Falkland Plateau). Stratigraphy of column 1 based on ammonites (Jeletzky 1983), of column 2 based on planktonic foraminifera (Krasheninnkow and Basov, 1983), of column 3 based on ammonites and calcareous nannofossils. The ranges of the planktonic foraminifera were adopted from Krasheninnkow and Basov (1983). Dotted line = inferred ranges.

has been proposed by Jakubowski (1987). In this study four nannofossil zones (NLK 9 to NLK 6) were recognised for the interval discussed.

The base of NLK 9 is defined by the last occurrence of *Nannoconus abundans.* This zone ranges from the latest Barremian into the Early Aptian. *Lithraphidites moray-firthensis* is common in NLK 9, but this taxon extends into the Late Barremian. NLK 8, which spans the latest Early Aptian and the earliest Late Aptian, is approximately equivalent to the 'middle' Aptian Gargasian.

NLK 8 can be subdivided into two subzones, 8 B (lower part of NLK 8) and 8 A. The base of NLK 8 B is characterised by the first consistent occurrence of *Eprolithus varolii,* the base of 8 A by the first appearance of *Rhagodiscus angustus* and *Eprolithus apertior,* and the top of 8 A by the end of the *R. asper* acme. The base of NLK 7 is defined by a lower abundance of *R. asper,* the top by the last occurrences of *Micrantholithus hoschulzii/obtusus* and *E. varolii.* This zone represents the latest Aptian. The north-western European sections, which have been studied in more detail, can all be assigned to NLK 8 B to NLK 6, apart from the Barremian part of the Gott section.

The Gargasian of the Gott section (samples 201–208) yields *R. angustus* and *E. varolii* and is thus correlated with NLK 8 A of Jakubowski (1987). This is supported by the *R. asper* acme, but no *M. hoschulzii/obtusus* have been observed.

These nannofossil data correspond to the macropalaeontological data which indicate a middle Late Aptian age for these samples.

The Rethmar section, including the Early– Late Aptian boundary, is assigned to the uppermost part of NLK 8 B and the lower part of NLK 8 A based on the first appearance of *R. angustus* in the lower part of the Gargasian.

The KMS section (Late Aptian–earliest Albian) extends from the top of NLK 8 A into the base of NLK 6, as indicated by the presence of *E. varolii,* the *R. asper* acme at the base and the absence of *Prediscosphaera.*

Finally, Site 511 can be dated by *Micrantholithus* sp. 1 (restricted to 511-61-5 and 511-61-4), *Conusphaera mexicana* (LAD: 511-61-3, 130–131 cm), *R. angustus* (FAD: 511-60-1, 30–31 cm), *M. hoschulzii/obtusus* (LAD): 511-57-3, 81–82 cm) and *Prediscosphaera* spp (FAD: 511-57-1, 63–64 cm).

The biostratigraphic data gathered from the north-western European sections suggest an *N. truittii* assemblage Zone, which is equivalent to the 'middle' Aptian Gargasian. A secondary index for this zone is the acme of *R. asper.* *Eprolithus varolii,* which extends throughout the Gargasian into the lowermost Albian, is first observed in the Bedoulian (Early Aptian), but up to now no complete Early Aptian sequences have been examined. *Rhagodiscus angustus* first occurs in the lowermost Gargasian (Rethmar sample 64239) and ranges throughout the Aptian. The

---

PLATE 7.1

Calcareous nannofossils of Tethyan origin from north-west Germany and the South Atlantic as indicators for warm water influences.

Scale of the bars: 1 μm.

Plate 7.1, Figs. 1, 2. *Rhagodiscus asper,* distal view; Zgl. Gott–Sarstedt (north-west Germany); bed no. 203, *nutfieldiensis* Zone (*inflexus*-Mergel), Late Aptian (Gargasian); sample no. Gott 203/1.

Plate 7.1, Fig. 3. *Rhagodiscus asper,* distal view; Rethmar section (north-west Germany); *ewaldi*-Mergel, uppermost Early Aptian (Gargasian); sample no. Rethmar 64240.

Plate 7.1, Fig. 4. *Rhagodiscus angustus,* distal view; Rethmar section (north-west Germany); *ewaldi*-Mergel, uppermost Lower Aptian (Gargasian); sample no. Rethmar 64240.

Plate 7.1, Fig. 5. *Nannoconus* sp.; Zgl. Gott–Sarstedt (north-west Germany); bed no. 203, *nutfieldiensis* Zone (*inflexus*-Mergel), Late Aptian (Gargasian); sample no. Gott 203/1.

Plate 7.1, Figs. 6, 7. *Nannoconus truittii*; Zgl. Gott–Sarstedt (north-west Germany); bed no. 203, *nutfieldiensis* Zone (*inflexus*-Mergel), Late Aptian (Gargasian); sample no. Gott 203/1.

Plate 7.1, Fig. 8. *Nannoconus* sp. cf. *N. truittii*; Zgl. Gott–Sarstedt (north-west Germany); bed no 203, *nutfieldiensis* Zone (*inflexus*-Mergel), Late Aptian (Gargasian); sample no. Gott 203/1.

Plate 7.1, Fig. 9. *Nannoconus* sp.; DSDP Site 511, 57–6, 18–20 cm (Falkland Plateau); Late Aptian; sample no. 57–6, 18–20 cm.

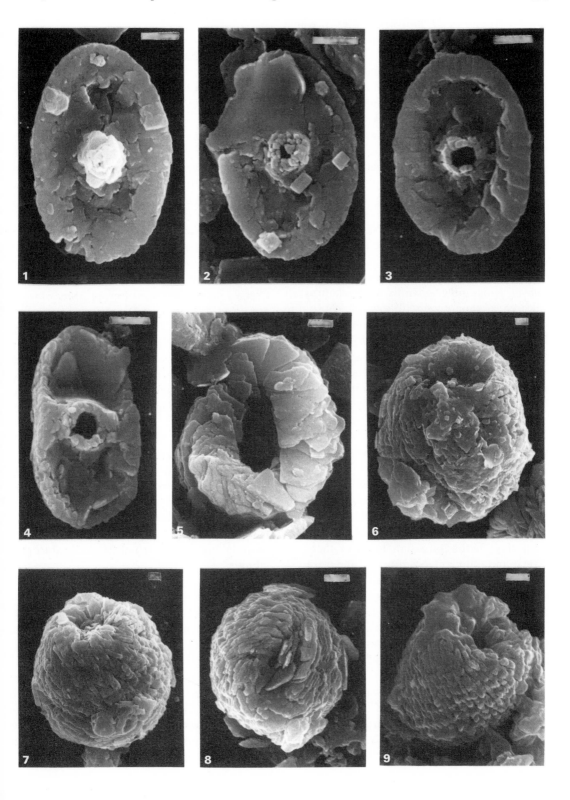

last occurrence of *M. hoschulzii/obtusus,* which has been noticed slightly above the Lower–Upper Aptian boundary, might be useful as well.

In addition to these species, the LAD of *C. mexicana* and *Micrantholithus* sp. 1, which until recently have been observed only in the South Atlantic, can be used to mark the lowermost Aptian. Examination of additional samples from north-west Europe will indicate whether these species are biostratigraphically useful in northern Boreal areas.

7.4 FACTORS CONTROLLING THE APTIAN NANNOFOSSIL DISTRIBUTION
(a) North-west Europe
Nannofossils: from the distribution described above, it is quite obvious that changes in calcareous nannofossils follow or cause the changes in lithology (Figs. 7.2 and 7.3). Both Bedoulian and Clansayesian consist of black clays, poor in $CaCO_3$. These beds are characterised by nannofossil assemblages poor in diversity and in most cases poor in abundance as well; *Nannoconus* is absent in these beds. The nannofossils of the intercalated marls of the Gargasian show assemblages rich in diversity and abundance, occurrences of *Nannoconus* spp., and an increase of *R. asper.*

Foraminifera: the above-described distribution pattern of the nannofossils is clearly mirrored in the distribution of the microfauna, which consists of impoverished assemblages in the black clays. In the Gargasian, however, there are diverse and abundant foraminiferal and ostracod assemblages, mass occurrences of *Hedbergella* spp. (5000 specimens/g sediment), rich calcareous foraminiferal assemblages, dominance of right-coiled foraminifera (up to 65%) restricted to certain beds, and large specimens of calcareous and agglutinated foraminifera.

Macrofossils: while the black clays are barren (Bedoulian) or nearly barren (Clansayesian) of belemnites, the intermediate Gargasian contains common *Neohibolites ewaldi, N. clava* and *N. inflexus* and *Duvalia grasiana (inflexus*-Mergel only).

Both *Neohibolites* and *Duvalia* are Tethyan-derived belemnite genera; however, while *Neohibolites* migrates to north-west Europe, *Duvalia* is an essentially Tethyan genus. In the Tethys *Duvalia* ranges from the Tithonian to the Aptian; from north-west Europe it is only known from the *inflexus*-Mergel. The north-west German Gargasian also yielded the following Tethyan derived ammonites: *Phylloceras morelli* was reported from the lowermost Upper Aptian and *Zürcherella zürcheri* is fairly common in these beds in the western part of the basin. Of special interest are the rare occurrences of the genus *Phylloceras*, which is quite common in the Lower Cretaceous of the Tethys. Up to now only six specimens are known from the Lower Cretaceous of north-west Europe, three of which were recorded from the Gargasian of north-west Germany. Koenen (1902) reported one specimen from the Bentheim area and one specimen from the Hanover area; very recently another specimen was found in the Bentheim area (Kaplan and Lehmann pers. commun.). These data have been compiled in Fig. 7.2.

(b) South Atlantic, Site 511
The Aptian sediments of this site show features similar to those in north-western Europe. In contrast to the over- and underlying black clays, the intercalated marls are characterised by the occurrence of *Nannoconus*, a highly diverse nannoflora and the presence of planktonic foraminifera (*T. roberti*) of tropical–subtropical affinities.

(c) Preservation
The obvious restriction of *Nannoconus* to horizons richer in calcium carbonate, which correlates with higher diversity and abundance of the nannofossil assemblages in these beds, may suggest a preservational factor explaining the poorer black clay assemblages by dissolution. Roth and Krumbach (1986), ranking the Aptian–Cenomanian coccoliths according to their dissolution resistance, pointed out that floral assemblages with *W. barnesae* >40% are

highly influenced by dissolution. In Site 511 *W. barnesae* makes up between 50% and 90% of the total assemblage in most samples including the black shales (Fig. 7.7) However, explaining this pattern by dissolution is contradicted by Parker *et al.* (1983), in the case of the Aptian black shales of Site 511. They demonstrated that the preservation and diversity of the nannoflora of the black shales did not produce the interbedded coccolith-rich and coccolith-poor layers, including the *Nannoconus*-rich horizons. The coccoliths are generally quite well preserved in the black shales and show little if any sign of etching. Even the most delicate forms such as species of *Corollithion* are quite well preserved in both limestone and clay facies. Thus the coccoliths of the black shales are apparently diverse and the floral assemblage has not been altered.

Also remarkable is the constitution of the floral assemblages in the lowermost 12 samples of the KMS section (Fig. 7.4). These samples, belonging to the uppermost part of the marl facies and the lowermost part of the clay facies, show a low diversity (2–14 species/sample), but an immense abundance (up to 1320 specimens/mm²; K5). These high abundance assemblages are correlated with peaks of *W. barnesae* (> 50%). If the same favourable conditions, which are typical for the Gargasian, had persisted into the black clay facies of the uppermost Aptian, abundances comparable with those of the Gott and Rethmar sections would have been expected, differing only by the absence of dissolution-susceptible forms such as *Corollithion*. Abundances for rich samples of the Gott and Rethmar sections range between 100–400 1/mm², never becoming as abundant as in the KMS section. The nannoconids, however, are likely to be solution resistant, for their walls do not show delicate elements. The idea that the constitution of the assemblages is not controlled by preservation is supported by the microfauna, which is clearly distinctive in the 'middle' Aptian. Both north-west Europe and Site 511 show Tethyan derived foraminifera at those intervals containing *Nannoconus*. Secondly, it is difficult

to explain the occurrence of right-coiled foraminifera by dissolution. Similarly the distribution pattern of the Tethyan belemnite genera *Neohibolites* and *Duvalia* (*Neohibolites* common throughout the marl facies, *Duvalia* only in the upper part, both genera absent in the black clays) cannot be explained by dissolution. Finally, the restricted occurrence of the Tethyan ammonite genera *Zürcherella* and *Phylloceras*, the latter in particular, in the Gargasian can only be explained by primary ecological changes.

Although dissolution cannot be totally excluded for the European sections, it is simpler to explain the nannofloral and faunal changes in the Aptian by major environmental and/or palaeogeographical alterations.

(d) Water depth
For the Late Aptian–Early Albian of the Atlantic, Roth and Krumbach (1986, p. 259) recognised a neritic factor with a high correlation of *Braarudosphaera* spp. and *Nannoconus* spp. This study, which is based on factor analysis, suggests that these forms are most abundant in shallow neritic settings. Similar statistics, which were performed by the author on Upper Hauterivian nannofossil assemblages from north-west Germany, show a *Micrantholithus* spp. – *Nannoconus* spp. factor restricted to thin marly horizons in the rhythmically bedded Upper Hauterivian. It is difficult to explain these numerous cycles of marl and clay horizons by a rhythmical change of the water depth. It seems more likely that this rhythmical bedding and the restriction of *Nannoconus* spp. to the thin marl beds are to be explained by warm water ingressions.

Most of the north-western European sections discussed above were probably deposited under shallow water inner shelf conditions all through the Aptian, although the water depth might have varied. The Gott and Skegness sections are nearshore sequences, situated about 30–40 km off the coast (Fig. 7.8). The KMS and the Rethmar sections display the basin facies of the Lower Saxony Basin and the North Sea. These data are corroborated by field observations, showing that the clay–marl–clay facies of the north-western European Aptian is to be found all through the eastern part of the Lower Saxony Basin. Although the sedimentation of the marl facies was partly controlled by the water depth, there are additional factors which influenced the distribution of *Nannoconus* spp.

For the Aptian of Site 511 Basov and Krasheninnikow (1983, p. 755) suggested a shallow water environment (cores 511–55 to 511–57) based on benthic foraminifera. They assumed water depths between 100 and 400 m for this interval. Shallow water conditions during the entire Aptian, however, do not explain the restriction of *Nannoconus* spp. to a 20 cm thin horizon, unless an uplift of the Falkland Plateau for a short period is postulated.

(e) Water temperature

Both in north-west Europe and the South Atlantic the occurrence of *Nannoconus* spp. in the marly parts of the 'middle' Aptian is linked to the appearance of Tethyan derived planktonic foraminifera (Figs. 7.2, 7.6 and 7.7). On the Falkland Plateau the tropical–subtropical indicator *T. roberti* is restricted to these horizons, indicating warm water conditions during this interval (Krasheninnikow and Basov 1983).

Lutze (1968) explained the composition of the foraminiferal assemblages of the Gargasian of north-west Germany by the incursion of warm to subtropical water into the Lower Saxony Basin. The assemblages are characterised by large sized benthic forams, high diversity and abundance of foraminifera and ostracods, mass occurrence of planktonic foraminifera, and dominance of right-coiled forms. The 'middle' Aptian *Hedbergella* event, represented by pale and varicoloured marls of a Gargasian age associated with rich–abundant *Hedbergella* faunas, is known from all over the North Sea (e.g. Lott *et al.* 1985, Rawson and Riley 1982) and even rare *Ticinella* sp. have been observed (D. Jutson pers. commun.).

Lutze's hypothesis for north-west Germany is strongly supported by the restriction of the stenothermal belemnite genus *Duvalia* to the upper part of the Gargasian, which coincides with the mass occurrence of the more common genus *Neohibolites* (Mutterlose 1987b), both of Tethyan origin. The occurrence of the Tethyan ammonite genera *Phylloceras* and *Zürcherella* can be explained in the same way.

The '*Nannoconus* event' of the Gargasian in the north-western European Aptian can be explained by the idea that *Nannoconus* is a Tethyan derived genus, whose palaeobiogeographical distribution is mainly controlled by the palaeogeographical setting, and a warm climate. While some of the species of *Nannoconus* in the Barremian are endemic to the Boreal area (*N. borealis* and *N. abundans*), the representatives of the '*N. truittii* assemblage' are likely to be of Tethyan origin. Warm water ingressions or warm periods could easily explain the synchronous occurrences of *Nannoconus* spp. in the Gargasian of both north-western Germany and eastern England.

Erba (1987) performed factor analysis with the calcareous nannofossils from the Aptian and Albian of the Umbrian–Marchaen basin and postulated for *R. asper* a warm water factor having a moderate primary productivity. Assuming that Erba is correct, the hypothesis linking nannoconids to warm water is supported by the fact that *R. asper* is most abundant in those beds which yielded *Nannoconus*. The nannofloras of those sites which have been evaluated in more detail (Gott, Rethmar, KMS, DSDP Site 511) clearly show that the occurrence of *Nannoconus* is related to relatively high abundances in *R. asper*. The Gott section which shows the most abundant *Nannoconus* assemblage yielded as well floras richest in *R. asper*. In north-west Europe the *R. asper* acme is restricted to those samples having the highest amount of $CaCO_3$, which corresponds to the 'middle' Aptian (Gargasian). As *R. asper* is common throughout the Lower Cretaceous (it has been observed in most samples), it seems likely that its distribution is primary ecologically controlled. Thus it is presumed that high abundances of *R. asper* and the presence of

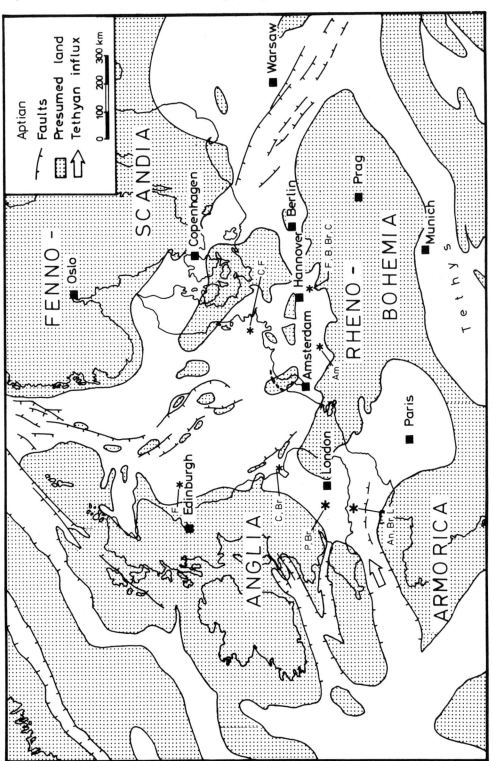

Fig. 7.8 — Palaeogeographical map for the Upper Aptian of north-west Europe, showing the presumed ingression of warm water faunas and floras from the west. The stars indicate areas in north-west Europe from which Tethyan derived faunas and floras have been described (Isle of Wight; south-west England, Farringdon sponge gravel; Skegness borehole; IGS borehole, western Central North Sea; Helgoland; Bentheim area; sections west and south-west of Hannover). The sizes of the stars indicate the amount of Tethyan influence. An = Anthozoa, B = Belemnites, Br. = Brachiopoda, C = Coccoliths, F = Foraminifera, L = Lamellibranchiata, P = Porifera. Palaeogeography modified after Ziegler (1982).

'Tethyan' nannoconids can be used as an indicator for warm surface water.

Generally speaking tropical waters are characterised by a higher number of taxa of all groups. In recent oceans there is an increase in diversity toward the equator (e.g. Strauch 1972, p. 115). In recent coccolithophores tropical and subtropical assemblages of the Atlantic contain over three times more species than the subarctic and subantarctic assemblages (Haq 1978, p. 87). These observations fit nicely into the above-described pattern of diversity for calcareous nannofossils. Highly diverse assemblages are associated with *Nannoconus* and *R. asper* in those beds, which are believed to have been deposited under a warm water regime. In contrast, floras of a low diversity are associated with cooler water. Okada and Honjo (1973), who studied the distribution of coccolithophores in the Pacific Ocean, pointed out that surface temperatures and distribution of surface currents are the most important factors limiting the biogeography of various species.

The idea of explaining the black clays and black shales of the Bedoulian and Clansayesian by a cooler or moderate surface water temperature and the intervening marls of the Gargasian by warmer water is supported by Kemper (1982), who postulated an upwelling of cold water for the black clays of the uppermost Aptian and Lower Albian of north-west Germany.

7.5 PALAEOBIOGEOGRAPHY AND MIGRATION PATTERNS

(a) Palaeobiogeography of the '*N. truittii* assemblage'

In 1955 Brönnimann described rich *Nannoconus* assemblages from Cuba, recognising for the Aptian–Albian interval the *N. truittii* association with *N. truittii, N. minutus, N. elongatus, N. bucheri* and *N. wassallii*. Since then the '*N. truittii* assemblage' has been described in detail from Mexico (Trejo 1960), the Southern Alps (Bouche 1963), Hungary (Baldi-Beke 1962), Spain (Geel 1966), Israel (Moshkovitz 1972) and France

(Deres and Achériteguy 1972). A more recent updated overview is given by Deres and Achériteguy (1980), including data from Spanish Sahara, Morocco, Algeria, Tunisia, Libya, Mozambique, Columbia, Surinam, Canada and offshore Ireland. Additional data are supplied by the DSDP reports. The sites which yielded Aptian *Nannoconus*-bearing sediments, however, are restricted to the Tethyan belt, except for those of the Falkland Plateau. Thus, apart from sites 327, 339 and 511, no essentially new data are supplied for the Aptian.

According to these data (compare Fig. 7.9) the '*N. truitti* assemblage' is mainly known from the Tethys, the northernmost occurrences being reported from the Irish Sea (loc. 4) and offshore east Canada (loc. 3). However, the '*N. truittii* assemblage' is not known further south than Surinam. *N. circularis* (Lower Aptian) and *N. boletus* ssp. *curtus* (Upper Aptian) only occur in Mozambique and the Falkland Plateau.

While various species of *Nannoconus* are common throughout the Aptian within the Tethys, in the more extreme settings of north-west Europe (locs. 1 and 2) and the Falkland Plateau (loc. 19) this genus is restricted to the 'middle' Aptian. This distribution pattern suggests a migration of some *Nannoconus* species into the north-western European basin and the Falkland Plateau during the 'middle' Aptian, when palaeoclimatic conditions favoured this migration. However, the pattern clearly contradicts an *in situ* evolution of *Nannoconus* at this time in these extreme settings.

(b) North-west Europe

The Lower Saxony Basin, about 280 km long and 80 km wide, can be regarded as the southernmost part of the North Sea, showing strong subsidence (see Fig. 7.8). This basin was separated from the Boreal seas during most of the Early Cretaceous by an archipelago, the Pompeckj' Swell, which, however, did not form a faunal–floral barrier. In the Early Aptian a major transgression and the opening of the northernmost Atlantic took place, causing a fundamental change in the palaeogeographical situation of both north-west

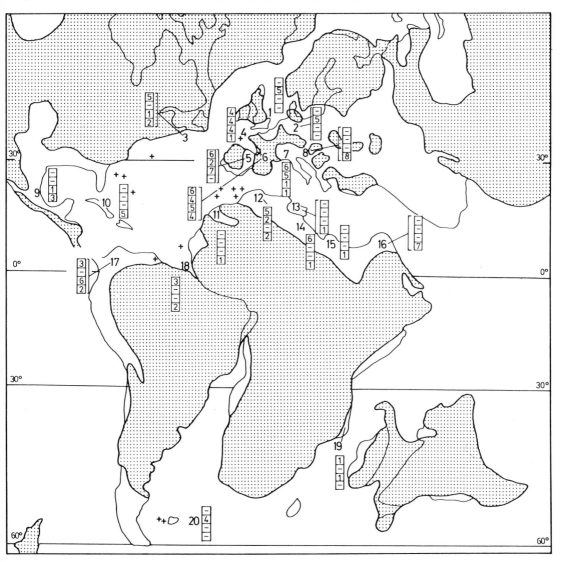

Species of *Nannoconus* in the

Upper Aptian
'mid' Aptian
Lower Aptian
Aptian
+   DSDP Sites

Fig. 7.9 — Palaeogeographical map for the Aptian showing the known distribution of the genus *Nannoconus*. Plate tectonics after Barron *et al.* (1981). *Nannoconus* assemblages based on Baldi-Beke (1962), Bouche (1963), Brönnimann (1955), Deres and Achéritéguy (1972, 1980), Dufour and Noël (1970), Geel (1966), Moshkovitz (1972), Taylor (1982), Trejo (1960) and DSDP reports (14, 36, 41, 44, 47, 48, 51B, 79, 89, 93).

Regions: 1 = eastern England, 2 = north-west Germany, 3 = offshore Canada, 4 = offshore Ireland, 5 = northern Spain, 6 = south-west France, 7 = south-east France, 8 = Hungary, 9 = Mexico, 10 = Cuba, 11 = Spanish Sahara, 12 = Morocco, 13 = Tunisia, 14 = Algeria, 15 = Libya, 16 = Israel, 17 = Columbia, 18 = Surinam, 19 = Mozambique, 20 = DSDP Site 511. + = DSDP Sites. For correlation of species and stratigraphy compare legend.

Germany and eastern England. During this time a major oceanic encroachment occurred in the North Sea, consequently establishing pelagic conditions. The North Sea had marine connections to the north of the Boreal Arctic Seas. In the latest Aptian cold water masses could penetrate into the Lower Saxony Basin, causing an upwelling-controlled regime (Kemper 1982). The middle Atlantic became oceanic, and new palaeocurrents developed.

Furthermore, an eastern and a western connection with the Tethys existed at least from the beginning of the Gargasian. Indeed, there is evidence that this connection was already in existence earlier, as indicated by the influx of Tethyan foraminifera in the 'Fischschiefer' (Bedoulian) of north-west Germany. A marine seaway via the English Proto-Channel existed to the middle part of the Protoatlantic, enabling Tethyan derived faunas and floras to immigrate into the North Sea.

This hypothesis of an influx of warm water faunas–floras from the west is strongly supported by Tethyan derived faunas in the 'middle' Aptian of England (Michael 1979, p. 310). Rich stenothermal benthic associations including corals, calcareous sponges (Farringdon sponge gravel), bryozoans, brachiopods and the bivalve genus *Toucasia*, are known from southern and south-eastern England. These warm water benthic associations were not able to migrate into the north-western German basin; only planktonic and nektonic organisms migrated that far east (Fig. 7.8).

This influx of floras and faunas from the west was interrupted when a new palaeogeographical regime influenced both north-west Germany and eastern England. An upwelling system, active during the latest Aptian (Clansayesian) and the Early Albian, was the main feature controlling the faunal and floral distribution.

(c) South Atlantic
While a shallow marine trans-Gondwana seaway between the Tethys and the Falkland Plateau–Antarctica via Madagascar already existed from the Tithonian onwards (Barron *et al.* 1981, Plate

3), the initial time of separation of the South American and African plates is placed in the Valanginian. The oldest marine sediments recorded on the margin of South America are Aptian, but it is only in the Albian that the increased separation of Africa and South America was associated with extensive marine transgressions in Africa (Barron *et al.* 1981), enabling faunas and floras to use either the African–South American or the African–Australian seaway. Thus floras and faunas could migrate from the Tethys to the Falkland Plateau in the Aptian only via a trans-Gondwana African-Australian seaway or via the Pacific. The former migration route is here preferred, for an '*N. truittii* assemblage' of Aptian age is known from Mozambique (Fig. 7.9).

If the suggested stratigraphy of Site 511 is correct, it is possible that the ingressions of warm surface water to both the north-western European and the Falkland Plateau sections occurred at the same interval in the 'middle' Aptian. This would indicate a synchronous warm water pulse. From the latest Aptian onwards the environmental setting of both areas was dictated by regional factors such as an upwelling system in north-west Europe.

7.6 CONCLUSIONS
The vertical distribution of calcareous nannofossils in the north-western European Aptian provides evidence for the temperature control of these assemblages. The following conclusions have been drawn from the present study.

(1) Relative high abundances of *Nannoconus* spp. of north-west Europe are considered to be an indicator for transgressive phases as well as an influx of warm water floras from the Tethys. These *Nannoconus*-bearing assemblages are restricted to the 'middle' Aptian (Gargasian); they are absent in the lower- and uppermost Aptian.

(2) Dominance of the cosmopolitan species *R. asper* in nannofossil assemblages indicates favourable conditions, enabling this species to

replace the most abundant species *W. barnesae*. The *R. asper* acme in the 'middle' Aptian might be used for biostratigraphic purposes.

(3) The 'middle' Aptian Tethyan influx is characterised by an overall increase in abundance and diversity of calcareous nannofossils.

(4) The occurrence of stenothermal, Tethyan derived fossil groups (ammonites, belemnites) and planktonic foraminifera, which is restricted to the 'middle' Aptian, is explained by a major change in the palaeocurrent system, enabling Tethyan faunas and floras to invade Boreal seas from the west.

(5) All the palaeontological data gained from various fossil groups support the idea of an influx of warm water faunas and floras for the 'middle' Aptian from the west via a Proto-Channel. A similar influx seems likely for the Falkland Plateau. It is possible that both of these occurred simultaneously, indicating either a major warm water pulse or a general trend of warmer climatic conditions.

7.7 ACKNOWLEDGEMENTS

I am grateful to Drs. P. H. Roth (Salt Lake City), D. K. Watkins (Lincoln) and S. W. Wise (Tallahassee) for helpful discussions and comments. Drs. U. Kaplan (Gütersloh) and J. Lehmann (Rheine) supplied additional information.

I acknowledge financial support by the Deutsche Forschungsgemeinschaft (Mu 667/2-1).

7.8 REFERENCES

Baldi-Beke, M. 1962. A. Magyarorszagi Nannoconuszok (Protozoa inc. sedis). *Geol. Hung., Paleontol.*, **30**, 109–148.

Barron, E. J., Harrison, C. G. A., Sloan J. L. and Hay, W. W. 1981. Paleogeography, 180 million years ago to the present. *Eclogae Geol. Helv.*, **74**, 443–470.

Basov, I. A. and Krasheninnikow, V. A. 1983. Benthic foraminifera in Mesozoic and Cenozoic sediments of the southwestern Atlantic as an indicator of paleoenvironment, Deep Sea Drilling Project Leg 71. *Init. Rep. DSDP*, **71**, 739–787.

Basov, I. A., Ciesielski, P. F., Krasheninnikow, V. A., Weaver, F.M. and Wise, S. W. 1983. Biostratigraphic and palaeontologic synthesis: Deep Sea Drilling Project Leg 71, Falkland Plateau and Argentine Basin. *Int. Rep. DSDP*, **71**, 445–460.

Bouche, P. M. 1963. État des connaissances sur les nannofossiles calcaires du Crétacé inférieur. *Coll. Crét. inf. (Lyon 1963)*, *Mém. Bur. Rech. Géol. Minières*, **34**, 452–459.

Brönnimann, P. 1955. Microfossils *incertae sedis* from the Upper Jurassic and Lower Cretaceous of Cuba. *Micropaleontology*, **11**, 28–51.

Čepek, P. 1981. Kalzitisches Nannoplankton. In F. Bender, *Angewandte Geowissenschaften*, **1**, Ferdinand Enke Verlag, Stuttgart, 407–414.

Čepek, P. 1982. Das kalzitische Nannoplankton des späten Apt und frühen Alb (*Parhabdolithus angustus*-Zone) des Gebietes von Hannover. *Geol. Jb.*, **A65**, 283–306.

Deres, F. and Archéritéguy, J. 1972. Contribution a l'étude des Nannoconides dans la Crétacé inférieur du Bassin d'Aquitaine. *Mém. Bur. Rech. Géol. Minières*, **77**, 153–159.

Deres, F. and Archéritéguy, J. 1980. Biostratigraphie des Nannoconides. *Bull. Centre Rech. SNEA*, **4** (1), 1–53.

Dufour, T. and Noël, D. 1970. Nannofossiles et constitution petrographique de la 'Majolica', des 'Schistes à Fucoides' et de la 'Scaglia Rossa' d'Ombrie (Italie). *Rev. de Micropaléontol.*, **13** (2), 107–114.

Erba, E. 1987. Mid-Cretaceous cyclic pelagic facies from the Umbrian–Marchaen Basin: what do nannofossils suggest? *INA Newsletter*, **9**, 52–53.

Gallois, R. W. 1975. A borehole section across the Barremian–Aptian boundary (Lower Cretaceous) at Skegness, Lincolnshire. *Proc. Yorks. Geol. Soc.*, **40**, 499–503.

Geel, T. 1966. Biostratigraphy of Upper Jurassic and Cretaceous sediments near Caravaca (SE Spain) with special emphasis on *Tintinnina* and *Nannoconus*. *Geol. Mijnbouw.*, **45**, 375–385.

Haq, B. U. 1978. Calcareous nannoplankton. In B. U. Haq and A. Boersma (Eds.), *Introduction to Marine Micropaleontology*, Elsevier, New York, 79–107.

Hay, W. W. 1965. Calcareous nannofossils. In B. Kummel and D. Raup, *Handbook of Palaeontological Techniques*, Freeman, San Francisco, 3–7.

Jakubowski, M. 1987. A proposed Lower Cretaceous calcareous nannofossil zonation scheme for the Moray Firth area of the North Sea. *Abh. Geol. B.-A.*, **39**, 99–119.

Jeletzky, J. A. 1983. Macroinvertebrate paleontology, biochronology, and paleoenvironments of Lower Cretaceous and Upper Jurassic rocks, Deep Sea Drilling Hole 511, Eastern Falkland Plateau. *Init. Rep. DSDP*, **71**, 951–975.

Kemper, E. 1982. Das späte Apt und frühe Alb Nordwestdeutschlands.-Versuch der umfassenden Analyse einer Schichtenfolge. *Geol. Jb.*, **A65**, 1–703.

Kemper, E. 1987. Das Klima der Kreide-Zeit. *Geol. Jb.*, **A96**, 5–185.

Koenen, A. v. 1902. *Die Ammonitiden des Norddeutschen Neocom*, Textbd. 451 S., Tafelbd. 55 Taf.

Krasheninnikow, V. A. and Basov, I. A. 1983. Stratigraphy of Cretaceous sediments of the Falkland Plateau based on planktonic foraminifers, Deep Sea Drilling Project,

Leg 71. *Init. Rep. DSDP*, **71**, 789–820.

Lott, G. K. Ball, K. C. and Wilkinson, I. P. 1985. Mid-Cretaceous stratigraphy of a cored borehole in the western part of the Central North Sea Basin. Proc. Yorks. Geol. Soc., **45** (4), 235–248.

Lutze, G. F. 1968. Ökoanalyse der Mikrofauna des Aptium von Sarstedt bei Hanover. *Ber. Naturhist. Ges., Beih.*, **5**, 427–443.

Michael, E. 1979. Mediterrane Fauneneinflüsse in den borealen Unterkreide-Becken Europas, besonders Nordwestdeutschlands. In *Aspekte der Kreide Europas, IUGS Series A*, **6**, 305–321.

Moshkovitz, S. 1972. Biostratigraphy of the genus *Nannoconus* in the Lower Cretaceous sediments of the subsurface: Ashqelon-Helez area, central Israel. *Isr. J. Earth-Sci.*, **21**, 1–28.

Mutterlose, J. 1984. Die Unterkreide Aufschlüsse (Valangin-Alb) im Raum Hanover-Braunschweig. *Mitt. Geol. Inst. Univ. Hanover*, **24**, 1–60.

Mutterlose, J. 1987a. Calcareous nannofossils and belemnites as warmwater indicators from the NW-German middle Aptian. *Geol. Jb.*, **A96**, 293–313

Mutterlose, J. 1987b. Migration and evolution patterns in Upper Jurassic and Lower Cretaceous belemnites. In J. Weidmann (Ed.), *Cephalopods—Present and Past*, Schweizerbart, Stuttgart, 525–537.

Okada, H. and Honjo S. 1973. The distribution of oceanic coccolithophorids in the Pacific. *Deep-Sea Res.*, **20**, 355–374.

Parker, M. E., Arthur, M. A., Wise S. W. and Wenkam, C. R. 1983. Carbonate and organic carbon cycles in Aptian-Albian black shales at Deep Sea Drilling Project Site 511, Falkland Plateau. *Init. Rep. DSDP*, **71**, 1051–1070.

Perch-Nielsen, K. 1979. Calcareous nannofossils from the Cretaceous between the North Sea and the Mediterranean. In *Aspekte der Kreide Europas, IUGS Series A*, **6**, 223–272.

Perch-Nielsen, K. 1985. Mesozoic calcareous nannofossils. In H. M. Bolli, J. B. Saunders and K. Perch-Nielsen (Eds.), *Plankton Stratigraphy*, Cambridge University Press, Cambridge, 329–426.

Rawson, P. F. and Riley L. A. 1982. Latest Jurassic–Early Cretaceous events and the 'Late' Cimmerian unconformity' in North Sea area. *Bull Am. Assoc. Pet. Geol.*, **66** (12), 2628–2648.

Roth, P. H. and Krumbach, K. R. 1986. Middle Cretaceous calcareous nannofossil biogeography and preservation in the Atlantic and Indian Oceans: implications for paleoceanography. *Mar. Micropaleont.*, **10**, 235–266.

Strauch, F. 1972. Zur Klimabindung mariner Organismen und ihre geologisch-paläontologische Bedeutung. *Neus Jb. Geol. Paläont., Abh.,*, **140**, 82–127.

Swinnerton, H. H. 1935. The rocks below the Red Chalk in Lincolnshire, and their cephalopod faunas. *Q. J. Geol. Soc. Lond.*, **91**, 1–46.

Taylor, R. J. 1982. Lower Cretaceous (Ryazanian to Albian) calcareous nannofossils. In A. R. Lord (Ed.), *A stratigraphical Index of Calcareous Nannofossils*, Ellis Horwood, Chichester, 40–80.

Trejo, M. H. 1960. La familia Nannoconidae y su Alcance estratigrafico en America (protozoa, incertae saedis). *Bol. Asoc. Mex. Geol. Petrol.*, **12**, 259–314.

Wise, S. W. 1983. Mesozoic and Cenozoic calcareous nannofossils recovered by Deep Sea Drilling Project Leg 71 in the Falkland Plateau Region, Southwest Atlantic Ocean. *Init. Rep. DSDP*. **71**, 481–550.

Ziegler, P. A. 1982. *Geological Atlas of Western and Central Europe*, Shell Internationale Petroleum Maatschappij B.V.

7.9 TAXONOMIC INDEX

Calcareous nannofossils: see general index

Foraminifera (in alphabetical order of generic names):

Clavihedbergella Banner and Blow 1959
Hedbergella Brönnimann and Brown 1958
H. infracretacea (Glaessner 1937)
Ticinella Reichel 1950
T. roberti (Gandolfi 1942)
Ticinella sp.

Belemnites (in alphabetical order of generic names):

Duvalia Bayle and Zeiller 1878
D. grasiana (Duval-Jouve 1841)
Neohibolites Stolley 1919
N. clava (Stolley 1911)
N. ewaldi (v. Strombeck 1857)
N. inflexus (Stolley 1911)

Ammonites (in alphabetical order of generic names):

Cheloniceras Hyatt 1903
Phylloceras Suess 1865
P. morelli (d'Orbigny 1846)
Zürcherella zürcheri (Jacob 1906)

Bivalves

Toucasia Munier-Chalmas 1873

8

Biostratigraphy and palaeogeographical applications of Lower Cretaceous nannofossils from north-western Europe

Jason A. Crux

Rapid changes within Lower Cretaceous nannofloral assemblages permit the division of Upper Ryazanian to Barremian strata of north-western Europe into 16 nannofossil zones. This degree of biostratigraphical refinement is approaching that achievable using ammonites and belemnites.

During certain periods of time, the nannofloras show strong Tethyan or Boreal–Arctic provincialism. The increase and decrease of Tethyan or Boreal–Arctic nannofloral elements are correlated to the pattern of transgressions, regressions and the opening–closing of seaways in north-western Europe. The following history is inferred from the nannofloral development in the NW European basins.

(a) A Late Ryazanian transgression with weak marine connections to the west of Britain and possibly beyond to Tethys.

(b) A transgression in the Early Valanginian, mainly from the Boreal Arctic Ocean.

(c) An extensive Late Valanginian transgression with good marine connections to Tethys through the Polish Trough for most of the time and a reduction of the Boreal–Arctic influence upon the north-western European area.

(d) A major and more extensive transgression in the Early Hauterivian bringing stronger Tethyan influences into the north-western European area.

(e) The nannofloras from the uppermost Lower Hauterivian (*inversum* Zone and *Aegocrioceras* Beds) are transitional between the Tethyan influenced assemblages below and those showing Boreal–Arctic influence above. This change occurred during a strongly transgressive interval. The arrival of Boreal–Arctic species in the area was possibly caused by an improvement of the marine connections to the north, or by a temporary closing of the Polish Trough, or by a change in the climate.

(f) There was a return of Tethyan influence to the nannofloras of the uppermost Hauterivian, possibly caused by a continuation of the transgression, opening seaways to the west or through the Polish Trough.

(g) A short, sharp regression in the Early Barremian caused partial closure of marine connections to the north between the Shetlands and Norway (the Polish Trough was by this time permanently closed). Although this regression was possibly followed by a transgression, the connections with Tethys remained poor. This period of sluggish circulation led to anoxic conditions in the bottom waters and the deposition of the blätterton organic-rich facies. The anoxic environment was eventually broken down by an influx of water from the north at the end of the Early Barremian. This sequence of events is reflected in the nannofloras by an abrupt change from strongly Tethyan influenced assemblages to those of a more endemic nature. With the re-establishment of good marine connections to the north, these were in turn replaced by assemblages showing a strong Boreal–Arctic influence.

(h) A minor regression in the Late Barremian once again allowed the establishment of endemic nannofloras.

8.1 INTRODUCTION

The Lower Cretaceous (Ryazanian–Barremian) strata of north-western Europe contain rich and distinctive nannofloras. These comprise a mixture of cosmopolitan, endemic, Tethyan and Boreal–Arctic taxa. The diversity of the nannofloras is greater than that of their contemporaries in the Tethyan Realm, with more than 40 species occurring in a single sample.

Previous studies have mainly concentrated on the description of new taxa and the biostratigraphical distribution of the nannofloras (Black 1971b, Thierstein 1973, Taylor 1978, 1982, Perch-Nielsen 1979, Köthe 1981, Jakubowski 1986, 1987, Crux 1987). More recently, Mutterlose and Harding (1987) and Kemper *et al.* (1987) have used the nannofossils to interpret the origin of distinctive lithological units such as the 'Hauptblätterton' of north-western Germany and the *Aegocrioceras* Beds.

The aims of the present study are twofold: firstly, to develop the nannofossil zonation schemes proposed by Taylor (1982) and Jakubowski (1987) and to integrate the results with the ammonite and belemnite zonations. Secondly, to use the Tethyan and Boreal–Arctic variations in the nannofloras to recognise regressions, transgressions and the opening and closing of seaways.

8.2 METHODS

For light microscope examination, the samples were prepared by crushing 1 cm³ of rock in a clean pestle and mortar. A little distilled water was added, the resulting slurry was poured onto a 20 μm sieve and washed through with further distilled water. The material that passed through the sieve was collected and allowed to settle in a glass vial. Most of the remaining water was decanted and the residue was agitated to homogenise it. A pipette was used to place two drops of the residue onto a microscope slide where it was smeared out into a thin layer, this was dried on a hot plate and a coverslip was attached using a permanent mounting medium.

For scanning electron microscope examination the samples were prepared using a centrifuge, as described by Taylor and Hamilton (1982).

8.3 SECTIONS STUDIED

A total of 231 samples from six sections of Ryazanian to Barremian age from England, Germany and offshore Norway have been examined for their calcareous nannofossil content (Fig. 8.1). The most complete of these sequences, although condensed in places, is the Speeton section from Filey Bay, Yorkshire, where Upper Ryazanian to Albian sediments are partially exposed. Four sections from the Hanover area make up a composite section of Upper Ryazanian to Aptian sediments. A short core from Sklinnabanken, offshore Norway, which comprises Upper Ryazanian to Lower Valanginian and Upper Hauterivian to Lower Barremian sediments, has also been studied. Brief details of these sections are given below, together with an account of their nannofloras and range charts showing the stratigraphical distribution of the species.

Generally the ages assigned in this work are taken from the most recently published accounts of each section. Exceptions to this are (1) the dating of the Sklinnabanken Core 7B, which is established in the present study by means of nannofossil biostratigraphy, and (2) the position of the Lower–Upper Hauterivian boundary in

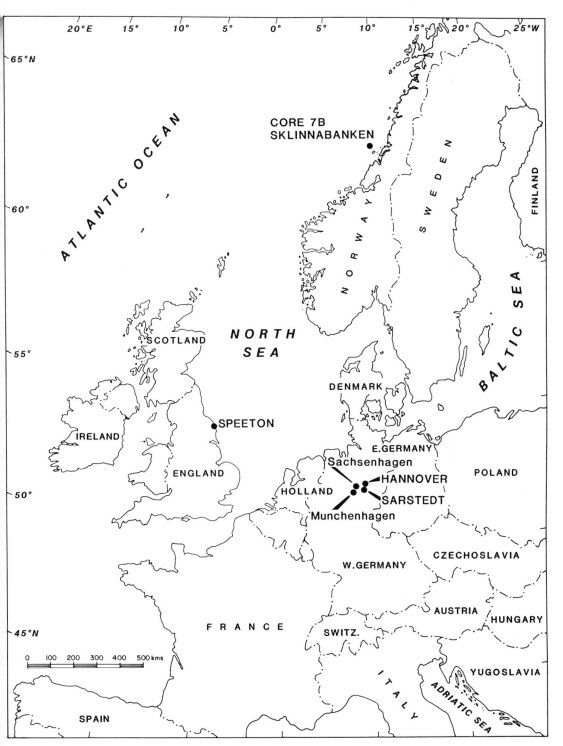

Fig. 8.1 — Location of studied sections.

the Moorberg, Sarstedt section, which is taken to lie above the *Aegocrioceras* Beds rather than below, where Mutterlose (1984) placed it. This results in the boundary lying in approximately the same horizon as it is placed in the Speeton section.

A remaining problem is the positioning of the Hauterivian–Barremian boundary between the various sections studied. The boundary is placed at the base of the *pugio* belemnite Zone in the Otto Gott, Sarstedt section, but at the base of the stratigraphically older *variabilis* ammonite Zone at Speeton. In comparative discussions of the two sections, the boundary defined by ammonites at Speeton is approximately applied to the Otto Gott section in order to compare like with like.

(a) Speeton, Filey Bay, Yorkshire (England)
Location: the type locality of the Speeton Clay lies at the southern end of Filey Bay, Yorkshire (Grid Ref. TA 145763 to 155754), where the clay forms low cliffs along a 1.2 km coastal section.
References: the Speeton section was first described in detail by Lamplugh (1889), who divided it into four major units, the A to D Beds, according to belemnite faunas. He further subdivided the D and C Beds on an essentially lithological basis. Further lithological subdivisions for different parts of the section have been made by Neale (1960, 1962), Kaye (1964), Fletcher (1969), Rawson (1971) and Rawson and Mutterlose (1983). The most recent descriptions of the section giving ammonite and/ or belemnite stratigraphies are Neale (1962) for the Ryazanian D Beds, Neale (1960) for the Valanginian to Hauterivian D Beds, Rawson (1971) for the Hauterivian to Barremian D and C Beds and Rawson and Mutterlose (1983) for the Barremian Lower B and basal Cement Beds. In addition, zones used for the Valanginian by Rawson *et al.* (1978) are applied.
Stratigraphical extent: the extent and completeness of the stratigraphical record available for sampling at the Speeton section is controlled by three factors. Firstly the original history of sedimentation has resulted in a major

gap in the succession with the absence of Upper Valanginian strata and condensed sequences in the Lower and Upper Ryazanian, Lower Valanginian and the Lower Hauterivian. The second factor is the state of the low cliffs which are constantly slumping and slipping, preventing collection of particular beds at any one time. The third factor is the deposition of beach sands on beds which are best exposed on the foreshore. The combination of the last two factors has prevented detailed sampling of the Upper Hauterivian C6–C4 Beds and the uppermost Barremian beds. It has been possible to relate all samples collected to ammonite biostratigraphy in the Speeton section.

23 samples have been collected from the uppermost Ryazanian *albidum* Zone to the Lower Valanginian *Polyptychites* spp. Zone. 27 samples have been collected from the Hauterivian strata; three from the condensed *amblygonium* and *noricum* Zones, the remainder from the *regale* to *marginatus* Zones. 31 samples were collected from the Barremian *variabilis* to *denckmanni* Zones.
Nannofloras: the distribution and composition of the nannofloras of the Speeton section are presented in Figs. 8.2–8.4. Samples from the lower part of the *albidum* Zone are barren or contain only rare *Watznaueria barnesae*. The upper part of the *albidum* Zone contains rich but poorly diversified nannofloras. The dominant species are divided equally between survivors from the Jurassic such as *W. barnesae*, *Staurolithites crux*, *Biscutum ellipticum* and *Diazomatolithus lehmanii* and new Cretaceous species such as *Rhagodiscus asper*, *Cretarhabdus crenulatus*, *Micrantholithus obtusus* and *Crucibiscutum salebrosum*. Distinctive species present in this interval are *Nannoconus* sp. (discs), *Sollasites arcuatus*, *Micrantholithus brevis* and *Perissocyclus fletcheri*; these species are all very short ranging. A dramatic drop in abundance and diversity of the nannofloras occurs at the top of the Ryazanian. The Lower Valanginian D4C to D2E Beds are highly condensed and have a low calcite content. The few nannofossils present are solution-resistant

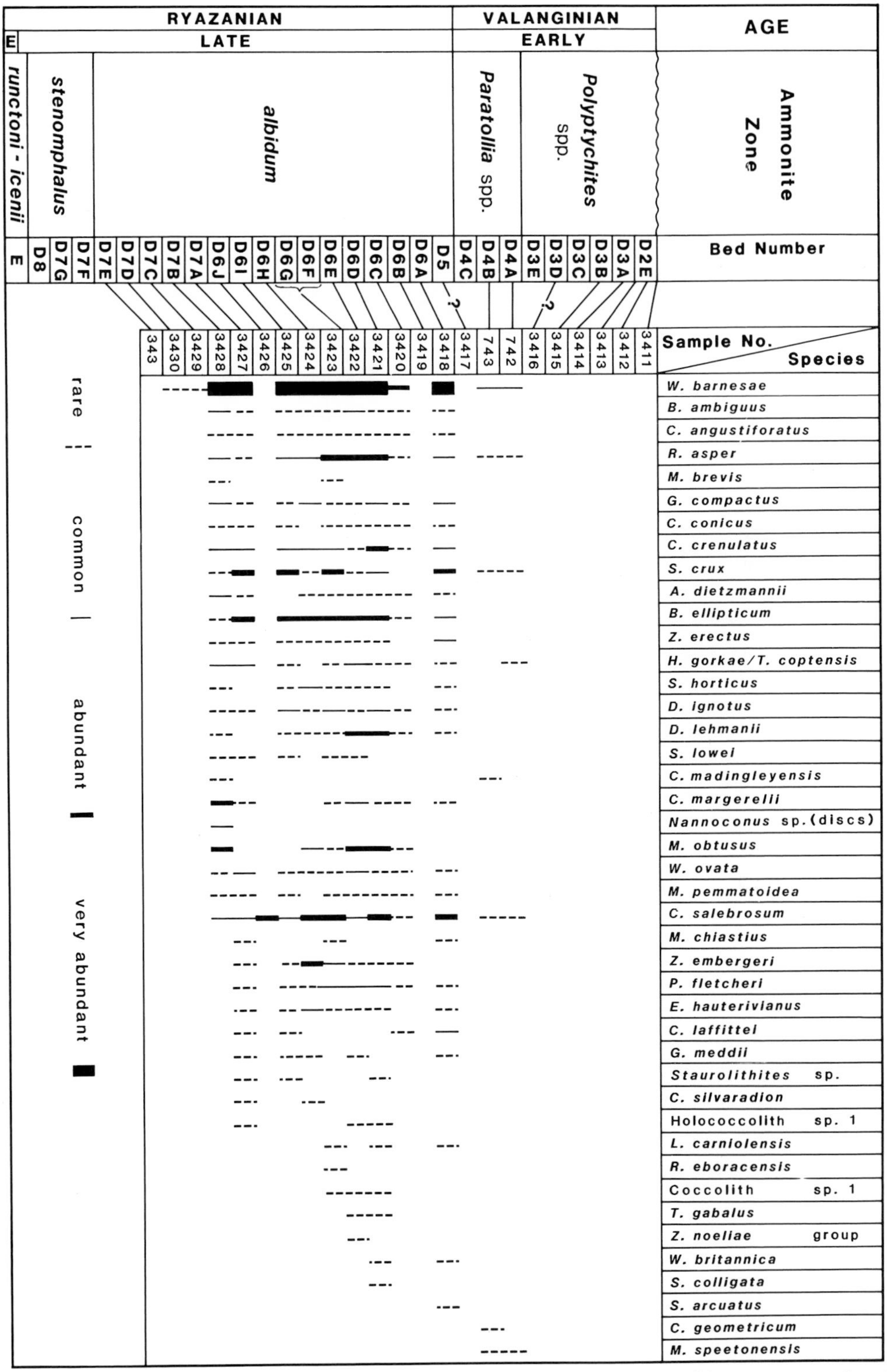

Fig. 8.2 — Stratigraphical distribution of nannofossils in the Ryazanian to Valanginian of Speeton: rare = less than 1 specimen/5 fields of view; common = more than 1 specimen/5 fields of view; abundant = more than 1 specimen/field of view; very abundant = more than 5 specimens/field of view. These nannofossil abundances were estimated at a magnification of × 1500

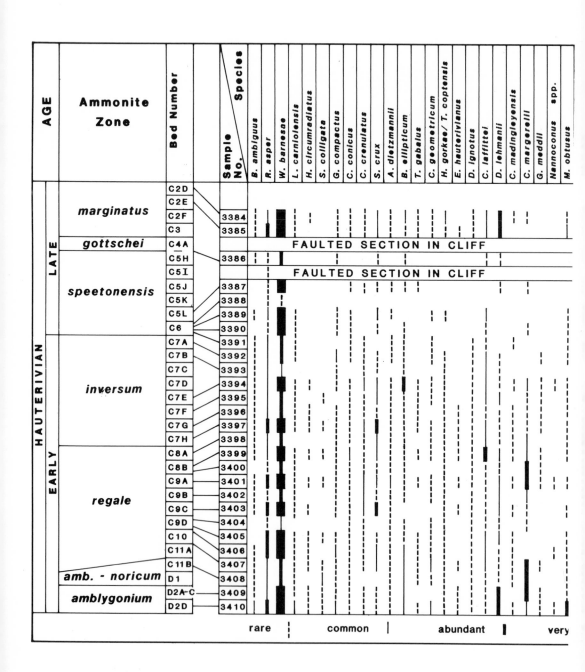

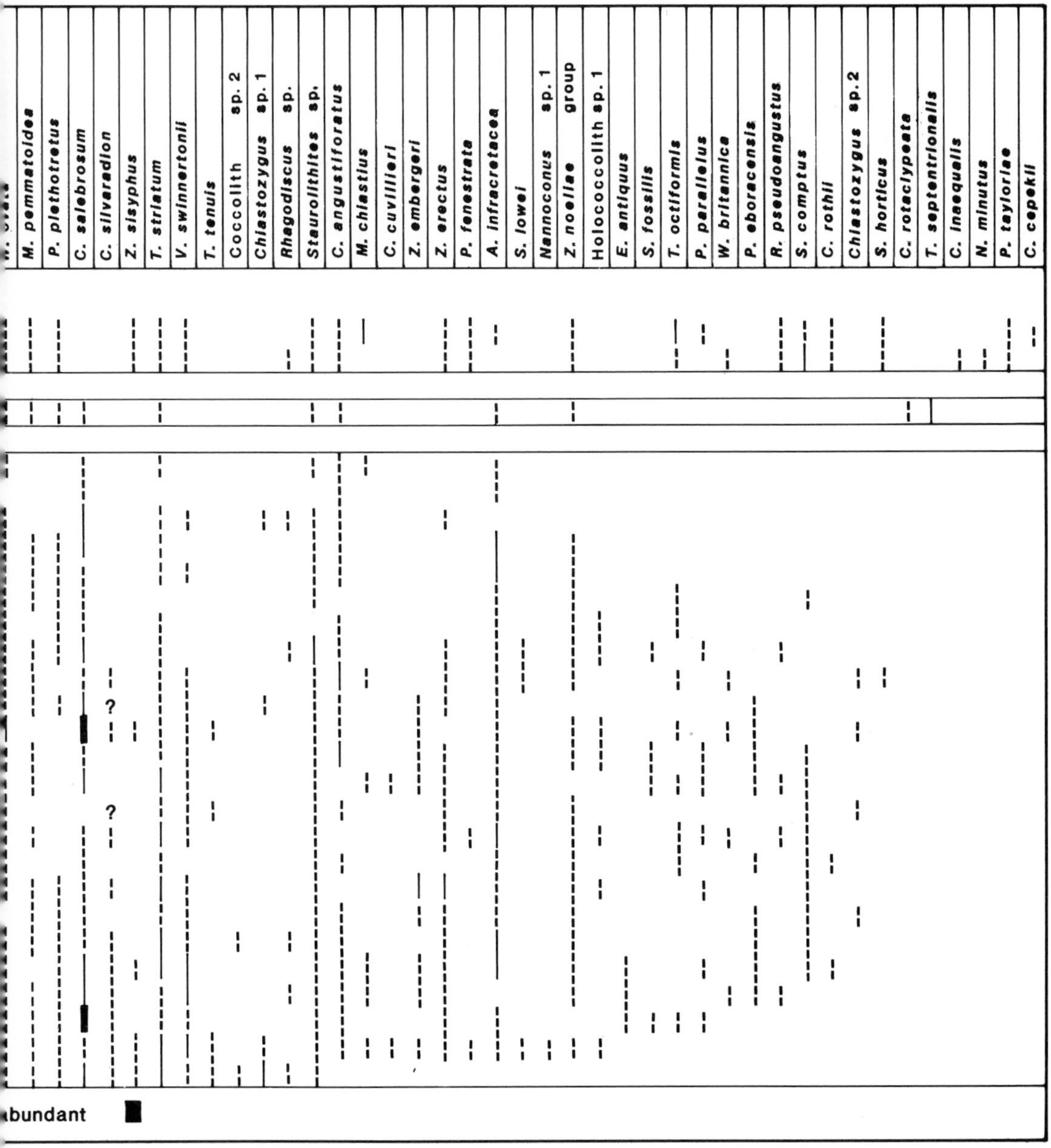

Fig. 8.3 — Stratigraphical distribution of nannofossils in the Hauterivian of Speeton.

AGE	Ammonite Zone	BED NUMBER AFTER RAWSON AND MUTTERLOSE (1983)	SAMPLE NO.
BARREMIAN	denckmanni	47	3353
		48	3354
	elegans	49	3355
		50	3356
	fissicostatum	LB1A	3357
		LB1B	3358
		LB1C	3359
		LB1D	3360
		LB1E	3361
		LB1F	3362
		LB2AI	3363
		LB2AII	3364
		LB2B	3365
		LB2CI	3366
		LB2CII	3367
		LB2D	3368
		LB3A	3369
	rarocinctum	LB3B	3370
		LB3C	3371
		LB3D	3372
		LB3E	3373
		LB4A	3374
		LB4B	3375
		LB4C	3376
		LB4D	3377
		LB5A	3378
		LB5B	3379
		LB5C	3380
		LB5D	3381
	variabilis	LB5E	3382
		LB6	3383
		CIA	
		CIB	
		C2A	
		C2B	
		C2C	

SPECIES: B. ambiguus, C. angustiforatus, R. asper, W. barnesae, L. carniolensis, M. chiastius, H. circumradiatus, G. compactus, S. comptus, C. conicus, C. crenulatus, S. crux, A. dietzmannii, B. ellipticum, Z. erectus, S. fossilis, T. gabalus, C. geometricum, H. gorkae/T. coptensis, S. horticus, D. ignotus, C. inaequalis, A. infracretacea, C. laffittei, D. lehmanii, S. lowei

Legend: rare ⋮   common | abundant ▮   very abundant

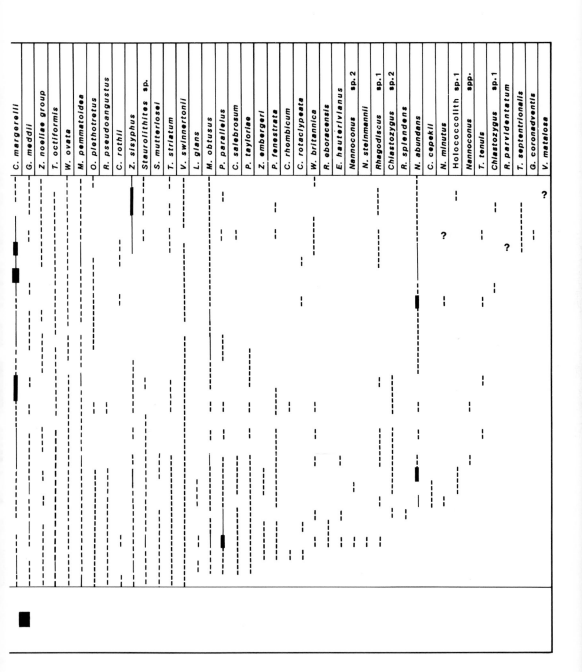

Fig. 8.4 — Stratigraphical distribution of nannofossils in the Barremian of Speeton.

species which are often poorly preserved. The presence of *Micrantholithus speetonensis*, however, makes these assemblages characteristic enough to be biostratigraphically useful.

Upper Valanginian strata are missing in the Speeton section. The hiatus is followed by a very condensed sequence of lowermost Hauterivian strata, rich in calcareous nannofossils. Although the nannofloras of the Lower Hauterivian are generally dominated by *W. barnesae*, there is an unusually high abundance of *Cyclagelosphaera margerelii* and *R. asper* in these beds. Several species which probably evolved in the Valanginian appear for the first time in these beds. These are *Haqius circumradiatus, Tegumentum striatum, T. octiformis* and *Eprolithus antiquus*. New species which evolved during the Early Hauterivian and are recorded at Speeton are *Scapholithus fossilis* and *Rhagodiscus pseudoangustus*. The extinctions of *Corollithion silvaradion* and *E. antiquus* are useful biostratigraphical events within this interval.

The Upper Hauterivian yielded relatively sparse nannofloras compared with those of the Lower Hauterivian. This pattern was also observed in the Moorberg, Sarstedt section. Species which evolved in the Late Hauterivian and are recorded at Speeton include *Tegulalithus septentrionalis, Cretarhabdus inaequalis* and *Chiastozygus cepekii*. The common occurrence of *T. septentrionalis* in the Upper Hauterivian is biostratigraphically useful.

The lowermost beds of the Barremian are relatively rich in nannofossils compared to the underlying Upper Hauterivian. The nannofloras are dominated by *W. barnesae*, with lesser numbers of *R. asper, S. crux, B. ellipticum, Corollithion geometricum, Assipetra infracretacea, Discorhabdus ignotus, Cylindralithus laffittei* and *D. lehmanii*. The nannofloras become less abundant in the upper part of the *fissicostatum* and lower *elegans* Zones (Beds LB2A–LB1A) and two species, *C. margerelii* and *Nannoconus abundans*, become relatively more abundant.

Above Bed LB1A two species, *T. septentrionalis* and *D. lehmanii*, return after being absent from the nannofloras of Beds C2E–LB1A and LB2CII–LB1A respectively. This return is associated with an increase in abundance of *Zeugrhabdotus sisyphus* and *Sollasites horticus*.

It is possible to recognise 15 of the 16 nannofossil zones proposed in this study in the Speeton section (Figs. 8.10 and 8.11).

(b) Sachsenhagen, near Hanover (Germany)
Location: west of Hanover, Map Ref. TK 25 Sachsenhagen, Nr. 3621 re: 35 17 300, h: 58 07 150.
Reference: Mutterlose (1984) gave a recent account of the section exposed in this large shallow pit. He divided the section into five parts, A–E from bottom to top. These divisions are faunal and reflect the progress of the depositional environment from brackish to fully marine.
Stratigraphical extent: Upper Ryazanian (Berriasian) to Lower Valanginian. 26 samples were collected through the succession.
Nannofloras: the samples collected are barren of nannofossils and poor in calcareous material in general.

(c) Münchenhagen, near Hanover (Germany)
Location: west of Hanover, 0.5 km east of the Loccum–Wiedensahl road and about 1 km north of the Zweigstrasse to Münchenhagen. Map Ref. TK 25 Schlusselburg, Nr. 3520, re: 35 09 775, h: 58 08 625.
Reference: Mutterlose (1984) gave a recent account of this section. He described the beds exposed, which he numbered 95 to 124, from bottom to top.
Stratigraphical extent: the 20 m section exposes the part of the Upper Valanginian (the upper part of the *Dichotomites* Beds). 12 samples were taken only from Beds 108 to 124, since the lower part of the section was covered by chemical waste.
Nannofloras: the stratigraphical distribution of the nannofossils recovered is shown in Fig. 8.5. The nannofloras are dominated by *W. barnesae* with lesser numbers of *C. margerelii*. This section is important for establishing the ranges of some

key taxa used in the biostratigraphical zonation scheme of this study. The presence of *T. striatum* in this section, but its absence in the Lower Valanginian strata of Speeton and Core 7B, Sklinnabanken, offshore Norway, shows it to have evolved in the early Late Valanginian. The total absence of *E. antiquus* from this section possibly indicates that it did not evolve until the latest Valanginian or earliest Hauterivian. The two highest samples taken from this section are barren of nannofossils, except for rare *W. barnesae*. The whole of the fossiliferous section from Münchenhagen can be assigned to the *T. striatum* Zone of this study.

(d) Moorberg, near Hanover (Germany)
Location: 30 km south of Hanover, east of the B6. Map Ref. TK 25 Sarstedt Nr. 3725, re: 35 59 880, h: 57 89 550.
Reference: Mutterlose (1984) gave a recent account of this brick pit, although much of the section described has since been covered by refuse. The section consists of strongly dipping Lower Cretaceous strata unconformably overlying Bajocian to Toarcian sediments. Mutterlose (1984) numbered the Lower Cretaceous beds 100 to 33 and then from − 1 up to − 50 upwards through the succession.
Stratigraphical extent: the section once extended from the Lower Hauterivian *amblygonium* Zone to the Lower Barremian *fissicostatum* Zone. The filling in of the pit with refuse, prevented the collection of samples above Bed − 3, which lies within the Upper Hauterivian *gottschei* Zone. 33 samples were taken from Beds 101 to − 3.
Nannofloras: the stratigraphical distribution of the nannofossils recovered in the 33 samples from this section is shown in Fig. 8.6. The nannofloras of the Lower Hauterivian are dominated by *W. barnesae*, with lesser numbers of *R. asper* and *C. margerelii*. There is a marked drop in abundance and diversity of the nannofloras in the upper part of the Lower Hauterivian (*Aegocrioceras* Beds, Bed 73). In the remainder of the Hauterivian section, the diversity and abundance of the nannofloras increase with the first *T. septentrionalis* and later

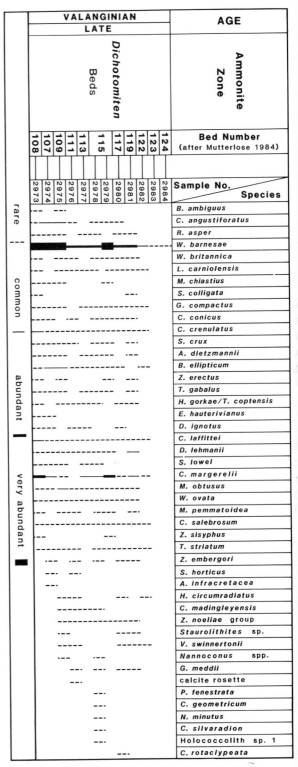

Fig. 8.5 — Stratigraphical distribution of nannofossils in the Münchenhagen clay pit.

AGE	Ammonite Zone	Bed Number (after Mutterlose 1984)	Sample No.	Species
LATE (HAUTERIVIAN)	gottschei	-3	43	
		33	42	
		38–34	2923	
	staffi	41	41	
		58	40	
		65 ?–	39	
		67	2922	
		72	36	
EARLY (HAUTERIVIAN)	Aegocrioceras Beds	73	35	
			2921	
			2920	
	regale	78	2919	
		80	2918	
			2917	
		81	2916	
		83	2915	
		86	2914	
		92	2913	
		95	2912	
		96	2911	
			2910	
	noricum	97	2909	
			2908	
		98	2907	
		99	2906	
			2905	
		100	2904	
	amblygonium	101	2903	
			2902	
			2901	
			2900	
			2899	
			2898	

Species (columns, left to right): *B. ambiguus*, *C. angustiforatus*, *R. asper*, *W. barnesae*, *L. carniolensis*, *G. compactus*, *C. concus*, *C. crenulatus*, *S. crux*, *A. dietzmannii*, *B. ellipticum*, *Z. embergeri*, *T. gabalus*, *C. geometricum*, *H. gorkae/T.coptensis*, *E. hauterivianus*, *D. ignotus*, *A. infracretacea*, *C. laffittei*, *D. lehmanii*, *S. lowei*, *C. madingleyensis*, *G. meddii*, *M. obtusus*, *W. ovata*, *Rhagodiscus sp.*, *C. salebrosum*

Legend: rare ┊   common │   abundant ▌   very abundant ▐

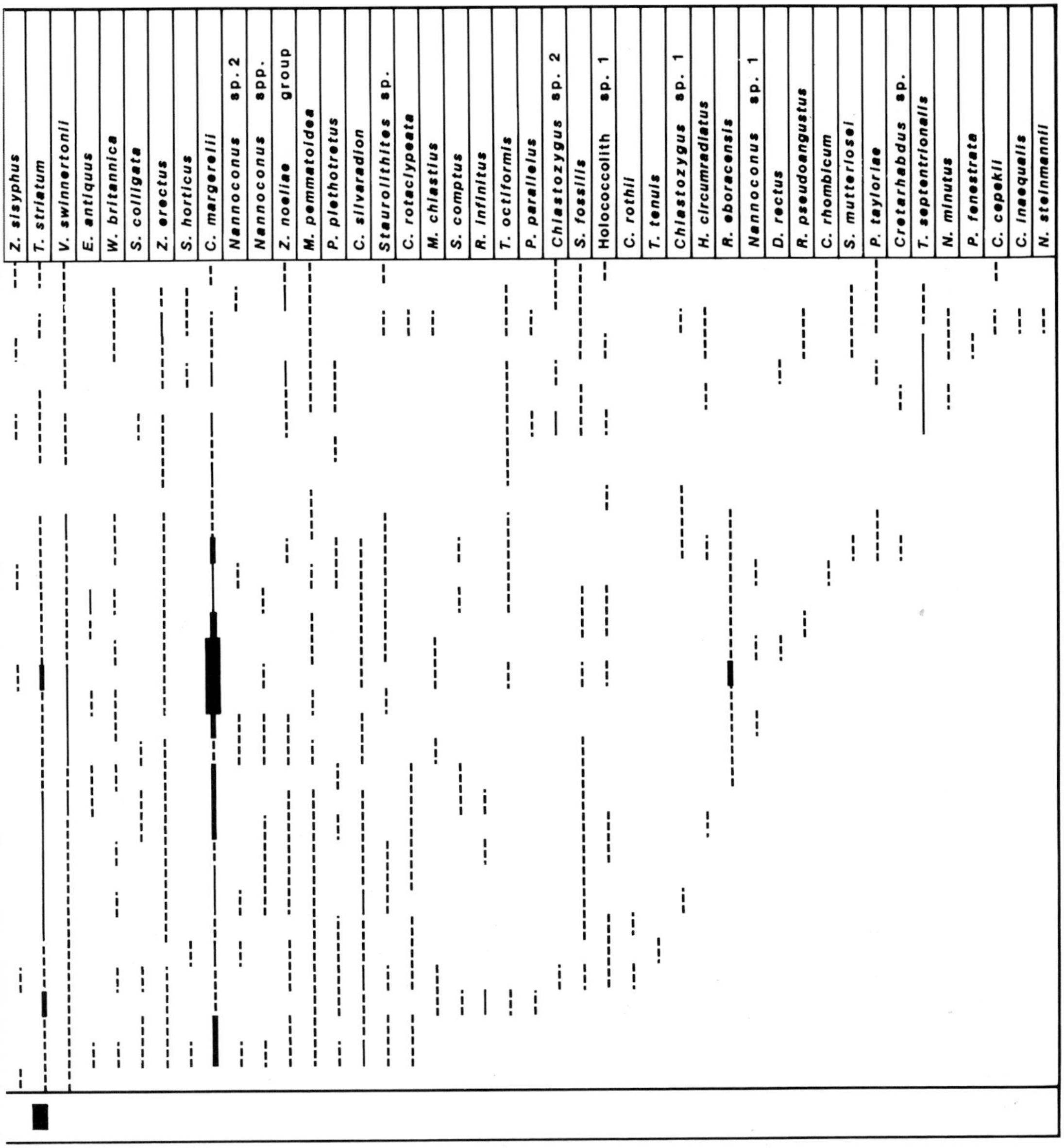

Fig. 8.6 — Stratigraphical distribution of nannofossils in the Moorberg, Sarstedt, brick pit.

Nannoconus spp. being relatively common. Useful biostratigraphical events within the Moorberg section include the first occurrences of *T. octiformis, T. septentrionalis* and *C. inaequalis* and the last occurrences of *E. antiquus, C. silvaradion*, common *T. septentrionalis* and large *T. striatum*.

The *E. antiquus, C. silvaradion, C. margerelii, T. septentrionalis* and *Stradnerlithus comptus* Zones, as defined in this study, were recognised in the Moorberg section.

(e) Otto Gott, Sarstedt, near Hanover (Germany)
Location: 30 km south of Hanover, 1 km east of the B6. Map Ref. TK 25 Sarstedt, Nr. 3725, re: 35 60 400, h: 57 90 650.
Reference: Mutterlose (1984) gave a recent account of the beds exposed in this actively worked brick pit. He numbered the beds from 50 to 215 upwards and related them to a composite ammonite and belemnite zonation scheme. No ammonite zonal assignments are available for the upper part of the section which is dated with belemnites.
Stratigraphical extent: the section ranges from the Upper Hauterivian *gottschei* ammonite Zone to the Upper Barremian *depressa* belemnite Zone. This is followed by an unconformity and an Upper Aptian section. 56 samples were collected from the *gottschei* to *depressa* Zones, Beds 50–199.
Nannofloras: the stratigraphical distribution of the nannofossils in the 56 samples from this section is presented in Fig. 8.7. The nannofloras of the Otto Gott section are more abundant and diverse in the Upper Hauterivian and Lower Barremian than in the Upper Barremian. *Watznaueria barnesae* dominates the nannofloras throughout the section with lesser numbers of *R. asper, S. crux* and *B. ellipticum*. The very lowest part of the section in the Upper Hauterivian has a relatively high abundance of nannoconids. These become less common in the Lower Barremian and are replaced by a single species, *N. abundans* (Mutterlose and Harding (1987) additionally record *Nannoconus borealis* and *N. elongatus*). *Tegulalithus septentrionalis* is

present in the Upper Hauterivian but becomes absent in the Lower Barremian and re-enters just below the Lower–Upper Barremian boundary. This return is associated with the abundant occurrence of *Z. sisyphus* and the return of *D. lehmanii* as was the case also in the Speeton section. Biostratigraphically important nannofossil events in this section are the last occurrences of *R. pseudoangustus* and *N. abundans*, the first occurrences of *N. abundans* and *Vagalapilla matalosa* and the re-entry of *T. septentrionalis*.

The *S. comptus–C. inaequalis, N. abundans, C. conicus, Z. sisyphus* and *V. matalosa* zones are present in the Otto Gott section.

(f) Core 7B, Sklinnabanken (offshore Norway)
Location: the core was taken close to the eastern margin of the Trøndelag Platform, offshore Helgeland, northern Norway (65° 06.0' N, 10° 17.3' E).
Reference: the core is described and related to seismic stratigraphy by Aarhus *et al.* (1986). They also provided an account of the macrofossils, palynomorphs and nannofossils of the core, but their findings do not totally agree with the present study. The dating of this section is possibly hampered by reworking of the palynomorphs (M. Partington pers. commun.).
Stratigraphical extent: the core ranges from the Upper Ryazanian to Lower Valanginian where a hiatus causes the remainder of the Valanginian and most of the Hauterivian to be absent. Sedimentation resumed in the latest Hauterivian to earliest Barremian. 22 smear slides (given to the author by Jacob Verdenius) from 4.50 to 10.95 m depth in the core were studied.
Nannofloras: the stratigraphical distribution of the nannofossils in this core is presented in Fig. 8.8. The Upper Ryazanian and Lower Valanginian nannofloras are dominated by *W. barnesae* and *C. salebrosum*. The lowermost two samples contain *S. arcuatus*, a species restricted to D6A at Speeton. Other biostratigraphically important species are *P. fletcheri* and the 'calcite rosettes' of Verdenius in Aarhus *et al.* (1986). The former appears to have a stratigraphically

restricted range from Upper Ryazanian to Lower Valanginian, while the latter appears to be restricted to the Valanginian.

The Upper Hauterivian to Lower Barremian strata contain nannofloras dominated by *W. barnesae* with common to abundant *R. asper*. This interval is dated by the presence of *T. septentrionalis*, *C. cepekii*, *C. inaequalis* and *S. comptus*.

The *S. arcuatus*, unnamed Zone and the *S. comptus* Zones are recognised in this core.

8.4 BIOZONATION SCHEME

Two previous biozonation schemes have been proposed specifically for Early Cretaceous nannofossils of the north-western European area. These are Taylor (1982) and Jakubowski (1987). The former was based on a study of the Speeton (England), Otto Gott and Moorberg, Sarstedt (Germany) sections. Taylor proposed five nannofossil zones integrated with the ammonite biostratigraphy for the Late Ryazanian to Barremian interval. Some of her zones are included in the present scheme, but further age resolution has been achieved by using previously undescribed species and knowledge gained from North Sea well studies.

The most recent zonation scheme proposed for the Early Cretaceous is that of Jakubowski (1987). This study is based mainly on the examination of well material from the Moray Firth area (North Sea) with some reference to the Nettleton (Lincolnshire) and Speeton sections onshore. Some of the zones proposed by Jakubowski have been found useful in the present study, but others have proved to be of only local significance in the Moray Firth area. In addition, the lack of ammonite dated reference material has led Jakubowski to interpret wrongly the age significance of some of his biostratigraphical events.

Further reference to the two previous zonation schemes is made in the comments sections for the individual zones. A comparison of the zonation proposed here and the two previous schemes is given in Fig. 8.9.

In this study, 16 interval zones are defined for the Late Ryazanian to Late Barremian interval. They are graphically compared with the ammonite and belemnite zones in Fig. 8.10 (Ryazanian–Valanginian interval) and 8.11 (Hauterivian–Barremian interval). Range charts showing the composite ranges for the biostratigraphically useful species in both Germany and England are presented in Figs. 8.12 and 8.13.

(a) *Nannoconus* sp. (discs) Zone
Definition: from the first to the last occurrence of *Nannoconus* sp. (discs).
Author: Crux, this study.
Assemblage characteristics: the nannoflora is dominated by *Watznaueria barnesae* with common to abundant *Bukrylithus ambiguus*, *Rhagodiscus asper*, *Glaukolithus compactus*, *Cretarhabdus crenulatus*, *Axopodorhabdus dietzmannii*, *Hemipodorhabdus gorkae–Tetrapodorhabdus coptensis*, *Cyclagelosphaera margerelii*, *Nannoconus* sp. (discs) and *Crucibiscutum salebrosum*. *Micrantholithus brevis* occurs for the first time in this zone.
Stratigraphical extent: this zone is an interval within the Upper Ryazanian, corresponding to part of the *albidum* ammonite Zone.
Type section: Speeton, Bed D7B.
Comments: Bed D7B is the oldest bed sampled at Speeton that yielded diverse nannofloras. In more basinal sections more complete Ryazanian sections may occur, allowing the extension of this zone into older strata. *Nannoconus* sp. (discs) has not been previously used in zonation schemes, but has been observed in the Norwegian sector of the North Sea and the Porcupine Basin, offshore Ireland. Thus its distribution is sufficiently widespread to be useful for correlation.

This zone corresponds to part of the *Retecapsa angustiforata* Zone of Taylor (1982) and of the *R. angustiforata* Subzone of Jakubowski (1987). This zone is defined by the first occurrence of *Cretarhabdus angustiforatus*; however, H. Dockerill (pers. commun.) reports this species from the Jurassic of France.

AGE		Ammonite / Belemnite Zone	Bed Number (after Muterlose 1984)	Sample No.
B A R R E M I A N	LATE	depressa	199 ?	2853
			197	2854
				2855
				2856
			196	2857
			192	2858
		germanica	187	2861
				2862
			?	2863
			185	2864
				2865
			184	2866
			183	2867
				2868
				2869
				2870
			138	2871
			137	2872
				2873
			133	2874
				2875
			131	2876
				2877
			130	2878
			127 ?	2879
				? 2880
			126 ?	2881
				? 2882
			124 ?	2883
		brunsvicensis	116	2884
				2885
			111	2886
			110	2887
	EARLY		109	2888
		Aulacoteuthis spp.	102 ?	2889
			?	2890
			101	2833
				2834
				2835
			100	2836
				2837
				2838
				2839
				2840
		pugio	92 ?	2841
			88 ?	2842
			83	2843
			79	2844
			76	2845
HAUTERIVIAN	LATE	discofalcatus	72	2846
			69	2847
			67	2848
			59	2849
			58	2850
			55	2851
		gottschei	50	2852

Species:
B. ambiguus, C. angustiforatus, R. asper, W. barnesae, L. carniolensis, M. chiastius, H. circumradiatus, G. compactus, C. conicus, C. crenulatus, S. crux, A. dietzmannii, B. ellipticum, Z. erectus, S. fossilis, C. geometricum, H. gorkae / T. coptensis, S. horticus, D. ignotus, C. laffittei, D. lehmanii, S. lowei, C. madingleyensis, C. margerelii, N. minutus, Nannoconus sp.2, Nannoconus spp., M. obtusus, T. octiformis, W. ovata, P. parallelus

rare ⋮   common | abundant ▮   very abundant ▉

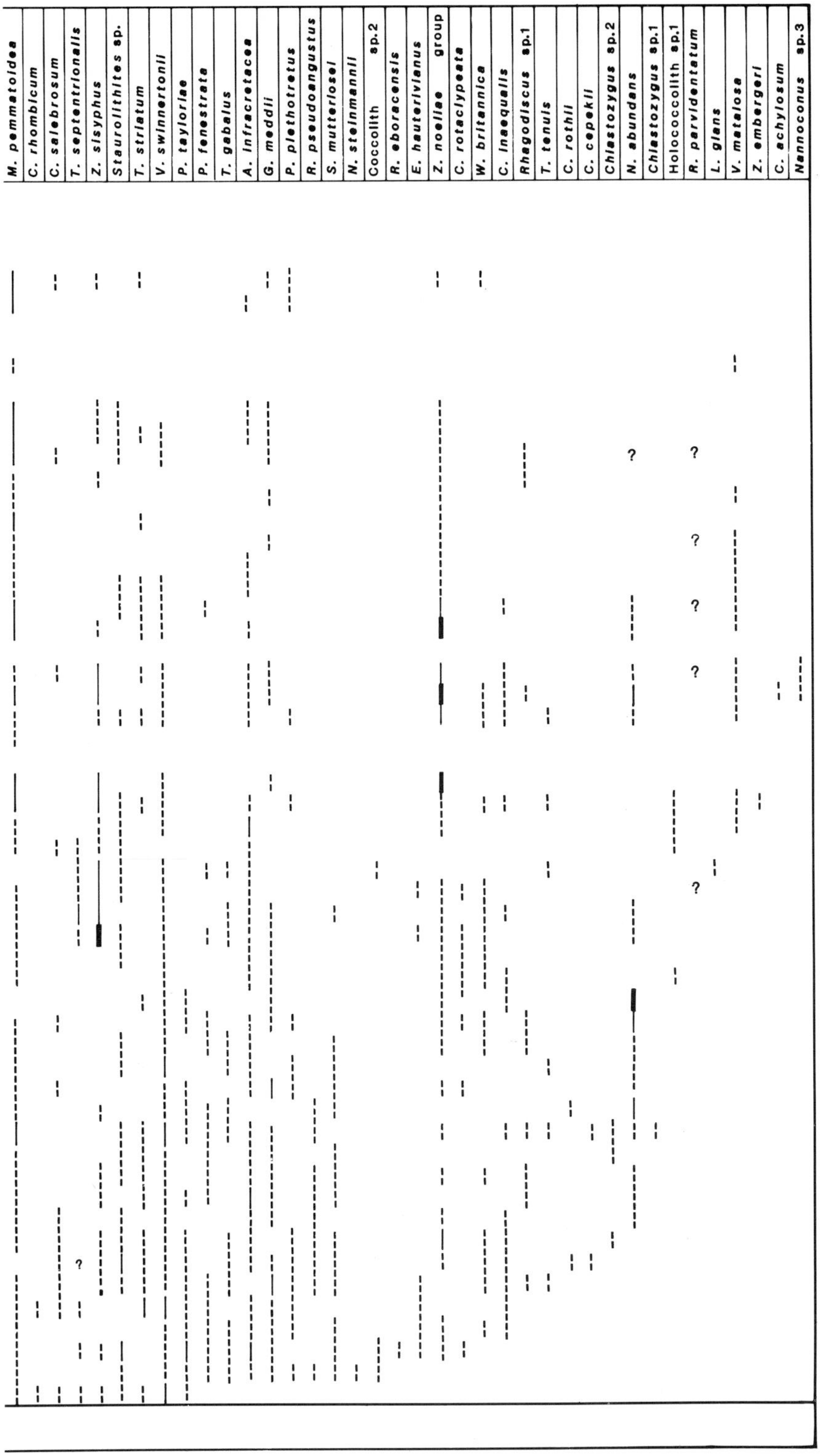

Fig. 8.7 — Stratigraphical distribution of nannofossils in the Otto Gott, Sarstedt, brick pit.

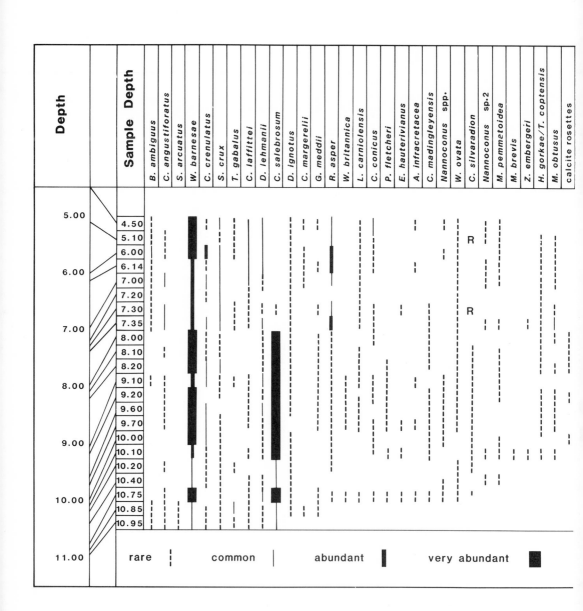

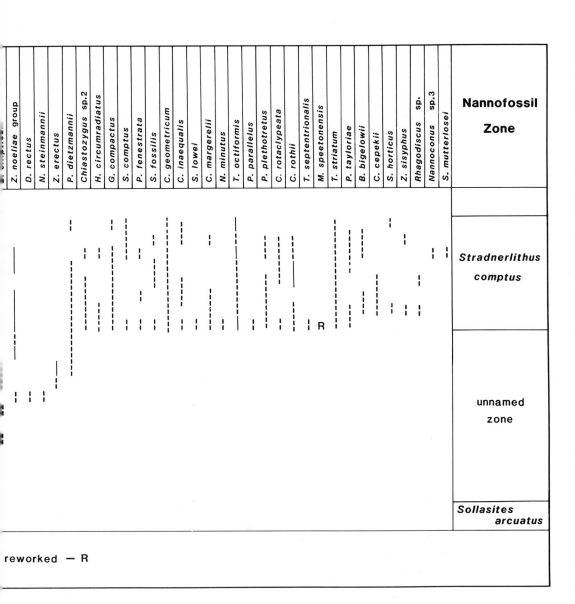

Fig. 8.8 — Stratigraphical distribution of nannofossils in Core 7B, Sklinnabanken.

AGE		Ammonite Zone	Taylor (1982)	Taylor (1982)
BARREMIAN		denckmanni	⌐⌐⌐⌐⌐⌐⌐⌐⌐ Nannoconus abundans	⏐ N. abundans
		elegans		
		fissicostatum		
		rarocinctum		
		variabilis		◿ N. abundans
HAUTERIVIAN	LATE	marginatus	Chiastozygus striatus	◿ "D Noelii"
		gottschei		⏐ S. colligata
		speetonensis		C. cuvillieri
	EARLY	inversum		
		regale		
		noricum		
		amblygonium		◿ C. striatus
VALANGINIAN	LATE	unnamed	Micrantholithus speetonensis	⏐ M. speetonensis
		pitrei		
		Dichotomites spp.		
	EARLY	Polyptychites spp.		◿ M speetonensis
		Paratollia spp.	C. salebrosus	S. colligata
RYAZA-NIAN	LATE	albidum	R. angustiforata	◿ C. salebrosus
		stenomphalus		◿ R. angustiforata

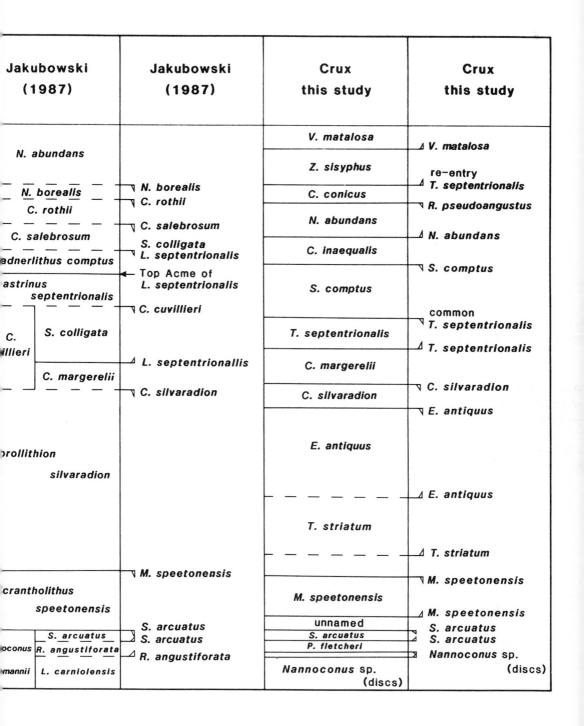

Fig. 8.9 — Nannofossil zonation scheme of the present study compared with previous schemes.

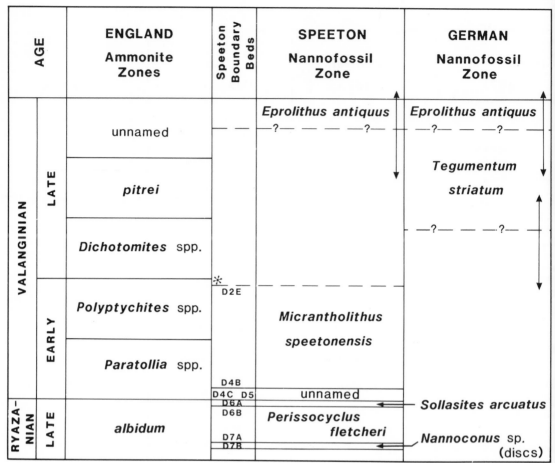

AGE		ENGLAND Ammonite Zones	Speeton Boundary Beds	SPEETON Nannofossil Zone	GERMAN Nannofossil Zone
VALANGINIAN	LATE	unnamed		Eprolithus antiquus —?— — — — —?—	Eprolithus antiquus —?— — — —?—
		pitrei			Tegumentum striatum
		Dichotomites spp.			—?— — — —?—
	EARLY	Polyptychites spp.	* D2E	Micrantholithus speetonensis	
		Paratollia spp.	D4B		
RYAZA-NIAN	LATE	albidum	D4C D5 D6A D6B D7A D7B	unnamed Perissocyclus fletcheri	Sollasites arcuatus Nannoconus sp. (discs)

After Perch – Nielsen 1979 *

Fig. 8.10 — Ammonite and nannofossil zones of the Ryazanian to Valanginian.

(b) *Perissocyclus fletcheri* Zone

Definition: from the last occurrence of *Nannoconus* sp. (discs) to the first occurrence of *Sollasites arcuatus*.

Author: Crux, this study.

Assemblage characteristics: the nannoflora is dominated by *W. barnesae* with common to abundant *R. asper*, *G. compactus*, *C. crenulatus*, *Staurolithites crux*, *Biscutum ellipticum*, *H. gorkae–T. coptensis*, *Discorhabdus ignotus*, *Diazomatolithus lehmanii*, *Micrantholithus obtusus*, *C. salebrosum*, *Zeugrhabdotus embergeri* and *Perissocyclus fletcheri*.

Stratigraphical extent: this zone lies entirely within the Upper Ryazanian; it is equivalent to part of the *albidum* ammonite Zone.

Lithostratigraphical equivalents: Beds D7A–D6B of the Speeton section.

Type section: Speeton, Beds D7A–D6B.

Comments: the presence of common *P. fletcheri* is characteristic of this zone. This is equivalent to part of the *R. angustiforata* Zone of Taylor (1982) and the *R. angustiforata* Subzone of Jakubowski (1987).

Fig. 8.11 is a biozonation chart (printed rotated) correlating the following schemes:

MOORBERG/OTTO GOTT, SARSTEDT — Nannofossil Zone
Vagalapilla matalosa · Zeugrhabdotus sisyphus · C. conicus · Nannoconus abundans · Cretarhabdus inaequalis · Stradnerlithus comptus · Tegulalithus septentrionalis · Cyclagelosphaera margerelii · Corollithion silvaradion · Eprolithus antiquus

Boundary Beds
OG116, OG116, OG102, OG102, OG100, OG100, OG79, OG76, MO38-34, MO41, MO67, MO72, MO73, MO78, MO80, MO80

MOORBERG/OTTO GOTT, SARSTEDT — Ammonite and Belemnite Zone
depressa · germanica · brunsvicensis · Aulacoteuthis spp. · pugio · disofalcatus · gottschei · staffii · ? – ? – ? · Aegocrioceras Beds · regale · noricum · amblygonium

SPEETON — Nannofossil Zone
? Vagalapilla matalosa · Zeugrhabdotus sisyphus · C. conicus · Nannoconus abundans · Cretarhabdus inaequalis · Stradnerlithus comptus · Tegulalithus septentrionalis · Cyclagelosphaera margerelii · Corollithion silvaradion · Eprolithus antiquus

Boundary Beds
48, 49, LB1A, LB1A, LB2D, LB3A, LB5B, LB5C, C2A, C2B, C7E, C7F, C9D, C10

SPEETON — Belemnite Zone
germanica · brunsvicensis · Aulacoteuthis spp. · pugio

SPEETON — Ammonite Zone
denckmanni · elegans · fissicostatum · rarocinctum · variabilis · marginatus · gottschei · speetonensis · inversum · regale · noricum · amblygonium

AGE
BARREMIAN (LATE / EARLY) · HAUTERIVIAN

Fig. 8.11 — Ammonite, belemnite and nannofossil zones of the Hauterivian to Barremian.

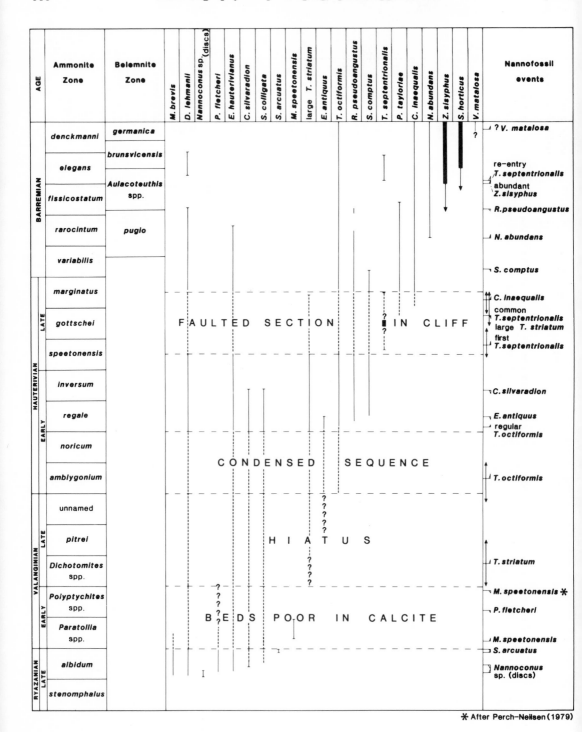

Fig. 8.12 — Composite range chart of biostratigraphically useful nannofossils in the Speeton section.

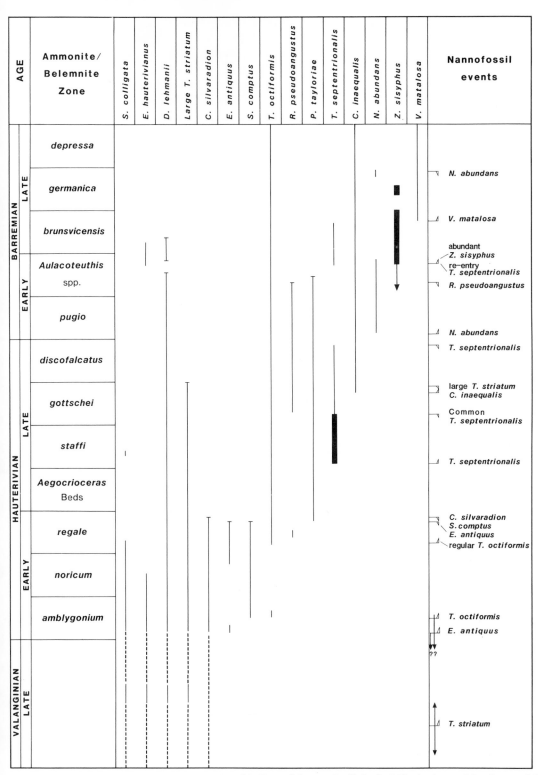

Fig. 8.13 — Composite range chart of biostratigraphically useful nannofossils in the Münchenhagen, Moorberg and Otto Gott, Sarstedt sections.

(c) *Sollasites arcuatus* Zone

Definition: from the first to the last occurrence of *S. arcuatus.*

Author: Jakubowski (1987).

Assemblage characteristics: the nannoflora is dominated by *W. barnesae* with common to abundant *R. asper, G. compactus, C. crenulatus, S. crux, B. ellipticum, Zeugrhabdotus erectus, C. salebrosum* and *Cylindralithus laffittei.* The nominate taxon is rare.

Stratigraphical extent: this biozone lies entirely within the Upper Ryazanian; it is equivalent to part of the *albidum* ammonite zone.

Lithostratigraphical equivalents: this biozone is restricted to a single bed of the Speeton section, Bed D6A. It is also present in the Core 7B from Sklinnabanken, offshore Norway, between 10.95 and 10.85 m.

Type section: Speeton, Bed D6A.

Comments: although *S. arcuatus* is only rare within this zone it is known to occur over a wide geographical area (Speeton, offshore Norway, Moray Firth) and is a reliable biostratigraphical marker.

This zone corresponds to parts of the *R. angustiforata* and the *Cruciplacolithus salebrosus* Zones of Taylor (1982).

(d) Unnamed Zone

Definition: from the last occurrence of *S. arcuatus* to the first occurrence of *Micrantholithus speetonensis.*

Author: Jakubowski emend. Crux this study.

Assemblage characteristics: the nannoflora is dominated by *W. barnesae* and *C. salebrosum. Perissocyclus fletcheri, Corollithion silvaradion, M. brevis, Diadorhombus rectus* and 'calcite rosettes' are among the biostratigraphically significant species.

Stratigraphical extent: uppermost Ryazanian to Lower Valanginian?; it is probably equivalent to the uppermost part of the *albidum* ammonite Zone and the lower part of the *Paratollia* spp. Zone.

Lithostratigraphical equivalents: D5–D4C Beds of the Speeton section and 10.75–8.00 m in Core 7B, Sklinnabanken, offshore Norway.

Type section: Sklinnabanken, Core 7B, 10.75–8.00 m.

Comments: this zone is barren in the Speeton section. The last occurrence of *P. fletcheri* and the inception of 'calcite rosettes' in this zone may be biostratigraphically useful in the future.

This zone corresponds to part of the *C. salebrosus* Zone of Taylor (1982) and to part of the *M. speetonensis* Zone of Jakubowski (1987).

(e) *Micrantholithus speetonensis* Zone

Definition: from the first to the last occurrence of *M. speetonensis.*

Author: Taylor (1982) emend. Crux this study.

Assemblage characteristics: only a very sparse nannoflora was recorded from this zone in the Speeton section. It includes rare specimens of *W. barnesae, R. asper, S. crux, H. gorkae–T. coptensis, Cretarhabdus madingleyensis, C. salebrosum, Corollithion geometricum* and *M. speetonensis.*

Stratigraphical extent: Lower Valanginian, equivalent to the upper part of the *Paratollia* spp. ammonite Zone and most of the *Polyptychites* spp. Zone?

Lithostratigraphical equivalents: Beds D4B–D2E of the Speeton section.

Type section: Speeton, Beds D4B–D2E.

Comments: the upper limit of this zone is defined by the extinction of *M. speetonensis.* This event was recorded in Bed D2E at Speeton by Perch-Nielsen (1979), but *M. speetonensis* was only recorded as high as Bed D4A in the present study. Perch-Nielsen's range is adopted here to define the duration of this zone. Taylor (1982) also followed Perch-Nielsen's range but misquoted it, placing the extinction of *M. speetonensis* in the Upper Valanginian. Jakubowski (1987) recorded this event at the Lower–Upper Valanginian boundary: his range was also based on Perch-Nielsen's (1979) record. A stratigraphical break occurs in the Speeton section with the uppermost Lower Valanginian and Upper Valanginian strata missing; it is thus possible that *M. speetonensis* could range into younger sediments elsewhere. A sharp reduction in the abundance of *C. salebrosum* in this zone may also be biostratigraphically useful.

This zone is equivalent to parts of the *M. speetonensis* Zones of Taylor (1982) and Jakubowski (1987).

(f) Unstudied Interval

Description: from the extinction of *M. speetonensis* to the first occurrence of *Tegumentum striatum*.

Stratigraphical extent: uppermost Lower Valanginian to Upper Valanginian, equivalent to the later part of the *Polyptychites* spp. ammonite Zone to the early part? of the *Dichotomites* spp. Zone.

Comments: this interval was not sampled. As noted above, it is possible that *M. speetonensis* ranges into the earlier part of the Late Valanginian and may thus have an overlapping range with *T. striatum*. Jakubowski (1987) noted the presence of *T. striatum* in the Lower Valanginian D3A Bed at Speeton. This range was not substantiated in the present study and thus this interval is reported.

(g) *Tegumentum striatum* Zone

Definition: from the first occurrence of *T. striatum* to the first occurrence of *Eprolithus antiquus*.

Author: Taylor (1982) emend. Crux this study.

Assemblage characteristics: the nannofloras are dominated by *W. barnesae* and *C. margerelii*. Other biostratigraphically useful species present include *T. striatum, Assipetra infracretacea, Speetonia colligata* and *C. silvaradion*.

Stratigraphical extent: Upper Valanginian to ?Lower Hauterivian, *Dichotomites* spp. Zone to the *?amblygonium* Zone. (Mutterlose (pers. commun.) reports *T. striatum* from the lowermost Upper Valanginian of Germany; thus this zone can be extended into the unstudied interval.)

Lithostratigraphical equivalents: a complete section through this zone was not sampled in the present study. A short interval within the zone occurs in the Münchenhagen clay pit, Beds 108–124.

Type section: Münchenhagen, Beds 108–124.

Comments: if Jakubowski (1987) is correct in claiming that *T. striatum* occurs in the Lower Valanginian D3A Bed at Speeton, then this zone would overlap with the underlying zone (*M. speetonensis* Zone). In the present study no such overlap was observed. Taylor (1982) did not record the first occurrence of *T. striatum* until the Lower Hauterivian. However, no Upper Valanginian sections were available to her study.

This zone corresponds to part of the *Chiastozygus striatus* Zone of Taylor (1982) and part of the *C. silvaradion* Zone of Jakubowski (1987).

(h) *Eprolithus antiquus* Zone

Definition: from the first to the last occurrence of *E. antiquus*.

Author: Crux this study.

Assemblage characteristics: the nannofloras are dominated by *W. barnesae* and *C. margerelii* with common to abundant *B. ambiguus, Lithraphidites carniolensis, C. crenulatus, C. laffittei, D. lehmanii, T. striatum, R. asper* and *C. salebrosum. Scapholithus fossilis* and *Tegumentum octiformis* have their inceptions in this zone.

Stratigraphical extent: the exact lower limit of this zone was not defined in the present study, but it must lie in the uppermost Valanginian to lowermost Hauterivian. The total range of this zone is Upper? Valanginian to Lower Hauterivian, unnamed ammonite Zone (*Eleuiceras* Beds) to the *regale* Zone.

Lithostratigraphical equivalents: Beds D2D–C10 of the Speeton section and Beds 101–80 of the Moorberg, Sarstedt section.

Type section: Moorberg, Sarstedt, Beds 101–80.

Comments: the exact level of the inception of *E. antiquus* is not known as no uppermost Valanginian sections were available for study. Perch-Nielsen (1985) assigned a Hauterivian to Albian range to *E. antiquus*. This range, which extends into much younger strata than that recorded in the present study, is probably explained by the inclusion in her species concept of *E. antiquus*, of forms attributable to *Eprolithus varolii*.

This zone corresponds to part of the *C. striatus* Zone of Taylor (1982) and part of the *C. silvaradion* Zone of Jakubowski (1987).

(i) *Corollithion silvaradion* Zone

Definition: from the last occurrence of *E. antiquus* to the last occurrence of *C. silvaradion*.
Author: Jakubowski (1987) emend. Crux, this study.
Assemblage characteristics: the nannofloras of this biozone are dominated by *W. barnesae* with common to abundant *R. asper*, *S. crux*, *C. laffittei*, *C. margerelii* and *C. salebrosum*. *Corollithion silvaradion* has its extinction at the top of this zone.
Stratigraphical extent: this zone lies completely within the Lower Hauterivian. In the Speeton section it is equivalent to the *regale* to *inversum* ammonite Zones, while in the Moorberg, Sarstedt section it lies entirely within the *regale* ammonite Zone. The ammonite correlation of this part of the two sections is rather tentative and the apparent difference in the extent of this zone in the two sections may not be real.
Lithostratigraphical equivalents: Beds C9D–C7F of the Speeton section and Beds 80–78 of the Moorberg, Sarstedt section.
Type section: Speeton, Beds C9D–C7F.
Comments: the top of this zone is placed at the extinction of *C. silvaradion*. The biostratigraphical usefulness of this species was first recognised by Köthe (1981), who noted its extinction in the *regale* ammonite Zone of the Moorberg, Sarstedt section. Jakubowski (1987) used the extinction of this species to define the upper limit of his *C. silvaradion* Zone. He placed this event in the Lower Hauterivian.

This zone is equivalent to part of the *C. striatus* Zone of Taylor (1982) and part of the *C. silvaradion* Zone of Jakubowski (1987).

(j) *Cyclagelosphaera margerelii* Zone

Definition: from the last occurrence of *C. silvaradion* to the first occurrence of *Tegulalithus septentrionalis*.
Author: Jakubowski (1987) emended to zonal status.
Assemblage characteristics: this zone is characterised by relatively sparse nannofloras compared with the overlying and underlying sections in both Speeton and Moorberg, Sarstedt. The nannofloras are dominated by *W. barnesae* with no other species particularly common.
Stratigraphical extent: Lower to Upper Hauterivian, equivalent to the inversum to *speetonensis* ammonite Zones of the Speeton section and the *regale* to *staffii* Zones of the Moorberg, Sarstedt section.
Lithostratigraphical equivalents: Beds C7E–C5L (and higher?) of the Speeton section and Beds 73–72 of the Moorberg, Sarstedt section.
Type section: Moorberg, Sarstedt, Beds 73–72.
Comments: the upper limit of this zone cannot be accurately identified in the Speeton section owing to faulting in the cliff.

This zone is equivalent to part of the *C. striatus* Zone of Taylor (1982).

(k) *Tegulalithus septentrionalis* Zone

Definition: from the first occurrence of *T. septentrionalis* to the top of its acme.
Author: Jakubowski (1987) emend. Crux, this study.
Assemblage characteristics: the nannofloras are dominated by *W. barnesae* with common *T. septentrionalis*. *Cretarhabdus inaequalis* has its inception and *Speetonia colligata* its last occurrence within this zone.
Stratigraphical extent: Upper Hauterivian, equivalent to the *staffii* to *gottschei* ammonite Zones of the Moorberg, Sarstedt section and the *speetonensis*? to *gottschei* Zones of the Speeton section.
Lithostratigraphical equivalents: Beds 67–34 of the Moorberg, Sarstedt section. The full extent of this zone is unknown at Speeton owing to faulting in the cliffs.
Type section: Moorberg, Sarstedt, Beds 67–34.
Comments: this zone may be diachronous over long distances as *T. septentrionalis* is known to

have an earlier first occurrence in high latitudes (Thierstein 1973). Jakubowski (1987) used the extinction of *Cruciellipsis cuvillieri* to divide this zone, which he extended to the Hauterivian–Barremian boundary. This biostratigraphical event is not considered useful in the present study owing to the extreme rarity of *C. cuvillieri*.

This zone corresponds to part of the *C. striatus* Zone of Taylor (1982) and is probably equivalent to the *S. colligata* Subzone and the *L. septentrionalis* Zone of Jakubowski (1987).

(l) *Stradnerlithus comptus* Zone

Definition: from the top of the *T. septentrionalis* acme to the last occurrence of *Stradnerlithus comptus*.

Author: Jakubowski (1987), emend. Crux, this study.

Assemblage characteristics: the nannofloras of this zone are dominated by *W. barnesae* with common to abundant *R. asper, S. crux, C. geometricum, D. ignotus, C. laffittei* and *D. lehmanii*. *Rhagodiscus pseudoangustus* and *C. inaequalis* are rare but characteristic of this zone. *S. comptus* has its last occurrence at the top of this zone. Various species of *Nannoconus* are common within this zone. There is a marked reduction in the diameter of the elliptical coccoliths of *T. striatum* within this zone.

Stratigraphical extent: uppermost Hauterivian to lowermost Barremian, equivalent to the *gottschei* to *variabilis* ammonite Zones.

Lithostratigraphical equivalents: Beds C2E–C2B of the Speeton section, 7.35–4.50 m in Core 7B, from Sklinnabanken, offshore Norway. It is not possible to differentiate this zone from the overlying *C. inaequalis* Zone in the Otto Gott, Sarstedt section, owing to the absence of *S. comptus*.

Type section: Speeton, Beds C2E–C2B.

Comments: the lower limit of this zone is poorly defined in the type section at Speeton owing to faulting in the cliffs, preventing accurate sampling over this interval. It is recognised in the Moorberg, Sarstedt section in Bed 33.

This zone corresponds to part of the *C. striatus* Zone of Taylor (1982). It is difficult to compare it

with the zones of Jakubowski (1987) as he records the top of the *T. septentrionalis* acme at the Hauterivian–Barremian boundary, while the findings of the present study suggest that it should lie much lower. The *S. comptus* Zone of the present study is probably equivalent to the *S. comptus* and part of the *C. salebrosum*? Zones of Jakubowski (1987).

(m) *Cretarhabdus inaequalis* Zone

Definition: from the last occurrence of *S. comptus* to the first occurrence of *Nannoconus abundans*.

Author: Crux, this study.

Assemblage characteristics: the nannofloras of this zone are dominated by *W. barnesae* with common to abundant *R. asper, S. crux, B. ellipticum, D. lehmanii, H. gorkae–T. coptensis* and *Pseudolithraphidites parallelus*.

Stratigraphical extent: Lower Barremian, equivalent to the *variabilis* to *rarocinctum* ammonite zones of the Speeton section and the *discofalcatus*? to *pugio* Zones of the Otto Gott, Sarstedt section.

Lithostratigraphical equivalents: Beds C2A–LB5C of the Speeton section and up to Bed 76 of the Otto Gott, Sarstedt section, where the base of the zone cannot be defined.

Type section: Speeton, Beds C2A–LB5C.

Comments: *C. inaequalis, Perissocyclus tayloriae* and *R. pseudoangustus* are rare but distinctive elements of the nannoflora of this zone. Jakubowski (1987) used the extinctions of *S. colligata* and *T. septentrionalis* to divide this interval. These events are not used here as the former species is very rare and the latter does not become extinct until the Upper Barremian or Albian?

This zone corresponds to parts of the *C. striatus* and *N. abundans* Zones of Taylor (1982) and parts of the *S. comptus* and *C. salebrosum* Zones of Jakubowski (1987).

(n) *Nannoconus abundans* Zone

Definition: from the first occurrence of *N. abundans* to the last occurrence of *R. pseudoangustus*.

Author: Taylor (1982) emend. Crux, this study.

Assemblage characteristics: the nannofloras are dominated by *W. barnesae* with common to abundant *R. asper*, *B. ambiguus*, *L. carniolensis*, *S. crux*, *B. ellipticum*, *A. infracretacea* and *N. abundans*. *Rhagodiscus pseudoangustus* becomes extinct at the top of this zone.

Stratigraphical extent: Lower Barremian, equivalent to the *rarocinctum* to *fissicostatum* ammonite Zones of the Speeton section and the *pugio* to *Aulacoteuthis* spp. belemnite Zones of the Speeton and Otto Gott, Sarstedt sections.

Lithostratigraphical equivalents: Beds LB5B–LB3A of the Speeton section and Beds 79–100 of the Otto Gott, Sarstedt section.

Type section: Speeton, Beds LB5B–LB3A.

Comments: Taylor (1982) recorded the first occurrence of *N. abundans* in the *variabilis* Zone; this record was not substantiated in the present study and my own record of its first occurrence is used. The last occurrence of *R. pseudoangustus* which marks the top of this zone may not be its true extinction. This species is probably ancestral to *Rhagodiscus angustus* which first appears in the Aptian; thus it is likely that somewhere the ranges of the two species overlap. Jakubowski (1987) uses the extinction of *C. salebrosum* to divide this zone. In the present study *C. salebrosum* was recorded throughout the Barremian and is known from strata as young as Cenomanian. However, Jakubowski may be applying a more restricted taxonomic concept, as small forms of this species are generally absent above this level.

This zone corresponds to part of the *N. abundans* Zone of Taylor (1982) and parts of the *C. salebrosum* and *Conusphaera rothii* Zones of Jakubowski (1987).

(o) *Cretarhabdus conicus* Zone

Definition: from the last occurrence of *R. pseudoangustus* to the re-entry of *T. septentrionalis*.

Author: Crux, this study.

Assemblage characteristics: the nannofloras are dominated by *W. barnesae* with common to abundant, *R. asper*, *L. carniolensis*, *Cretarhabdus conicus*, *S. crux*, *A. dietzmannii*, *B. ellipticum*, *C.*

laffittei, *C. margerelii*, *C. crenulatus* and *N. abundans*.

Stratigraphical extent: Lower Barremian, equivalent to the *fissicostatum* to *elegans* ammonite Zones of the Speeton section and within the *Aulacoteuthis* belemnite Zone of Otto Gott, Sarstedt.

Lithostratigraphical equivalents: Beds LB2D–LB1A of the Speeton section and Beds 100–102 of Otto Gott, Sarstedt.

Type section: Otto Gott, Sarstedt, Beds 100–102.

Comments: the extinction of *C. rothii*, used by Jakubowski (1987) to divide this zone, is of only local significance in the Moray Firth area. This species is too rare over most of the North Sea and surrounding areas.

This zone corresponds to part of the *N. abundans* Zone of Taylor (1982) and part of the *C. rothii* and *Nannoconus borealis* Zones of Jakubowski (1987).

(p) *Zeugrhabdotus sisyphus* Zone

Definition: from the re-entry of *T. septentrionalis* and/or the abundant occurrence of *Zeugrhabdotus sisyphus* to the first occurrence of *Vagalapilla matalosa*.

Author: Crux, this study.

Assemblage characteristics: the nannofloras are dominated by *W. barnesae* with common *R. asper*, *S. crux*, *B. ellipticum*, *D. lehmanii*, *C. margerelii*, *Z. sisyphus* and *S. horticus*. This zone is characterised by the unusual occurrence of common to abundant *Z. sisyphus* together with species such as *T. septentrionalis*, *Ethmorhabdus hauterivianus* and *D. lehmanii*, which became rare or absent during the deposition of 'Hauptblätterton'.

Stratigraphical extent: Lower to Upper Barremian, *elegans* to *denckmanni* ammonite Zones in the Speeton section and the *Aulacoteuthis* spp. to *brunsvicensis* belemnite Zones in the Speeton and Otto Gott, Sarstedt sections.

Lithostratigraphical equivalents: Beds LB1A to 49 of the Speeton section and Beds 102 to 116 of the Otto Gott, Sarstedt section.

Type section: Otto Gott, Sarstedt, Beds 102–116.

Comments: the absence and re-appearance of *T. septentrionalis* is possibly caused by the closing and reopening of seaways to the north. It is thus possible that this biozone is slightly diachronous. However, this is not apparent when it is compared with the belemnite zones in the Speeton and Otto Gott sections. Mutterlose and Harding's (1987) record of *T. septentrionalis* within the 'Hauptblätterton' is probably incorrect as they illustrate (Plate 2, Figs. 19–21) forms attributable to *N. abundans*.

This zone corresponds to part of the *N. abundans* to Taylor (1982) and part of the *N. borealis*? and *N. abundans* Zones of Jakubowski (1987).

(q) *Vagalapilla matalosa* Zone
Definition: from the first occurrence of *V. matalosa* to the top of the studied interval.
Author: Crux, this study.
Assemblage characteristics: the nannofloras from the lower part of this zone are dominated by *W. barnesae*: they become less abundant in the upper part of the zone and this dominance decreases. Other species which are rare to common within this zone are *R. asper*, *B. ellipticum*, *C. laffittei* and the *Zeugrhabdotus noeliae* group.
Stratigraphical extent: Upper Barremian, equivalent to the *germanica* to *depressa* belemnite Zones (top not seen).
Lithostratigraphical equivalents: Beds 116–199 of the Otto Gott, Sarstedt section and Beds 48 and 47 of the Speeton section.
Type section: Otto Gott, Sarstedt, Beds 116–199.
Comments: the upper limit of this zone remains undefined. Further study of the uppermost Barremian to Lower Albian sections may reveal a suitable biostratigraphical event. The extinction of *N. abundans* may be useful; this event probably occurs within the Otto Gott section, possibly in Bed 183. However, poor preservation above this level may cause its absence from the higher beds. Jakubowski (1987) considered *N. abundans* to range to the top of the Barremian. However, his work is based on well material with no ammonite or belemnite stratigraphy over this interval.

This zone corresponds to part of the *N. abundans* Zones of Taylor (1982) and Jakubowski (1987).

8.5 THE NANNOFLORAL DEVELOPMENT AND THE PALAEOGEOGRAPHICAL RECORD IN THE NORTH-WESTERN EUROPEAN EARLY CRETACEOUS
The Early Cretaceous basins of eastern England, the North Sea region and northern Germany formed a southern shelf sea extension of the Boreal–Arctic Ocean. There were intermittent marine connections to the south-east with Tethys through the Polish Trough during parts of the Early Cretaceous. A more enduring and stronger marine connection between north-western Europe and the Boreal–Arctic Ocean existed to the north between the Shetlands and Norway. Other connections with Tethys also existed around the north west of Britain.

The ammonite zonal sequences for north-western Europe are based on groups of genera which dominated the faunas at successive periods of time (Kemper *et al.* 1981). The change from one genus to another was sometimes a gradual evolutionary progression, but more often immigration of a new genus caused a rapid change in the fauna. These immigrations came from both the Boreal and Tethyan regions. Rawson (1973) and Kemper *et al.* (1981) distinguished three groups of migrants from Tethys to north-western Europe: (a) isolated strays, (b) mass migrations and (c) inter-regional mass occurrences. A relationship between the occurrence of the last two categories and the timing of transgressions was noticed by Kemper *et al.* (1981). Influxes of Tethyan derived genera are found to coincide with periods of transgression, while endemic and Boreal faunas prevail in intervals between transgressions. In the present study an attempt is made to see whether nannofloras exhibit a similar pattern and whether they can be used to contribute to our knowledge of the geological history of this period.

Our knowledge of Tethyan nannofloras is

relatively good when compared with the records from the Arctic regions. In this study, I have used the nannofossil species abundances from south-eastern France, Switzerland and the Atlantic given in Thierstein (1973) as Tethyan standards. No such extensive record is available from the Arctic: only a single publication on Valanginian nannofloras from Svalbard (Verdenius 1978). Further work is needed to define Boreal and Arctic nannofloral characteristics, and thus to enable their recognition in the north-western European area.

Wind and Čepek (1979) and Covington and Wise (1987) have reported rich Lower Cretaceous nannofloras from offshore Morocco and Carolina (USA) respectively. These assemblages are similar to those from Tethys recorded by Thierstein (1973), but in addition several rare occurrences of new species were recorded. Further studies are needed to determine whether these new species are geographically restricted, or simply rare delicate members of the nannoflora which are not always preserved and have previously been overlooked in other areas. Two of these species, *Watznaueria fasciata* and *Rhagodiscus pseudoangustus*, were recorded in the present study. This supports the idea that the geographical ranges of the other species will in time be proved more extensive than known at present. Similarly rare species such as *Clepsilithus polystreptus*, *Chiastozygus cepekii*, *Calculites sarstedtensis* and *Watznaueria rawsonii* described by Crux (1987) may not be geographically restricted to the north-western European area and they will probably be recorded from other regions in future studies. The nannofloras from the Proto-Atlantic Ocean are included in the broad term Tethyan in the present study. Table 8.1 lists species considered in this study to have Tethyan, endemic or Boreal–Arctic origins together with species indicative of certain environments.

In the following section, the nannofloral record from the localities studied is compared with the records of other fossil groups, the sediment distribution within the north-western European basins and the sea level changes

Table 8.1 — Summary of Tethyan, endemic and Boreal–Arctic nannofossils considered in this study, together with species indicative of certain environments.

Tethyan (including Proto-Atlantic)	Endemic	Boreal–Arctic (Austral?)
L. bollii	N. abundans	E. antiquus?
S. colligata	M. brevis?	S. arcuatus
S. comptus?	Nannoconus sp. (discs)	P. fletcheri?
C. cuvillieri	M. speetonensis?	S. horticus
N. minutus		C. salebrosum
Nannoconus sp.1		T. septentrionalis
Nannoconus sp.2		C. silvaradion
Nannoconus sp.3		Z. sisyphus
C. oblongata		T. striatum
C. rothii		
N. steinmannii		
T. verenae		
Transgressive intervals		Periods of good circulation
R. asper		E. hauterivianus?
C. margerelii		D. lehmanii
D. lehmanii		Z. sisyphus?

previously proposed for the Early Cretaceous. For these comparisons information is drawn from Kemper *et al.* (1981) on ammonite migrations, from Mutterlose *et al.* (1983) on belemnite migrations, and from Tyson and Funnell (1987) and Rawson and Riley (1982) on palaeogeography and sea level changes. A summary of the nannofloral record in north-western Europe and the comparisons drawn is given in Fig. 8.14.

(a) Preservation
The preservation of the nannofloras throughout the studied interval is generally good, with the nannofossils showing only slight signs of dissolution. Exceptions to this are as follows: dissolution and almost complete removal of the nannofloras in the Lower Valanginian at Speeton; dissolution of the nannofossils of the *Aegocrioceras* Beds, Moorberg, Sarstedt, which results in low diversities of the nannofloras; dissolution and low diversity in the nannofloras of the Upper Barremian, Otto Gott, Sarstedt; dissolution and overgrowth of secondary calcite

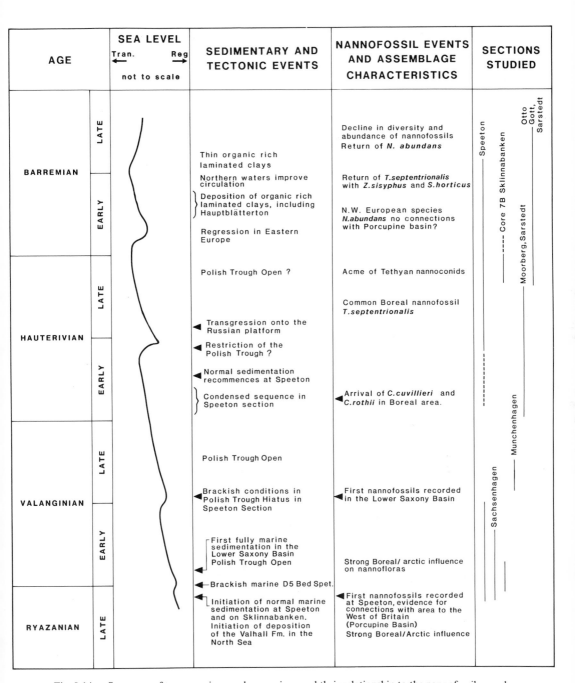

Fig. 8.14 — Summary of transgressions and regressions and their relationship to the nannofossil record.

in the Upper Ryazanian to Lower Valanginian interval of Core 7B, Sklinnabanken.

(b) Ryazanian

The early Ryazanian north-western European basins were areas of brackish and lacustrine limestone and shale deposition of the Purbecko-Wealden facies. Near active sediment source areas, deltaic and fluvio-lacustrine clastics were deposited. In the North Sea, marine black shales were deposited (Kimmeridge Clay Formation).

A well-documented Late Ryazanian regional transgression initiated deposition of the Speeton Clay Formation in Yorkshire, the Valhall Formation in the North Sea, and the Upper Spilsby Sandstone Member over the East Midlands shelf (England). This transgressive phase is also recorded as far north as Sklinnabanken, where uppermost Ryazanian to lowest Valanginian sediments were deposited.

The oldest nannofloras examined in the present study were deposited during this transgressive phase. They come from the D7D to D6A Beds at Speeton and approximately the lowest metre of the Sklinnabanken Core 7B. The first diverse nannoflora recorded in the Speeton section is from the D7B Bed (24 species). For the remainder of the Ryazanian similar assemblages occur. These diversities are comparable with those recorded over the same interval in the Tethyan region by Thierstein (1973). The species assemblages of the two areas are, however, markedly different. The assemblages recorded in the Core 7B from Sklinnabanken are less diverse than those from Speeton, but this is probably a result of poorer preservation. The nannofloras of the Speeton section are dominated by *W. barnesae* with common to abundant, *R. asper*, *C. crenulatus*, *S. crux*, *B. ellipticum*, *D. lehmanii*, *C. margerelii*, *C. salebrosum* and *Z. embergeri*. Some of these species are also common to abundant in the Tethyan sections (*W. barnesae*, *Z. embergeri*, *C. margerelii*, *D. lehmanii* and *C. crenulatus*).

Species abundant in the Tethyan sections but rare or absent at Speeton are *Nannoconus steinmannii*, *M. obtusus* and *Braarudosphaera*

bigelowii. Less common species found only in Speeton include *P. fletcheri*, *Nannoconus* sp. (discs) and *S. arcuatus*. Rare species found only in Tethys include *Nannoconus bronnimannii*, *Nannoconus truittii*, *Rucinolithus wisei* and *C. rothii*.

Thus quite substantial differences existed between the nannofloras of Speeton and Tethys, although they did have some elements in common. The relative high abundances of *C. margerelii* and *D. lehmanii* in the Speeton section possibly indicate transgressive conditions rather than marine connections with Tethys. These two species, together with *R. asper*, are found to be abundant in other transgressive intervals. The presence of a single nannoconid species, *Nannoconus* sp. (discs), suggests marine connections with areas to the west of Britain and possibly Tethys as this species occurs in the Porcupine Basin, offshore Ireland. The absence of *N. steinmannii*, *B. bigelowii*, *R. wisei* and *C. rothii* from Speeton at this time, together with the abundant presence of the Boreal species *C. salebrosum* and the occurrence of *P. fletcheri* and *S. arcuatus*, points to a strong Boreal influence on the nannofloras.

In conclusion, a weak Tethyan influence is seen on the essentially Boreal nannofloras of the Late Ryazanian at Speeton and Sklinnabanken. This influence is not so pronounced as in the later transgressions. There is also tentative evidence to suggest that these influences entered the area by way of a western seaway, rather than from the south east.

(c) Early Valanginian

The Early Valanginian was a period of further transgression in most of the north-western European basins. Marine sedimentation returned to the Lower Saxony Basin, and is seen in the Sachsenhagen pit. Fully marine conditions also returned to Speeton after a brief brackish interval (Bed D5), and further onlap occurred on the Sklinnabanken. In other areas, the transgression is also shown by the appearance of marine sediments in the Polish Trough and Sweden and the abrupt change from sands to clays and ironstone deposition in Lincolnshire. This Early Valanginian transgression was accompanied by the immigration of the ammonite genus *Platylenticeras*, which evolved rapidly to occupy the whole of the Lower Saxony Basin and the Speeton area (Kemper *et al*. 1981).

The nannofossil record from the Early Valanginian is disappointing. Very sparse nannofloras are present in the Early Valanginian D4B to D2E Beds from Speeton. The samples collected from the Sachsenhagen pit are barren and only a limited stratigraphical interval is represented in Core 7B from Sklinnabanken. The scarcity of nannofossils in the Speeton section is due to poor preservation. There is a general lack of calcite in these beds and the nannofloras present are strongly etched. Only limited conclusions can be drawn from such an impoverished assemblage. The only point of note is the presence of *M. speetonensis*. This species is probably endemic to the north-western European area, although it may also have originated from the Arctic area.

The nannofloras of the Core 7B from Sklinnabanken are dominated by *W. barnesae* and *C. salebrosum*. The abundance of the latter and the presence of the high latitude species *C. silvaradion* and *P. fletcheri* show a strong Arctic–Boreal influence, which is hardly surprising for the geographical location of this core. Some nannoconids are present but the Tethyan influence is weak.

In conclusion, the Tethyan influence in the Early Valanginian transgressive phase is much weaker than that seen in later phases, especially in comparison to the Tethyan influences of the latest Hauterivian to Barremian, where common *C. rothii* and several nannoconid species occur on the Sklinnabanken.

(d) Late Valanginian

A short regressive phase occurred in the 'mid' Valanginian, with brackish conditions returning to the Polish Trough (Tyson and Funnell 1987). A major transgressive interval which is evident over much of Europe succeeded this phase, with an expansion of the Saxony Basin in Germany.

The first nannofossil-rich sediments occur in the Upper Valanginian of the Hanover area which were sampled in the Münchenhagen pit. At Speeton the Upper Valanginian is absent, but the presence of reworked ammonites in the basal Hauterivian attests to the marine conditions that prevailed during the Late Valanginian.

Although endemic ammonite genera continued to dominate the faunas of north-western Europe until nearly the end of the Valanginian, some changes did occur. One of these was the increased migration in both directions between western Tethys and north-western Europe (Kemper *et al*. 1981).

The only Late Valanginian nannofloras available for comparison with those from Tethys are from Münchenhagen near Hanover. The strata sampled there only represent a short interval in the Late Valanginian. The nannofloras are dominated by *W. barnesae* with common to abundant *C. margerelii*. Other species which commonly occur in some samples are *C. angustiforatus*, *C. crenulatus*, *C. laffittei*, *B. ellipticum* and *D. lehmanii*. There is a marked decrease in the abundance of *C. salebrosum* which dominated the Lower Valanginian nannofloras of Sklinnabanken. *Tegumentum striatum* is recorded for the first time in this section. This species is thought to occur earlier at high latitudes (Verdenius 1978). Species that were common to abundant in the Tethyan area at this time include *W. barnesae*, *Z. embergeri*, *Cretarhabdus* sp., *C. rothii*, *C. margerelii*, *Nannoconus* spp., *D. lehmanii*, *Watznaueria britannica*, *L. carniolensis*, *C. laffittei*, *B. ellipticum*, *R. asper*, *C. cuvillieri*, *Micrantholithus* spp., *S. crux*, *B. bigelowii* and *Calcicalithina oblongata*. These two assemblages show that significant differences still existed between the nannofloras of the Tethyan and north-western European areas. Tethyan species such as *C. cuvillieri*, *C. oblongata* and *C. rothii* were excluded from the Saxony Basin (and *T. striatum* was excluded from the Tethyan area?). *Tubodiscus verenae* has not been recorded from either the Lower or Upper Valanginian of the north-western European area. This rare species, however, has a sporadic and stratigraphically restricted occurrence in Tethyan sections. Its apparent absence from north-western Europe may be due to insufficient sampling of the Valanginian (Taylor's (1982) record of *T. verenae* from the Hauterivian to Albian is not considered to be of true *T. verenae*).

In conclusion, the limited evidence provided by the nannofloras of the Münchenhagen pit points towards relatively better connections between Tethys and the north-western European area. This is indicated by the decline in the dominance of the Boreal–Arctic species *C. salebrosum* and the increased similarities of the nannofloral assemblages of the two areas. The marine connections were, however, inferior to those of the Hauterivian when the Tethyan species *C. cuvillieri* and *C. rothii* were able to migrate to north-western Europe for the first time.

(e) Early Hauterivian
The Early Hauterivian saw the general continuation of the Early Cretaceous transgression. There was a further expansion of marine sedimentation in the Lower Saxony Basin, the deposition recommenced at Speeton and in central Lincolnshire.

Amongst ammonite faunas, the Early Hauterivian was a period of significant Tethyan influence (Kemper *et al.* 1981). It is also a time when the belemnite *Hibolites* migrated northwards in large numbers to replace the Boreal–Arctic *Acroteuthis* (Mutterlose *et al.* 1983).

In the present study, a thick sequence of Lower Hauterivian strata has been sampled from Moorberg, Sarstedt. The succession sampled at Speeton is less complete with the lowermost two ammonite zones being condensed. The nannofloras of both sections are diverse and abundant, dominated by *W. barnesae* but with particularly common *C. margerelii* and *R. asper*. Other species which are sporadically common to abundant are *C. crenulatus, L. carniolensis, G. compactus, B. ambiguus, C. laffittei* and *T. striatum*. For the first time in the north-western

European area, *C. rothii* and *C. cuvillieri* are recorded. Nannoconids are also consistently present throughout the Lower Hauterivian. Species that were common to abundant in the Tethyan area during the Early Hauterivian include *W. barnesae, Z. embergeri, Z. sisyphus, C. margerelii, Nannoconus* spp., *L. carniolensis, C. conicus, B. ellipticum, D. ignotus, R. asper, Micrantholithus* spp., *C. crenulatus* and *C. oblongata. Calcicalithina oblongata* has not been recorded in the present study in the Speeton and German sections; Perch-Nielsen (1979), however, reported it to be present in the northern North Sea and Jakubowski (1987) reported its rare presence in the North Sea; it is also present in the Porcupine Basin. This suggests that this species migrated northwards around the west of the British Isles but failed to reach the southern North Sea and the Saxony Basin. *Lithraphidites bollii*, a species that evolved in the Hauterivian and is biostratigraphically useful in the Tethyan area, also appears to be excluded from much of the north-western European area.

Possible Boreal–Arctic influences on the Lower Hauterivian nannofloras are the abundant occurrences of *C. salebrosum* in certain beds at Speeton (C11B and C7H) and the occurrence of *E. antiquus,* although little is known about the palaeobiogeographical distribution of the latter species.

Thus the nannofloras of the Tethyan and north-western European areas were by this time quite similar to one another, although still with distinctive characteristics. The transgression of the Early Hauterivian was probably more extensive than previous transgressions. It caused a greater similarity between the nannofloral assemblages in the two areas as well as introducing for the first time previously exclusively Tethyan elements to the nannoflora of the north-western European area.

(f) Late Hauterivian
Sea level changes in the Late Hauterivian and in the Barremian are not so clearly defined as the transgressions that occurred from the Late Ryazanian to Early Hauterivian. Rawson and

Riley (1982) reported a major transgression in the mid-Hauterivian, which Kemper *et al.* (1981) also recorded in the *inversum* Zone. In support of this, they cite local coastal onlap in north Germany and a sharp facies change and minor erosion in England, with clays locally resting on highly condensed Upper Valanginian and Lower Hauterivian sediments. There was also a dramatic increase in the depositional area on the Russian Platform. In offshore areas, the mid-Hauterivian is marked on some highs by a change from condensed calcareous beds of Late Ryazanian to Early Hauterivian age (basal Valhall limestone) to shales of Late Hauterivian age. Throughout the remainder of the Late Hauterivian to Early Barremian, Rawson and Riley (1982) showed a mild regression. An important factor in this development is the marine connection through the Polish Trough. Tyson and Funnell (1987) summarised the available data: 'In the Polish Basin the Early Hauterivian basin is similar to that in the Late Valanginian but undergoes progressive shallowing that culminates in brackish episodes in the early Late Hauterivian but the southern link with Tethys apparently became broken or very restricted at this time.' In contrast to this Mutterlose (pers. commun.) considers the Polish Trough to have remained open.

Palaeobiogeographical data on the Late Hauterivian ammonites and belemnites suggest that there were poor marine connections between Tethys and the north-western European area. Ammonite faunas are almost exclusively Boreal, with the exception of some heteromorphs (Kemper *et al.* 1981), while the belemnite genus *Hibolites* underwent differential speciation in the two areas (Mutterlose *et al.* 1983).

The sediments of the uppermost Lower Hauterivian in the Moorberg section (*Aegocrioceras* Beds) yield only poorly preserved low diversity nannofloras. The succeeding beds from the *staffi* Zone contain common *T. septentrionalis*, which is considered to be a Boreal–Arctic species. It has its first occurrence in the Boreal area in the north-western European area in the Late Hauterivian but *T.*

septentrionalis, or closely related forms, does not reach the Tethyan area until the Aptian (Thierstein 1973). *Stradnerlithus comptus* is absent from the *Aegocrioceras* Beds and above in the Moorberg, Sarstedt section. It is only present in a single sample from the *inversum* Zone and is absent from the *speetonensis* and *gottschei* Zones at Speeton, although this interval is poorly sampled. This absence coincides with the strongly transgressive interval and the arrival of *T. septentrionalis. Tegulalithus septentrionalis* becomes less common together with a reduction in size of specimens of *T. striatum* in the upper part of the *gottschei* Zone of the Moorberg and Otto Gott, Sarstedt sections. At the same time there is an increase in the number of nannoconids, *R. asper* and *C. margerelii*. The nannoconids are most abundant in the lower part of the *discofalcatus* Zone. This pattern is not so easy to observe at Speeton, because faulting in the cliff only allowed a single sample to be taken from the upper part of the *speetonensis* and *gottschei* Zones. In this sample *T. septentrionalis* is common. In the succeeding samples, from the *marginatus* Zone, there is a marked increase in the numbers of *R. asper* and *D. lehmanii*.

This pattern of nannofloral development suggests a changeover from Tethyan influenced nannofloras of the Lower Hauterivian, below the *Aegocrioceras* Beds, to Boreal–Arctic influenced nannofloras above. These are in turn followed by a return of Tethyan influenced assemblages above the *gottschei* Zone. These changes are at least in part related to regional transgressions and regressions. The Early Hauterivian transgression introduced Tethyan influence into the nannofloras. This was followed by a second major transgression in the late Early Hauterivian (*Aegocrioceras* Beds–*inversum* Zone). In the Moorberg, Sarstedt section this resulted in a slightly condensed sequence which contains only sparse nannofloras. The sparse assemblages are possibly the result of dissolution due to the nannofossils' protracted stay at the sediment–surface water interface. The transgression may also have introduced deeper, colder and more corrosive waters into the area. At Speeton the dissolution of the nannofloras is not seen. The change from a diverse Tethyan influenced nannoflora to a less diverse Boreal–Arctic influenced assemblage is more gentle. The section is possibly more complete, but as mentioned above sampling of the beds above Bed C5L was poor.

The arrival of *T. septentrionalis* and the less diverse nannofloras in the north-western European area in the Late Hauterivian (*staffi–?speetonensis* Zone) probably indicates the presence of colder, more northerly waters. Their presence in the area could be due to one of the following: (a) the transgression that began in the late Early Hauterivian had by this time improved marine connections to the north so that these waters were temporarily able to displace the Tethyan influenced waters; (b) a temporary local restriction in the Polish Trough caused the connections to Tethys to be broken (Tyson and Funnell 1987); (c) a climatic change allowed the northerly waters to extend further south.

The first two explanations are more probable, but recently Kemper *et al.* (1987) have proposed a climatic change corresponding to the base of the *Aegocrioceras* Beds, based on changes in the nannofloras, microfaunas, ammonites and calcium carbonate content in the Moorberg, Sarstedt section and the well Konrad 101.

(g) Barremian
Rawson and Riley (1982) discussed the conflicting evidence for regression and transgression in the Barremian and concluded that it was essentially a regressive interval with only minor transgressive pulses. The sea had withdrawn from the Polish Trough and the Russian Platform by the Early Barremian. In contrast, the deposition of the blätterton beds in Germany and other similarly laminated clays at Speeton, together with the worldwide development of organic-rich sediments in the Barremian to Aptian, suggests a transgressive phase.

Mutterlose and Harding (1987) discussed the distribution of the blätterton facies in the Early Barremian. They found the greatest thickness of

the 'Hauptblätterton' to occur in areas furthest away from the northern seaways in the eastern part of the Lower Saxony Basin. From this they concluded that the facies represented a regressive phase.

Two Barremian sections were sampled in the present study. The Speeton section ranges from the Lower to the lower Upper Barremian, and the Otto Gott, Sarstedt, section from Lower to Upper Barremian. In both sections it was not possible to sample the uppermost Barremian. The nannofloras of the two sections are dominated by *W. barnesae* with common to abundant *B. ambiguus*, *R. asper*, *L. carniolensis*, *S. crux*, *B. ellipticum*, *C. geometricum*, *C. conicus*, *S. horticus*, *D. ignotus*, *A. infracretacea*, *D. lehmanii*, *C. margerelii*, *Z. sisyphus*, *Z. noeliae* group, *P. parallelus*, *C. laffittei* and *N. abundans*. The abundance patterns of these species show quite marked variations through the Lower and lower Upper Barremian. Significant features of these patterns are as follows: (1) *N. abundans*, an endemic north-western European species, dominates the nannoconid assemblages; (2) common to abundant *R. asper*; this is more pronounced in the Speeton section; (3) common to abundant *D. lehmanii* in the lowest Barremian (*variabilis–rarocinctum* Zones), dropping off to total absence in the lower part of the *fissicostatum* Zone and returning at the end of that zone; (4) *Cyclagelosphaera margerelii* common to abundant in the upper part of the Lower Barremian; (5) the absence of *T. septentrionalis* during most of the Lower Barremian, and its return in the uppermost beds together with abundant *Z. sisyphus* and, in Speeton only, *S. horticus*; (6) a general decline in the diversity and abundance of the nannofloras in the Upper Barremian; (7) the absence of *N. abundans* during the return of *T. septentrionalis*, and its own return in the Upper Barremian of Otto Gott, Sarstedt.

These patterns are interpreted as follows: a restriction of the northern seaways, approximately at the base of the *Aulacoteuthis* spp. belemnite Zone, possibly associated with a regression, is followed by a transgressive interval. The nannofloras reflect this by the change from Tethyan nannoconids to the common to abundant occurrence of *N. abundans,* an endemic species which evolved from Tethyan ancestors but does not occur outside north-western Europe. *Nannoconus abundans* is absent from the Porcupine Basin, offshore Ireland, possibly indicating poor marine connections between this area and the North Sea area at this time. The abundant occurrence of *R. asper* is clearly linked in the Otto Gott, Sarstedt section to the interval of the 'Hauptblätterton' deposition. Organic-rich sediments such as these are commonly associated with strongly transgressive intervals (Hallam and Bradshaw 1979). It has already been noted in this study that common to abundant *R. asper* is also associated with transgressive periods, possibly suggesting that the 'Hauptblätterton' was deposited during a transgressive interval. Erba (1987) has recently reported *R. asper* to indicate moderate primary productivity in periods of warmer waters in Aptian–Albian sections from central Italy. The return of the Boreal–Arctic species, *T. septentrionalis* in Bed LB1A of Speeton and Bed 102 of Otto Gott, Sarstedt indicates that the transgressive interval had improved marine connections to the north, allowing an influx of Boreal–Arctic species. Associated with the return of *T. septentrionalis* in both Speeton and Otto Gott is *Z. sisyphus* and, in Speeton only, *S. horticus.* This influx of northern waters caused basin flushing, the destruction of stratified anoxic water and cessation of black shale deposition. Basin flushing after the deposition of the 'Hauptblätterton' was possibly caused by colder and/or more saline Boreal–Arctic waters sinking under the warmer–more brackish waters of the north-western European area. This influx of northern waters also caused a decline in the abundance of *N. abundans.* However, this species (after a short absence in the Otto Gott section) was able to re-establish itself in the Late Barremian Beds (127–137). This may indicate a second period of restriction in the marine connections to the north.

The greater thickness of the 'Hauptblätterton'

away from the northern seaways, noted by Mutterlose and Harding (1987), shows that some oxygenated bottom waters were entering the north-western European area, but these were insufficient to oxygenate the whole area. These oxygenated bottom waters prevented the thick sections of organic-rich laminated clays (as seen in Germany) being deposited near the opening to the northern ocean.

Mutterlose and Harding (1987) and Mutterlose (pers. commun.) considered influxes of warm surface waters to be of prime importance amongst the causes of the 'Hauptblätterton' deposition. In support of this they cite the presence of *Nannoconus* spp. *Diadorhombus rectus* and *C. rothii (C. mexicana)*, as well as evidence from the palynofloras and other fossil groups. The nannoconid species present, however, are endemic north-western European species, with the exception of *N. elongatus.* The specimen of this species that they illustrate is possibly more closely related to *N. borealis,* than to true *N. elongatus.* The presence of *D. rectus* is probably the result of reworking; this species is usually considered to be restricted to the Valanginian (Perch-Nielsen 1985). Reworked palynomorphs are also common in the 'Hauptblätterton' (M. Partington pers. commun.).

The idea that influxes of warm surface waters could enter the restricted north-western European area is difficult to accept. Mutterlose and Harding (1987) agree that the only marine connection to other areas lay to the north. Influxes of surface waters through this northerly opening would be more likely to be of colder water that would sink and oxygenate the bottom waters (as postulated above), rather than warm waters that would remain on the surface.

8.6 TAXONOMY

Genera and species referred to in this paper are listed alphabetically. Citations not found in the references to this paper may be found in Perch-Nielsen (1985).

Genus: *Assipetra* Roth 1973

Assipetra infracretacea (Thierstein 1973) Roth 1973
Plate 8.14, Figs. 3, 4

Genus: *Axopodorhabdus* Wind & Wise in Wise & Wind 1977

Axopodorhabdus dietzmannii (Reinhardt 1965) Wind & Wise 1983
Plate 8.3, Fig. 8; Plate 8.11, Figs. 18–20

Genus: *Biscutum* Black in Black & Barnes 1959

Biscutum ellipticum (Górka 1957) Grün in Grün & Allemann 1975
Plate 8.2, Figs. 11, 12; Plate 8.10, Figs. 21–24

Genus: *Braarudosphaera* Deflandre 1947

Braarudosphaera bigelowii (Gran & Braarud 1935) Deflandre 1947

Genus: *Bukrylithus* Black 1971

Bukrylithus ambiguus Black 1971
Plate 8.6, Figs. 1, 2; Plate 8.12, Fig. 15

Genus: *Calcicalithina* Thierstein 1971

Calcicalithina oblongata (Worsley 1971) Thierstein 1971

Genus: *Calculites* Prins & Sissingh in Sissingh 1977

Calculites sarstedtensis Crux 1987
Plate 8.6, Figs. 9–12; Plate 8.13, Fig. 20

Comments: this rare species is easily overlooked under the light microscope and has not been included in the estimates of abundancies.

Genus: *Chiastozygus* Gartner 1968

Chiastozygus cepekii Crux 1987
Plate 8.5, Figs. 4–7, Plate 8.12, Figs. 10–14

Chiastozygus sp. 1

Plate 8.12, Figs. 8, 9

Description: a large species of *Chiastozygus* with a narrow rim. The central cross bars support a spine.

Chiastozygus sp. 2

Description: species of *Chiastozygus* whose central bars are divided into two parts by a median line running the length of the bar.

Genus: *Clepsilithus* Crux 1987

Clepsilithus polystreptus Crux 1987
Plate 8.5, Figs. 8, 9

Comments: this small species was not identified under the light microscope.

Genus *Conusphaera* Trejo 1969

Conusphaera rothii (Thierstein 1971) Jakubowski 1986
Plate 8.13, Figs. 31, 32

PLATE 8.1
The bars at the top of the photographs = 1 μm.

Plate 8.1, Figs. 1, 2. *Watznaueria barnesae.* Fig. 1, proximal view, Albian, Mundays Hill, Bedfordshire. Fig. 2, distal view, Albian, Mundays Hill, Bedfordshire.

Plate 8.1, Figs. 3, 7. *Cyclagelosphaera margerelii.* Fig. 3, distal view, Albian, Mundays Hill, Bedfordshire. Fig. 7, distal view, Hauterivian, Speeton, PAL 3408.

Plate 8.1, Figs. 4–6. *Watznaueria rawsonii.* Fig. 4, distal view, Barremian, Speeton, PAL 3378. Fig. 5, distal view, Hauterivian?, Otto Gott, PAL 2846. Fig. 6, coccosphere, Barremian?, Speeton.

Plate 8.1, Fig. 8. *Watznaueria fasciata.* Distal view, Hauterivian?, Otto Gott, PAL 2846.

Plate 8.1, Fig. 9. *Watznaueria ovata.* Distal view, Barremian, Otto Gott, PAL 2888.

Plate 8.1, Fig 10. *Watznaueria britannica.* Distal view, Barremian, Otto Gott, PAL 2888.
Plate 8.1, Figs. 11, 12. *Manivitella pemmatoidea.* Fig. 11, proximal view, Hauterivian, Otto Gott, PAL 2852. Fig. 12, proximal view, Hauterivian, Otto Gott, PAL 2852.

Pl. 8.1]   **Biostratigraphy and palaeogeographical applications**   183

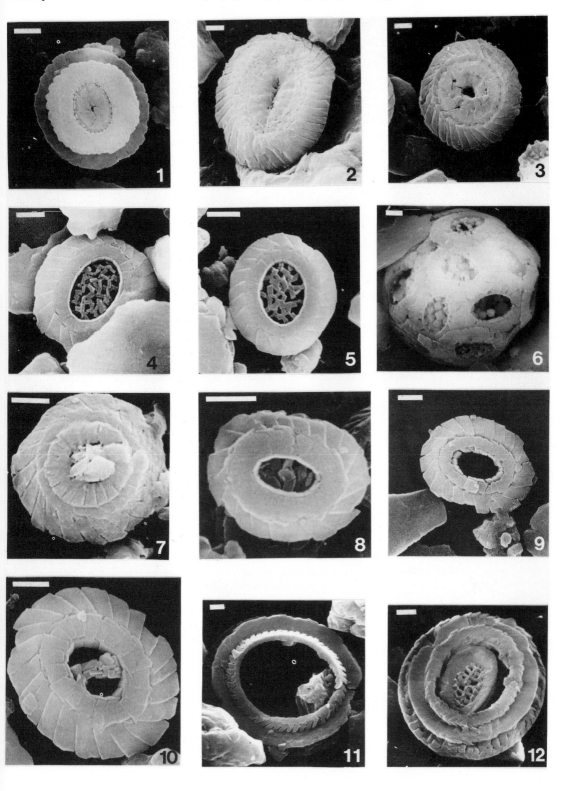

Genus: *Corollithion* Stradner 1961
Corollithion achylosum (Stover 1966) Thierstein 1971

Corollithion geometricum (Górka 1957) Manivit 1971
Plate 8.5, Fig. 12; Plate 8.13, Figs. 21, 22

Corollithion rhombicum (Stradner & Adamiker 1966) Bukry 1969
Plate 8.5, Fig. 11; Plate 8.13, Fig. 25

Corollithion silvaradion Filewicz *et al.* in Wind & Wise 1977
Plate 8.5, Fig. 10; Plate 8.13, Figs. 23, 24

Genus: *Cretarhabdus* Bramlette & Martini 1964
Cretarhabdus angustiforatus (Black 1971) Bukry 1973
Plate 8.4, Figs. 4, 5; Plate 8.11, Figs. 5–7

Cretarhabdus conicus Bramlette & Martini 1964
Plate 8.4, Fig. 1; Plate 8.11, Figs. 12, 13

Cretarhabdus crenulatus Bramlette & Martini 1964
Plate 8.4, Fig. 2; Plate 8.11, Fig. 8

Cretarhabdus inaequalis Crux 1987
Plate 8.4, Figs. 7–10; Plate 8.11, Figs. 14, 15

Cretarhabdus madingleyensis (Black 1971) nov. comb.
Plate 8.4, Fig. 3; Plate 8.11, Figs. 9–11

1968 *Polypodorhabdus madingleyensis* Black, p. 806, Plate 150, Fig. 2 (invalid)
1971a *Polypodorhabdus madingleyensis* Black, p. 619, Plate 45.4, Fig. 37.
Comments: no differences in the rim constructions of the genera *Cretarhabdus* Bramlette and Martini 1964 and *Polypodorhabdus* Nöel 1965 were seen in the present study so this species is placed in the senior genus.

Cretarhabdus sp.

Description: a species of *Cretarhabdus*, with six central windows–perforations, two of which are at each end of the elliptical coccolith.

Genus: *Crucibiscutum* Jakubowski 1986
Crucibiscutum salebrosum (Black 1971) Jakubowski 1986
Plate 8.2, Figs. 8, 9; Plate 8.10, Figs. 28–32

Genus: *Cruciellipsis* Thierstein 1971
Cruciellipsis cuvillieri (Manivit 1966) Thierstein 1971
Plate 8.11, Figs. 1, 2

PLATE 8.2
The bars at the top of the photographs = 1 μm.

Plate 8.2, Figs. 1–3. *Discorhabdus ignotus*. Fig. 1, distal view, Hauterivian?, Otto Gott, PAL 2846. Fig. 2, proximal view, Hauterivian?, Otto Gott, PAL 2846. Fig. 3, proximal view, Hauterivian?, Otto Gott, PAL 2846.

Plate 8.2, Fig. 4. *Sollasites arcuatus*. Distal view, Ryazanian, Speeton, PAL 3418.

Plate 8.2, Fig. 5. *Sollasites lowei*. Proximal view, Hauterivian?, Otto Gott, PAL 2846.

Plate 8.2, Fig. 6. *Sollasites horticus*. Distal view, Ryazanian, Speeton, PAL 3424.

Plate 8.2, Figs. 7, 10. *Diazomatolithus lehmanii*. Fig. 7, proximal view, Barremian, Speeton, PAL 2888. Fig. 10, proximal view, Hauterivian?, Otto Gott, PAL 2846.

Plate 8.2, Figs. 8, 9. *Crucibiscutum salebrosum*. Fig. 8, distal view, Barremian, Speeton, PAL 3378. Fig. 9, distal view, Hauterivian?, Otto Gott, PAL 2846.

Plate 8.2, Figs. 11, 12. *Biscutum ellipticum*. Fig. 11, proximal view, Hauterivian?, Otto Gott, PAL 2846. Fig. 12, distal view, Barremian, Speeton, PAL 3378.

Pl. 8.2]   **Biostratigraphy and palaeogeographical applications**   185

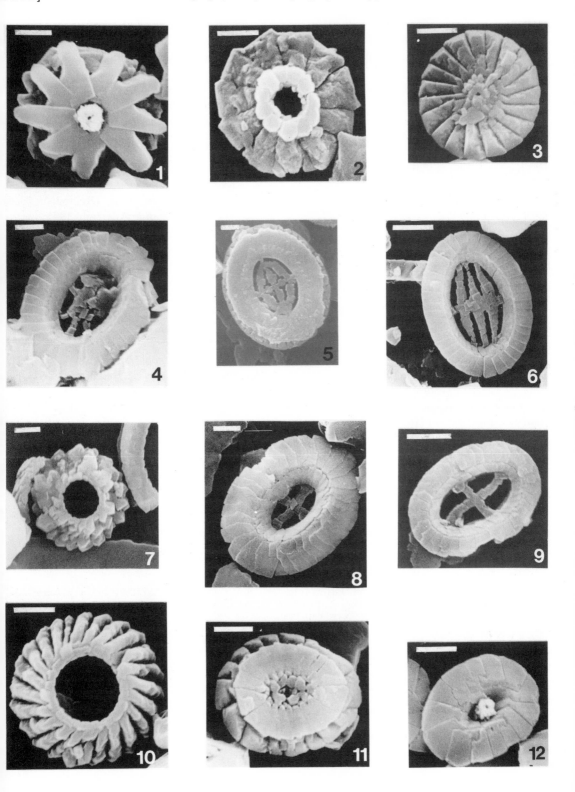

Genus: *Cyclagelosphaera* Noël 1965
Cyclagelosphaera margerelii Noël 1965
Plate 8.1, Figs. 3, 7; Plate 8.10, Figs. 13, 14

Cyclagelosphaera rotaclypeata Bukry 1969
Plate 8.10, Fig. 18

Genus: *Cylindralithus* Bramlette & Martini 1964
Cylindralithus laffittei (Noël 1957) Black 1973
Plate 8.9, Fig. 8; Plate 8.13, Figs. 29, 30

Genus: *Diadorhombus* Worsley 1971
Diadorhombus rectus Worsley 1971

Genus: *Diazomatolithus* Noël 1965
Diazomatolithus lehmanii Noël 1965
Plate 8.2, Figs. 7, 10; Plate 8.10, Figs. 1–5

Genus: *Discorhabdus* Noël 1965
Discorhabdus ignotus (Górka 1957) Perch-
Nielsen 1968
Plate 8.2, Figs. 1–3; Plate 8.10, Fig. 25

Genus: *Eprolithus* Stover 1966
Eprolithus antiquus Perch-Nielsen 1979
Plate 8.9, Figs. 9, 12; Plate 8.14, Figs. 13, 18, 23

Eprolithus varolii Jakubowski 1986

Genus: *Ethmorhabdus* Noël 1965
Ethmorhabdus hauterivianus (Black 1971)
Applegate *et al.* in Covington & Wise 1987
Plate 8.3, Fig. 10; Plate 8.11, Figs. 31–34

Genus: *Glaukolithus* Reinhardt 1964
Glaukolithus compactus (Bukry 1969) Perch-
Nielsen 1984
Plate 8.12, Figs. 25–27

Genus: *Grantarhabdus* Black 1971
Grantarhabdus coronadventis (Reinhardt 1966)
Grün in Grün & Allemann 1975

Grantarhabdus meddii Black 1971
Plate 8.4, Figs. 11, 12; Plate 8.11, Figs. 16, 17

Genus: *Haqius* Roth 1978
Haqius circumradiatus (Stover 1966) Roth 1978
Plate 8.10, Figs. 15–17

Genus: *Hemipodorhabdus* Black 1971
Hemipodorhabdus gorkae (Reinhardt 1969) Grün
in Grün & Allemann 1975
Plate 8.3, Fig. 12; Plate 8.11, Fig. 21

Genus: *Lapideacassis* Black 1971
Lapideacassis glans Black 1971
Plate 8.9, Fig. 11

PLATE 8.3
The bars at the top of the photographs = 1 μ m.

Plate 8.3, Fig. 1. *Perissocyclus fletcheri*. Distal view, Ryazanian, Speeton, PAL 3418.

Plate 8.3, Figs. 2–6. *Perissocyclus tayloriae*. Fig. 2, distal view, Aptian, Otto Gott, PAL 2891. Fig. 3, proximal view, Barremian, Speeton, PAL 3378. Fig. 4, holotype, distal view, Hauterivian, Otto Gott, PAL 2852. Fig. 5, proximal view, Barremian, Speeton, PAL 3379. Fig. 6, distal view, Barremian, Speeton, PAL 3378.

Plate 8.3, Fig. 7. *Speetonia colligata*. Proximal view, Hauterivian, Speeton, PAL 3408.

Plate 8.3, Fig. 8. *Axopodorhabdus dietzmannii*. Distal view, Barremian, Otto Gott, PAL 2888.

Plate 8.3, Fig. 9. *Repagulum parvidentatum*. Distal view, Albian, Mundays Hill, Bedfordshire.

Plate 8.3, Fig. 10. *Ethmorhabdus hauterivianus*. Proximal view, Ryazanian, Speeton, PAL 3424.

Plate 8.3, Fig. 11. *Tetrapodorhabdus coptensis*. Distal view, Ryazanian, Speeton, PAL 3424.

Plate 8.3, Fig. 12. *Hemipodorhabdus gorkae*. Distal view, Hauterivian, Otto Gott, PAL 2852.

Pl. 8.3] **Biostratigraphy and palaeogeographical applications** 187

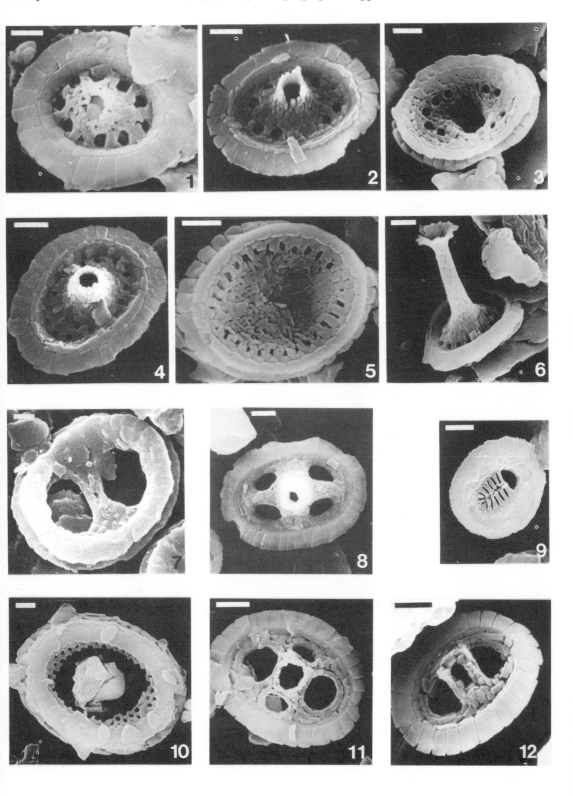

Genus: *Lithraphidites* Deflandre 1963
Lithraphidites bollii (Thierstein 1971) Thierstein
1973

Lithraphidites carniolensis Deflandre 1963
Plate 8.7, Fig. 12; Plate 8.13, Figs. 34, 35

Genus: *Manivitella* Thierstein 1971
Manivitella pemmatoidea (Deflandre in Manivit
1965) Thierstein 1971
Plate 8.1, Figs. 11, 12; Plate 8.10 Figs. 6, 7

Genus: *Micrantholithus* Deflandre in Deflandre
& Fert 1954
Micrantholithus brevis Jakubowski 1986

Micrantholithus obtusus Stradner 1963
Plate 8.14, Figs. 1, 2

Micrantholithus speetonensis Perch-Nielsen 1979
Plate 8.14, Fig. 5

Genus: *Microstaurus* Black 1971
Microstaurus chiastius (Worsley 1971) Grün in
Grün & Allemann 1975
Plate 8.4, Fig. 6; Plate 8.11, Figs. 3, 4

Genus: *Nannoconus* Kamptner 1931
Nannoconus abundans Stradner & Grün 1973
Plate 8.9, Figs. 4, 7, 10; Plate 8.14, Figs. 9–11

Nannoconus borealis Perch-Nielsen 1979

Nannoconus bronnimannii Trejo 1959

Nannoconus elongatus Brönnimann 1955

Nannoconus minutus Brönnimann 1955

Nannoconus steinmannii Kamptner 1931
Plate 8.14, Fig. 15

Nannoconus truittii Brönnimann 1955

Nannoconus sp. (discs)
Plate 8.14, Figs. 6, 7

Description: a flat species of *Nannoconus*, which
has dark birefringence colours between crossed-
nicols under the light microscope.

Comments: this biostratigraphically useful
species has not been observed under the scanning
electron microscope in this study. Perch-Nielsen
(pers. commun.) considers it to be similar to
Nannoconus abundans, from her own
observations of it. Under the light microscope
the two species can be distinguished by the
darker birefringence colours of *Nannoconus* sp.
(discs).

Nannoconus sp. 1
Plate 8.14, Fig. 16

PLATE 8.4
The bars at the top of the photographs = 1 μ m.

Plate 8.4, Fig. 1. *Cretarhabdus conicus*. Distal view, Barremian, Otto Gott, PAL 2888.

Plate 8.4, Fig. 2. *Cretarhabdus crenulatus*. Distal view, Aptian, Otto Gott, PAL 2891.

Plate 8.4, Fig. 3. *Cretarhabdus madingleyensis*. Distal view, Hauterivian?, Otto Gott, PAL 2846.

Plate 8.4, Figs. 4, 5. *Cretarhabdus angustiforatus*. Fig. 4, distal view, Barremian, Otto Gott, PAL 2888. Fig. 5, distal view, Hauterivian, Otto Gott, PAL 2852.

Plate 8.4, Fig. 6. *Microstaurus chiastius*. Distal view, Hauterivian, Otto Gott, PAL 2852.

Plate 8.4, Figs. 7–10. *Cretarhabdus inaequalis*. Fig. 7, distal view, Hauterivian?, Otto Gott, PAL 2846. Fig. 8, distal view, Barremian, Speeton, PAL 3378. Fig. 9, proximal view, Barremian, Speeton, PAL 3378. Fig. 10, distal view, Hauterivian?, Otto Gott, PAL 2846.

Plate 8.4, Figs. 11, 12. *Grantarhabdus meddii*. Fig. 11, distal view, Ryazanian, Speeton, PAL 3424, Fig. 12, distal view, Barremian, Speeton, PAL 3378.

Pl. 8.4]   **Biostratigraphy and palaeogeographical applications**   189

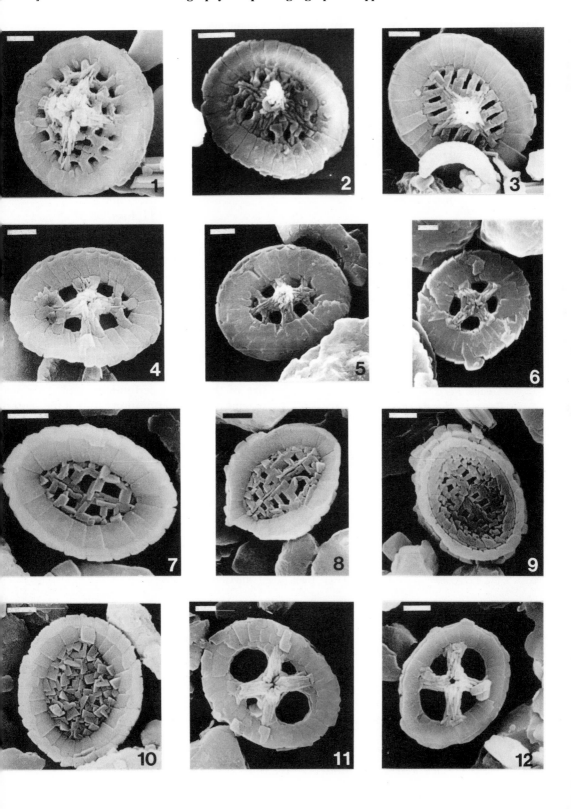

Description: a species of *Nannoconus* which is square shaped in side view under the light microscope.

Nannoconus sp. 2
Plate 8.14, Figs. 8, 12, 17

Description: a globular species of *Nannoconus* whose axial canal opens at one end.

Nannoconus sp. 3
not illustrated

Comments: a *Nannoconus* species similar to *Nannoconus multicadus* Deflandre & Deflandre 1959, it has a cylindrical shape with a central constriction which gives it the appearance of being two smaller nannoconids joined together.

Genus: *Percivalia* Bukry 1969
Percivalia fenestrata (Worsley 1971) Wise 1983
Plate 8.13, Figs. 6–9

Genus: *Perissocyclus* Black 1971
Perissocyclus fletcheri Black 1971
Plate 8.3, Fig. 1

Comments: a distinctive species of *Perissocyclus* with 10 central perforations.

Perissocyclus noeliae Black 1971
Plate 8.11, Figs. 22, 23

Comments: species of *Perissocyclus* with five to nine central perforations, these perforations are large and similar to those of *P. plethotretus*. *Perissocyclus plethotretus* is grouped with *P. noeliae* in the light microscope abundance estimates, as there is a continuous transition from the one to the other with *P. noeliae* having less than 10 perforations and *P. plethotretus* 10 or more.

Perissocyclus plethotretus (Wind & Čepek 1979)
nov. comb
Plate 8.11, Figs. 24–27

1979 *Octopodorhabdus plethotretus* Wind & Čepek, pp. 230–231, Plate 4, Figs. 1–5

Comments: this species is differentiated from *P. noeliae* by the greater number of central perforations, from *P. fletcheri* by its larger size and generally greater number of central perforations and from *P. tayloriae* by the large size of the central perforations and the equal spacing of these perforations around the central area.

Perissocyclus tayloriae nov. sp.
Plate 8.3, Figs. 2–6; Plate 8.11, Fig. 28

Derivation of name: after R.J. Taylor
Holotype: Plate 8.3, Fig. 4

PLATE 8.5
The bars at the top of the photographs = 1 μ m.

Plate 8.5, Fig. 1. *Tegumentum striatum*. Distal view, Hauterivian, Otto Gott, PAL 2825.

Plate 8.5, Figs. 2, 3. *Tegumentum octiformis*. Fig. 2, distal view, Barremian, Speeton, PAL 3379. Fig. 3, proximal view, Barremian, Speeton, PAL 3378.

Plate 8.5, Figs. 4–7. *Chiastozygus cepekii*. Fig. 4, distal view, Hauterivian?, Otto Gott, PAL 2846. Fig. 5, distal view, Barremian, Speeton, PAL 3379. Fig. 6, proximal view, Hauterivian?, Otto Gott, PAL 2846. Fig. 7, proximal view, Hauterivian?, Otto Gott, PAL 2846.

Plate 8.5, Figs. 8, 9. *Clepsilithus polystreptus*. Fig. 8, distal view, Hauterivian?, Otto Gott, PAL 2846. Fig. 9, proximal view, Hauterivian?, Otto Gott, PAL 2846.

Plate 8.5, Fig. 10. *Corollithion silvaradion*. Proximal view, Ryazanian, Speeton, PAL 3424.

Plate 8.5, Fig. 11. *Corollithion rhombicum*. Proximal view, Hauterivian, Speeton, PAL 3384.

Plate 8.5, Fig. 12. *Corollithion geometricum*. Distal view, Hauterivian, Speeton, PAL 3384.

Pl. 8.5] **Biostratigraphy and palaeogeographical applications** 191

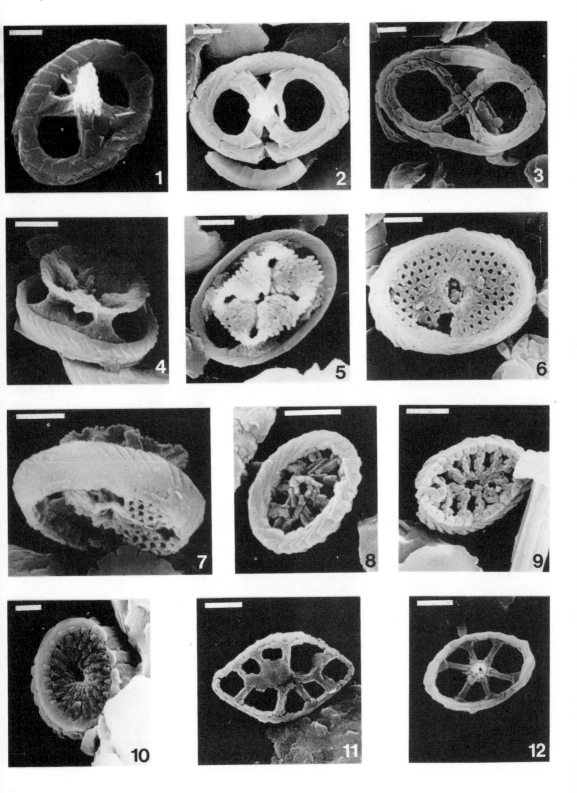

Diagnosis: large species of *Perissocyclus* with 8 to 22 small central perforations in one or two cycles. These perforations appear in some specimens to be divided into four quadrants and separated by large spine supporting buttresses.
Type level; Bed 50, Otto Gott, Sarstedt.
Range: Hauterivian–Aptian.

Genus: *Pseudolithraphidites* Keupp 1976

Diagnosis: this genus is emended to include spines of four parallel bars with small coccoliths attached. These coccoliths appear to have a central cross and a slightly imbricating rim of calcite elements
Pseudolithraphidites parallelus (Wind & Čepek 1979) nov. comb.
Plate 8.7, Figs. 10, 11; Plate 8.13, Fig. 33

1979 *Rhabdolekiskus parallelus* Wind & Čepek, p. 232, Plate 3, Figs. 3–6

Comments: Perch-Nielsen (1985) has pointed out that the genus *Rhabdolekiskus* Hill 1976 is a junior synonym of *Lithraphidites* Deflandre 1963. The similarity in the construction of the spines illustrated by Keupp 1976, (Figs. 25–27) and those of *Rhabdolekiskus parallelus* leads to the latter's inclusion in the genus *Pseudolithraphidites*. The very small size of the coccoliths of this species differentiate it from

members of the genus *Staurolithites*.
Genus: *Repagulum* Forchheimer 1972
Repagulum parvidentatum (Deflandre & Fert 1954) Forchheimer 1972
Plate 8.3, Fig. 9; Plate 8.10, Figs. 19, 20

Genus: *Rhagodiscus* Reinhardt 1967
Rhagodiscus angustus (Stradner 1963) Reinhardt 1971

Rhagodiscus asper (Stradner 1963) Reinhardt 1967
Plate 8.8, Figs. 1–5, Plate 8.13, Figs. 1, 2

Rhagodiscus eboracensis Black 1971

Rhagodiscus infinitus (Worsley 1971) Applegate *et al.* in Covington & Wise 1987
Plate 8.13, Figs. 11–14

Rhagodiscus pseudoangustus Crux 1987
Plate 8.8, Figs. 10, 11; Plate 8.13. Figs. 16–19

Rhagodiscus reightonensis (Taylor 1978) Watkins in Watkins & Bowdler 1984
Plate 8.8, Figs. 7, 8

Comments: I cannot differentiate this species from *R. asper* under the light microscope; it is thus included in *R. asper* in the estimates of abundancies.

PLATE 8.6
The bars at the top of the photographs = 1 μm.

Plate 8.6, Figs. 1, 2. *Bukrylithus ambiguus*. Fig. 1, distal view, Barremian, Speeton, PAL 3378. Fig. 2, proximal view, Barremian, Speeton, PAL 3378.

Plate 8.6, Figs. 3–6. *Staurolithites crux*. Fig. 3, distal view, Hauterivian, Otto Gott, PAL 2852. Fig. 4, distal view, Hauterivian, Otto Gott, PAL 2852. Fig. 5, proximal view, Hauterivian, Otto Gott, PAL 2849. Fig. 6, distal view, Hauterivian?, Otto Gott, PAL 2846.

Plate 8.6, Figs. 7, 8. *Staurolithites mutterlosei*. Fig. 7, holotype, distal view, Barremian, Speeton, PAL 3378. Fig. 8, proximal view, Barremian, Speeton, PAL 3378.

Plate 8.6, Figs. 9–12. *Calculites sarstedtensis*. Fig. 8, distal view, Hauterivian?, Otto Gott, PAL 2846. Fig. 10, distal view, Hauterivian?, Otto Gott, PAL 2846. Fig, 11, side view, Hauterivian?, Otto Gott, PAL 2846. Fig. 12, proximal view, Hauterivian?, Otto Gott, PAL 2846.

Pl. 8.6] **Biostratigraphy and palaeogeographical applications** 193

Rhagodiscus splendens (Deflandre 1953) Verbeek
1977
Plate 8.8, Fig. 12

Rhagodiscus sp.

Comments: this species, which was only seen in side view under the light microscope, has a short granular spine which flares at the distal end.

Genus: *Rucinolithus* Stover 1966
Rucinolithus wisei Thierstein 1971

Genus: *Scapholithus* Deflandre in Deflandre & Fert 1954
Scapholithus fossilis Deflandre in Deflandre & Fert 1954
Plate 8.7, Fig. 9

Genus: *Sollasites* Black 1967
Sollasites arcuatus Black 1971
Plate 8.2, Fig. 4

Sollasites horticus (Stradner *et al.* in Stradner & Adamiker 1966) Čepek & Hay 1969
Plate 8.2, Fig. 6; Plate 8.10, Fig. 26

Sollasites lowei (Bukry 1969) Roth 1970
Plate 8.2, Fig. 5; Plate 8.10, Fig. 27

Genus: *Speetonia* Black 1971
Speetonia colligata Black 1971
Plate 8.3, Fig. 7; Plate 8.11, Figs. 29, 30

Genus: *Staurolithites* Caratini 1963
Staurolithites crux (Deflandre in Deflandre & Fert 1954) Caratini 1963
Plate 8.6, Figs. 3–6; Plate 8.12, Figs. 17, 18

Staurolithites mutterlosei nov. sp.
Plate 8.6, Figs. 7, 8; Plate 8.12, Figs. 16, 21, 22

Derivation of name: after J. Mutterlose
Holotype: Plate 8.6, Fig. 7
Diagnosis: a species of *Staurolithites* whose central cross is slightly offset from the long and short axes of the coccolith. The longer bars of the cross are slightly curved. The rim has a complex structure composed of two superimposed cycles of imbricating elements. The two parts of the rim can be seen under the light microscope between crossed-nicols.
Type level: Bed LB 6, Speeton
Range: Hauterivian–Barremian
Staurolithites sp.
Plate 8.12, Figs. 19, 20

Description: a species of *Staurolithites* which shows a double rim structure (an inner rim).

PLATE 8.7
The bars at the top of the photographs = 1 μm.

Plate 8.7, Fig. 1. *Zeugrhabdotus sisyphus*. Distal view, Barremian, Otto Gott, PAL 2888.

Plate 8.7, Figs. 2, 3. *Zeugrhabdotus noeliae*. Fig. 2, distal view, Hauterivian, Speeton, PAL 3384. Fig. 3, distal view, Aptian, Otto Gott, PAL 2891.

Plate 8.7, Figs, 4, 5. *Zeugrhabdotus embergeri*. Fig. 4, distal view, Ryazanian, Speeton, PAL 3424. Fig. 5, proximal view, Barremian, Otto Gott, PAL 2888.

Plate 8.7, Fig. 6. *Zeugrhabdotus* sp. 1. Distal view, Hauterivian?, Otto Gott, PAL 2846.

Plate 8.7, Fig. 7. *Tranolithus gabalus*. Distal view, Hauterivian?, Otto Gott, PAL 2846.

Plate 8.7, Fig. 8. *Zeugrhabdotus* sp. 2. Distal view, Hauterivian, Otto Gott, PAL 2846.

Plate 8.7, Fig. 9. *Scapholithus fossilis*. Proximal view?, Hauterivian?, Otto Gott, PAL 2846.

Plate 8.7, Figs. 10, 11. *Pseudolithraphidites parallelus*. Fig. 10, side view, Hauterivian, Otto Gott, PAL 2852. Fig. 11, side view, proximal view, Hauterivian?, PAL 2846.

Plate 8.7, Fig. 12. *Lithraphidites carniolensis*. Side view, Aptian, Otto Gott, PAL 2891.

Pl. 8.7]   **Biostratigraphy and palaeogeographical applications**   195

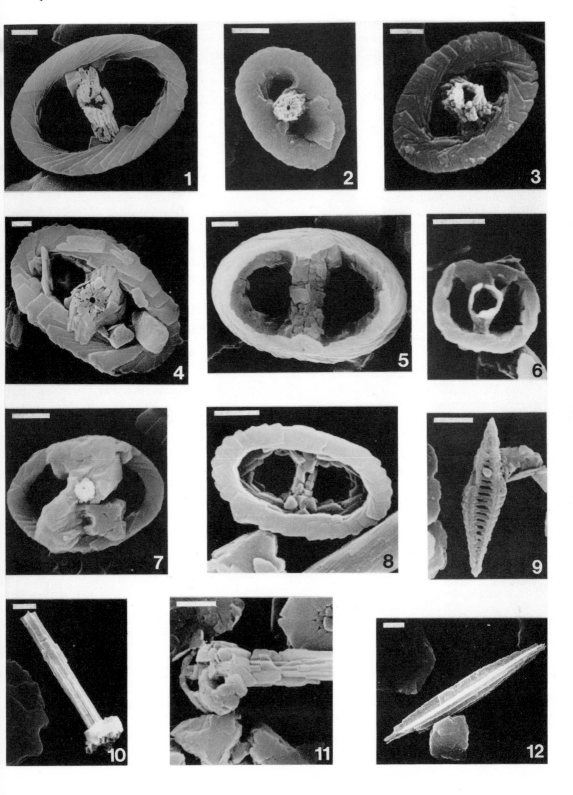

Genus: *Stradnerlithus* Black 1971
Stradnerlithus comptus Black 1971
Plate 8.13, Figs. 20–28

Genus: *Tegulalithus* Crux 1986
Tegulalithus septentrionalis (Stradner 1963) Crux
1986
Plate 8.9, Figs. 1–3, 5, 6; Plate 8.14, Figs. 19–22,
24–27

Comments: this species has an erratic occurrence throughout the Hauterivian–Albian. Many authors consider it to be synonymous with *Tegulalithus tessellatus* (Stradner, Adamiker & Maresch 1968) Crux 1986. *Tegulalithus septentrionalis* first occurs in the Upper Hauterivian in both Speeton and the Moorberg, Sarstedt section, it has an acme in the Upper Hauterivian and disappears at the Hauterivian–Barremian boundary in Germany. It re-occurs in both Speeton and the Otto Gott, Sarstedt section in the *Aulacoteuthis* Zone and disappears again in the *brunsvicensis* Zone. The records of *T. tesselatus* in the Aptian and Albian may simply be further re-occurrences. The appearance of the specimens from the Barremian is slightly different from that of the specimens from the Hauterivian. These differences are thought to be due to better preservation in the Hauterivian.

Genus: *Tegumentum* Thierstein in Roth & Thierstein 1972
Tegumentum octiformis (Köthe 1981) nov. comb.
Plate 8.5, Figs. 2, 3; Plate 8.12, Figs. 6, 7

1981 *Chiastozygus octiformis* Köthe, pp. 20–21, Plate 2, Fig. 2.
Comments: the rim structure of this species is that of the genus *Tegumentum*.

Tegumentum striatum (Black 1971) nov. comb.
Plate 8.5, Fig. 1; Plate 8.12, Figs. 1–5

1971b *Chiastozygus striatus* Black, p. 416, Plate 34, Fig. 7.
1978 *Tegumentum striatum* (Black) Taylor, p. 199 (invalid)
Comments: early forms of this species (Valanginian–Hauterivian) tend to be larger than later forms (Hauterivian–Barremian). The change-over from large forms to smaller forms occurs in the Upper Hauterivian at the top of the *gottschei* Zone. Some specialists consider the two sizes of *T. striatum* to be two different species. In the present study no difference in the structures of the large and small forms was observed under the scanning electron microscope.

Tegumentum tenuis (Black 1971) nov. comb.

1971b *Chiastozygus tenuis* Black, p. 416, Plate 34, Fig. 8.
Comments: this 'species' is possibly synonymous with *T. striatum*. It is differentiated in the present study by the smaller angle that the two cross bars make with one another when they meet in the centre of the coccolith. It is differentiated from *T. octiformis* by the presence of the distinctive 8-shaped cross-structure in the latter.

PLATE 8.8
The bars at the top of the photographs = 1 μ m.

Plate 8.8, Figs. 1–5. *Rhagodiscus asper*. Fig. 1, distal view, Aptian, Otto Gott, PAL 2891. Fig 2, distal view, Aptian, Otto Gott, PAL 2891. Fig, 3, distal view, Hauterivian, Otto Gott, PAL 2852. Fig. 4, distal view, Aptian, Otto Gott, PAL 2891. Fig. 5, proximal view, Hauterivian?, Otto Gott, PAL 2846.

Plate 8.8, Figs. 6, 9. *Viminites swinnertonii*. Fig. 6, distal view, Hauterivian, Otto Gott, PAL 2852. Fig. 9, distal view, Hauterivian, Otto Gott, PAL 2852.

Plate 8.8, Figs, 7, 8. *Rhagodiscus reightonensis*. Fig. 7, distal view, Barremian, Speeton, PAL 3378. Fig. 8, distal view, Barremian, Speeton, PAL 3379.

Plate 8.8, Figs. 10, 11. *Rhagodiscus pseudoangustus*. Fig. 10, distal view, Barremian, Speeton, PAL 3378. Fig. 11, proximal view, Barremian, Speeton, PAL 3378.

Plate 8.8, Fig. 12. *Rhagodiscus splendens*. Distal view, Barremian, Otto Gott, PAL 2886.

[Pl. 8.8]   **Biostratigraphy and palaeogeographical applications**   197

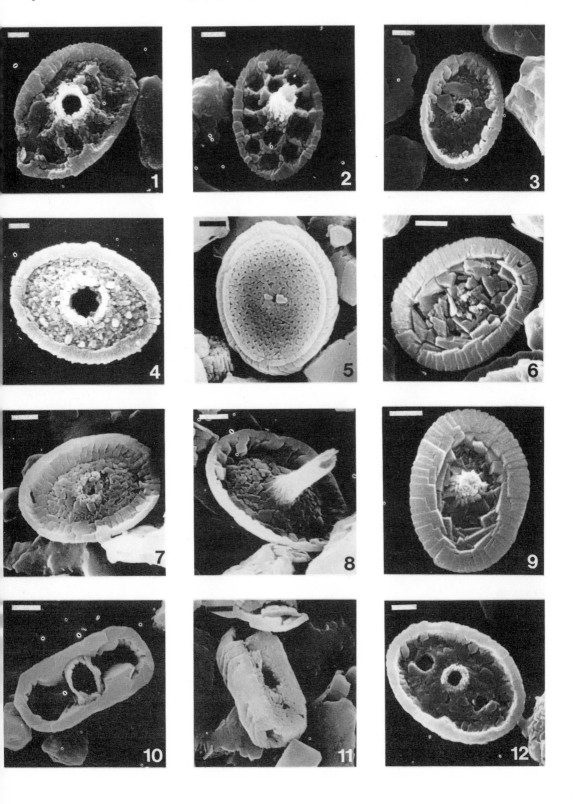

Genus: *Tetrapodorhabdus* Black 1971
Tetrapodorhabdus coptensis Black 1971
Plate 8.3, Fig. 11

Comments: this species is grouped with *Hemipodorhabdus gorkae* for the abundance estimates shown on the range charts.

Genus: *Tranolithus* Stover 1966
Tranolithus gabalus Stover 1966
Plate 8.7, Fig. 7; Plate 8.12, Figs. 32, 33

Genus: *Tubodiscus* Thierstein 1973
Tubodiscus verenae Thierstein 1973

Genus: *Vagalapilla* Bukry 1969
Vagalapilla matalosa (Stover 1966) Thierstein 1973
Plate 8.12, Figs. 23, 24

Genus: *Viminites* Black 1975
Viminites swinnertonii (Black 1971) Black 1975
Plate 8.8, Figs. 6, 9; Plate 8.13, Figs. 3–5

Genus: *Watznaueria* Reinhardt 1964
Watznaueria barnesae (Black in Black & Barnes 1959) Perch-Nielsen 1968
Plate 8.1, Figs. 1, 2; Plate 8.10, Fig. 11

Watznaueria britannica (Stradner 1963) Reinhardt 1964
Plate 8.1, Fig. 10; Plate 8.10, Fig. 12

Watznaueria fasciata Wind & Čepek 1979
Plate 8.1, Fig. 8

Comments: this species was not identified under the light microscope.

Watznaueria ovata Bukry 1969
Plate 8.1, Fig. 9; Plate 8.10, Figs. 8–10

Watznaueria rawsonii Crux 1987
Plate 8.1, Figs. 4–6

Comments: this species was not identified under the light microscope.

Genus: *Zeugrhabdotus* Reinhardt 1965
Zeugrhabdotus embergeri (Noël 1959) Perch-Nielsen 1984
Plate 8.7, Figs. 4, 5; Plate 8.12, Figs. 34, 35

Zeugrhabdotus erectus (Deflandre in Deflandre & Fert 1954) (Reinhardt 1965)
Plate 8.12, Figs. 28, 29

Zeugrhabdotus noeliae Rood *et al.* 1971
Plate 8.7, Figs. 2, 3; Plate 8.12, Fig. 31

Zeugrhabdotus sisyphus (Gartner 1968) nov. comb.
Plate 8.7, Fig. 1; Plate 8.12, Fig. 30

1968 *Zygodiscus sisyphus* Gartner, p. 34, Plate 14, Fig. 19; Plate 18, Figs. 17–19; Plate 21, Fig. 6; Plate 22, Figs. 5, 6; Plate 23, Figs. 17, 18; Plate 25, Figs. 19–22; Plate 26, Fig. 6.

Zeugrhabdotus sp. 1
Plate 8.7, Fig. 6

PLATE 8.9
The bars at the top of the photographs = 1 μm.

Plate 8.9, Figs. 1–3, 5, 6. *Tegulalithus septentrionalis*. Fig. 1, plan view, Hauterivian, Otto Gott, PAL 2852. Fig, 2, side view, Hauterivian, Otto Gott, PAL 2852. Fig. 3, plan view, Hauterivian, Otto Gott, PAL 2852. Fig. 5, plan view, Barremian, Otto Gott, PAL 2888. Fig. 6, plan view, Barremian, Otto Gott, PAL 2888.

Plate 8.9, Figs. 4, 7, 10. *Nannoconus abundans*. Fig. 4, side view, Barremian?, Speeton. Fig. 7, plan view, Barremian?, Speeton. Fig. 10, plan view, Barremian?, Speeton.

Plate 8.9, Fig. 8. *Cylindralithus laffittei*. Proximal view, Albian, Mundays Hill, Bedfordshire.

Plate 8.9, Figs. 9, 12. *Eprolithus antiquus*. Fig. 8, plan view, Hauterivian, Speeton, PAL 3408. Fig. 12, side view, Hauterivian, Speeton, PAL 3408.

Plate 8.9, Fig. 11. *Lapideacassis glans*. Side view, Hauterivian?, Otto Gott, PAL 2846.

Pl. 8.9] **Biostratigraphy and palaeogeographical applications** 199

Description: a small almost square species of *Zeugrhabdotus* whose central bar supports a large spine base.

Comments: this species was included in the *Z. noeliae* group in the light microscope abundance estimates.

Zeugrhabdotus sp. 2
Plate 8.7, Fig. 8

Description: a small species of *Zeugrhabdotus*, with a high strongly imbricating rim. The central bar does not have a spine or spine base.

Comments: this species was included in the *Z. noeliae* group in the light microscope abundance estimates.

Unnamed species

Coccolith sp. 1

Description: an elliptical coccolith with a bright golden elliptical rim when seen between crossed-nicols under the light microscope

Coccolith sp. 2

Description: a large elliptical darkly birefringent coccolith with a diffuse aspect.

Holococcolith sp. 1
Plate 8.10, Figs. 33–35

PLATE 8.10
All magnifications × 2430.

Plate 8.10, Figs. 1–5. *Diazomatolithus lehmanii*. Fig. 1, crossed-nicols, Hauterivian, Moorberg, PAL 2904. Fig. 2, crossed-nicols, Hauterivian, Otto Gott, PAL 2852. Fig. 3, bright field, same specimen. Fig. 4, bright field, Hauterivian, Speeton, PAL 3410. Fig. 5, crossed-nicols, same specimen.

Plate 8.10, Figs. 6, 7. *Manivitella pemmatoidea*. Fig. 6, crossed-nicols, Hauterivian, Otto Gott, PAL 2852. Fig. 7, bright field, same specimen.

Plate 8.10, Figs. 8–10. *Watznaueria ovata*. Fig. 8, crossed-nicols, Hauterivian, Otto Gott, PAL 2852. Fig. 9, bright field, same specimen. Fig. 10, crossed-nicols, Hauterivian, Otto Gott, PAL 2850.

Plate 8.10, Fig. 11. *Watznaueria barnesae*. Crossed-nicols, Valanginian, Speeton, PAL 743.

Plate 8.10, Fig. 12. *Watznaueria britannica*. Crossed-nicols, Albian, Mundays Hill, Bedfordshire.

Plate 8.10, Figs. 13, 14. *Cyclagelosphaera margerelii*. Fig. 13, crossed-nicols, Hauterivian, Otto Gott, PAL 2851. Fig. 14, bright field, same specimen.

Plate 8.10, Figs. 15–17. *Haqius circumradiatus*. Fig. 15, bright field, Barremian, Otto Gott, PAL 2843. Fig. 16, bright field, Hauterivian, Otto Gott, PAL 2849. Fig. 17, bright field, Hauterivian, Otto Gott, PAL 2849.

Plate 8.10, Fig. 18. *Cyclagelosphaera rotaclypeata*. Crossed-nicols, Albian, Mundays Hill, Bedfordshire.

Plate 8.10, Figs. 19, 20. *Repagulum parvidentatum*. Fig. 19, bright field, Mundays Hill, Bedfordshire. Fig. 20, crossed-nicols, Albian, same specimen.

Plate 8.10, Figs. 21–24. *Biscutum ellipticum*. Fig. 21, bright field, Hauterivian, Moorberg, PAL 2898. Fig. 22, crossed-nicols, same specimen. Fig. 23, bright field, Otto Gott, PAL 2852. Fig. 24, crossed-nicols, Hauterivian, same specimen.

Plate 8.10, Fig. 25. *Discorhabdus ignotus*. Bright field, Hauterivian, Otto Gott, PAL 2852.

Plate 8.10, Fig. 26. *Sollasites horticus*. Bright field, Hauterivian, Otto Gott, PAL 2852.

Plate 8.10, Fig. 27. *Sollasites lowei*. Bright field, Hauterivian, Moorberg, PAL 2899.

Plate 8.10, Figs. 28–30. *Crucibiscutum salebrosum*. Fig. 28, crossed-nicols, Ryazanian, Speeton, PAL 3422. Fig. 29, bright field, Hauterivian, Moorberg, PAL 2898. Fig. 30, crossed-nicols, same specimen.

Plate 8.10, Figs. 31, 32. Large *Crucibiscutum salebrosum*. Fig. 31, bright field, Hauterivian, Moorberg, PAL 2922. Fig. 32, crossed-nicols, same specimen.

Plate 8.10, Figs. 33–35. Holococcolith sp. 1. Fig. 33, bright field, Ryazanian, Speeton, PAL 3427. Fig. 34, crossed-nicols, same specimen. Fig. 35, crossed-nicols, Hauterivian, Moorberg, PAL 2903.

Pl. 8.10] **Biostratigraphy and palaeogeographical applications** 201

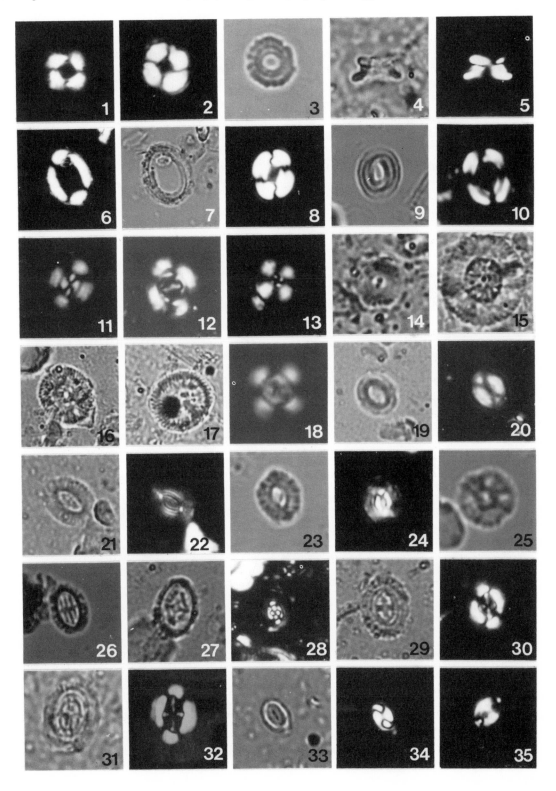

Description: a small brightly birefringent holococcolith? with a solid centre cut by a single slit parallel to the long axis of the coccolith. A narrow outer rim can also be seen between crossed-nicols.

'Calcite rosettes'
Plate 8.14, Fig. 14

Description: irregular rosettes of six to ten calcite elements.

Comments: this large species was illustrated by Kok (1985, Plate 1, Figs. 11–13), who quoted a Berriasian–Valanginian age for it in the North Sea area.

8.7 CONCLUSIONS

The rapid changes within the Lower Cretaceous nannofloral assemblages enable us to divide the Upper Ryazanian to Barremian into 16 nannofossil zones. This degree of biostratigraphical refinement is approaching

PLATE 8.11
All magnifications × 2430.

Plate 8.11, Figs. 1, 2 *Cruciellipsis cuvillieri*. Fig. 1, bright field, Hauterivian, Speeton, PAL 3399. Fig. 2, crossed-nicols, same specimen.

Plate 8.11, Figs. 3, 4. *Microstaurus chiastius*. Fig. 3, bright field, Hauterivian, Otto Gott, PAL 2852. Fig. 4, crossed-nicols, same specimen.

Plate 8.11, Figs. 5–7. *Cretarhabdus angustiforatus*. Fig. 5, bright field, Barremian, Otto Gott, PAL 2842. Fig. 6, bright field, Hauterivian, Otto Gott, PAL 2852. Fig. 7, crossed-nicols, same specimen.

Plate 8.11, Fig. 8. *Cretarhabdus crenulatus*. Bright field, Hauterivian?, Otto Gott, PAL 2849.

Plate 8.11, Figs. 9–11. *Cretarhabdus madingleyensis*. Fig. 9, crossed-nicols, Hauterivian, Moorberg, PAL 2908. Fig. 10, bright field, same specimen. Fig. 11, bright field, Barremian, Otto Gott, PAL 2842.

Plate 8.11, Figs. 12, 13. *Cretarhabdus conicus*. Fig. 12, crossed-nicols, Hauterivian, Otto Gott, PAL 2852. Fig. 13, bright field, same specimen.

Plate 8.11, Figs. 14, 15. *Cretarhabdus inaequalis*. Fig. 14, crossed-nicols, Barremian, Otto Gott, PAL 2884. Fig. 15, bright field, same specimen.

Plate 8.11, Figs. 16, 17. *Grantarhabdus meddii*. Fig. 16, bright field, Hauterivian, Otto Gott, PAL 2852. Fig. 17, crossed-nicols, same specimen.

Plate 8.11, Figs. 18–20. *Axopodorhabdus dietzmannii*. Fig. 18, bright field, Hauterivian?, Otto Gott, PAL 2848. Fig. 19, crossed-nicols, Hauterivian, Otto Gott, PAL 2852. Fig. 20, bright field, same specimen.

Plate 8.11, Fig. 21. *Hemipodorhabdus gorkae*. Bright field, Hauterivian, Otto Gott, PAL 2852.

Plate 8.11, Figs. 22, 23. *Perissocyclus noeliae*. Fig. 22, bright field, Hauterivian, Moorberg, PAL 2919. Fig. 23, bright field, Hauterivian, Otto Gott, PAL 2852.

Plate 8.11, Figs. 24–27. *Perissocyclus plethotretus*. Fig. 24, bright field, Hauterivian, Otto Gott, PAL 2852. Fig. 25, bright field, Hauterivian?, Otto Gott, PAL 2849. Fig. 26, bright field, Barremian, Otto Gott, PAL 2845. Fig. 27, bright field, Barremian, Otto Gott, PAL 2837.

Plate 8.11, Fig. 28. *Perissocyclus tayloriae*. Bright field, Hauterivian?, Otto Gott, PAL 2847.

Plate 8.11, Figs. 29, 30. *Speetonia colligata*. Fig. 29, crossed-nicols, Hauterivian, Moorberg, PAL 2911. Fig. 30, bright field, same specimen.

Plate 8.11, Figs. 31–34. *Ethmorhabdus hauterivianus*. Fig. 31, bright field, Hauterivian, Moorberg, PAL 2912. Fig. 32, crossed-nicols, same specimen. Fig. 33, bright field, Hauterivian, Moorberg, PAL 2905. Fig. 34, crossed-nicols, same specimen.

Pl. 8.11] Biostratigraphy and palaeogeographical applications 203

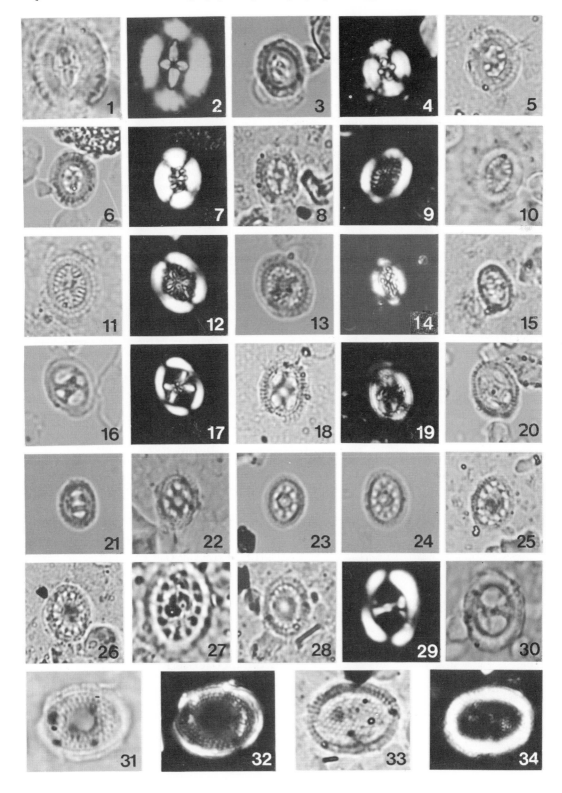

that achieved by the zonation schemes of ammonites and belemnites.

Analysis of the nannofossil assemblages throughout the lower Cretaceous provides evidence for transgressions, regressions and the opening and closing of seaways. The following conclusions have been drawn from the present study:

(a) There was only a weak Tethyan influence on the nannofloras of the Upper Ryazanian. This suggests that the Late Ryazanian

transgression was less extensive than later transgressions. The presence of *Nannoconus* sp. (discs) in these sediments suggests marine connections from these areas to the west of Britain, including the Porcupine Basin.

(b) The Boreal–Arctic influence and endemic character of the nannofloras of the Lower Valanginian remained strong with abundant *C. salebrosum* together with rare *M.*

PLATE 8.12
All magnifications × 2430.

Plate 8.12, Figs. 1–5. *Tegumentum striatum.* Fig. 1, bright field, Hauterivian, Otto Gott, PAL 2852. Fig. 2, crossed-nicols, same specimen. Fig. 3, bright field, Hauterivian?, Otto Gott, PAL 2849. Fig. 4, crossed-nicols, same specimen. Fig. 5, bright field, Hauterivian, Otto Gott, PAL 2852.

Plate 8.12, Figs. 6, 7. *Tegumentum octiformis.* Fig. 6, bright field, Hauterivian, Otto Gott, PAL 2852. Fig. 7, crossed-nicols, same specimen.

Plate 8.12, Figs. 8, 9. *Chiastozygus* sp. 1. Fig. 8, bright field, Barremian, Otto Gott, PAL 2840. Fig. 9, crossed-nicols, same specimen.

Plate 8.12, Figs. 10–14. *Chiastozygus cepekii.* Fig. 10, bright field, Hauterivian, Otto Gott, PAL 2852. Fig. 11, bright field, Hauterivian?, Otto Gott, PAL 2846. Fig. 12, crossed-nicols, same specimen. Fig. 13, bright field, Hauterivian, Moorberg, PAL 2923. Fig. 14, crossed-nicols, same specimen.

Plate 8.12, Fig. 15. *Bukrylithus ambiguus.* Crossed-nicols, Hauterivian, Moorberg, PAL 2922.

Plate 8.12, Figs. 16, 21, 22. *Staurolithites mutterlosei.* Fig. 16, bright field, Hauterivian?, Otto Gott, PAL 2846. Fig. 21, bright field, Hauterivian?, Otto Gott, PAL 2846. Fig. 22, crossed-nicols, same specimen.

Plate 8.12, Figs. 17, 18. *Staurolithites crux.* Fig. 17, crossed-nicols, Hauterivian, Otto Gott, PAL 2852. Fig. 18, bright field, same specimen.

Plate 8.12, Figs. 19, 20. *Staurolithites* sp. Fig. 19, crossed-nicols, Hauterivian, Otto Gott, PAL 2852. Fig. 20, bright field, same specimen.

Plate 8.12, Figs. 23, 24. *Vagalapilla matalosa.* Fig. 23, bright field, Barremian, Otto Gott, PAL 2883. Fig. 24, crossed-nicols, same specimen.

Plate 8.12, Figs. 25–27. *Glaukolithus compactus.* Fig. 25, bright field, Barremian, Otto Gott, PAL 2845. Fig. 26, bright field, Barremian, Speeton, PAL 3378. Fig. 27, crossed-nicols, same specimen.

Plate 8.12, Figs. 28, 29. *Zeugrhabdotus erectus.* Fig. 28, bright field, Hauterivian, Otto Gott, PAL 2852. Fig. 29, crossed-nicols, same specimen.

Plate 8.12, Fig. 30. *Zeugrhabdotus sisyphus.* Crossed-nicols, Hauterivian, Otto Gott, PAL 2852.

Plate 8.12, Fig. 31. *Zeugrhabdotus noeliae.* Crossed-nicols, Albian, Mundays Hill, Bedfordshire.

Plate 8.12, Figs. 32, 33. *Tranolithus gabalus.* Fig. 32, crossed-nicols, Hauterivian, Moorberg, PAL 2907. Fig. 33, bright field, same specimen.

Plate 8.12, Figs. 34, 35. *Zeugrhabdotus embergeri.* Fig. 34, crossed-nicols, Hauterivian, Speeton, PAL 3409. Fig. 35, bright field, same specimen.

Pl. 8.12]   **Biostratigraphy and palaeogeographical applications**   205

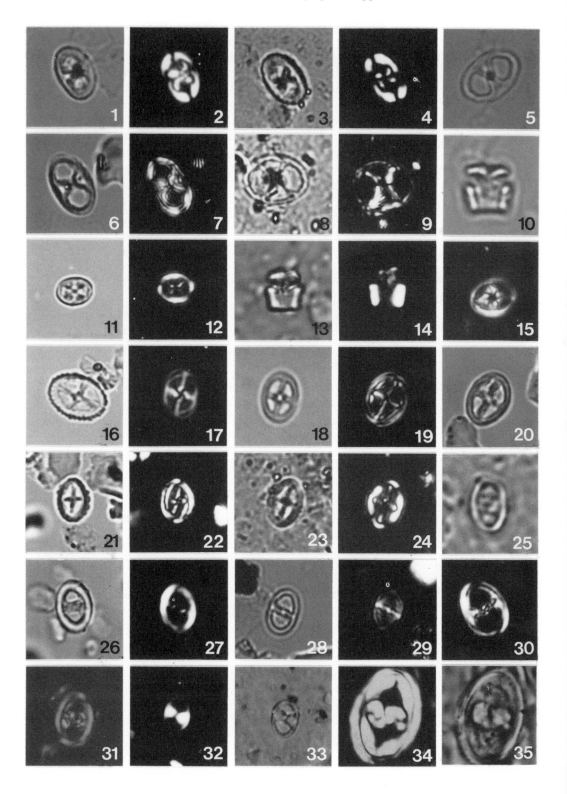

speetonensis, P. fletcheri and 'calcite rosettes'. The Tethyan influence on these nannofloras was weak with several Tethyan species (*C. rothii* and *C. cuvillieri*) excluded from the north-western European area.

(c) The limited nannofloras available from the Upper Valanginian indicate a reduction of Boreal–Arctic influence upon the north-western European area.

(d) The transgression of the Early Hauterivian was probably more extensive than the previous transgressions and caused greater similarity in the nannofloras of the Tethys and north-western European areas. It introduced the Tethyan species *C. rothii* and *C. cuvillieri* into the latter area for the first time. A possible restriction of the marine connections with Tethys, or an improvement of connections with the Boreal-Arctic area,

PLATE 8.13
All magnifications × 2430, except Figs. 34 and 35 which are × 700.

Plate 8.13, Figs. 1, 2. *Rhagodiscus asper*. Fig. 1, bright field, Hauterivian, Otto Gott, PAL 2852. Fig. 2, crossed-nicols, same specimen.

Plate 8.13, Figs. 3–5. *Viminites swinnertonii*. Fig. 3, bright field, Hauterivian, Otto Gott, PAL 2852. Fig. 4, crossed-nicols, same specimen. Fig. 5, crossed-nicols, Hauterivian, Otto Gott, PAL 2852.

Plate 8.13, Figs. 6–9. *Percivalia fenestrata*. Fig. 6, bright field, Hauterivian, Speeton, PAL 3409. Fig. 7, crossed-nicols, same specimen. Fig. 8, bright field, Hauterivian, Otto Gott, PAL 2848. Fig. 9, crossed-nicols, same specimen.

Plate 8.13, Figs. 10, 15. *Rhagodiscus splendens*. Fig. 10, bright field, Albian, Mundays Hill, Bedfordshire. Fig. 15, crossed-nicols, same specimen.

Plate 8.13, Figs. 11–14. *Rhagodiscus infinitus*. Fig. 11, bright field, Hauterivian, Moorberg, PAL 2901. Fig. 12, crossed-nicols, same specimen. Fig. 13, bright field, Hauterivian?, Moorberg, PAL 2848. Fig. 14, crossed-nicols, same specimen.

Plate 8.13, Figs. 16–19. *Rhagodiscus pseudoangustus*. Fig. 16, bright field, Hauterivian?, Otto Gott, PAL 2846. Fig. 17, crossed-nicols, same specimen. Fig. 18, bright field, Barremian, Speeton, PAL 3378. Fig. 19, crossed-nicols, same specimen.

Plate 8.13, Fig. 20. *Calculites sarstedtensis*. Bright field, Hauterivian?, Otto Gott, PAL 2846.

Plate 8.13, Figs. 21, 22. *Corollithion geometricum*. Fig. 21, bright field, Hauterivian, Moorberg, PAL 2922. Fig. 22, bright field, Hauterivian, Speeton, PAL 3401.

Plate 8.13, Figs. 23, 24. *Corollithion silvaradion*. Fig. 23, bright field, Ryazanian, Speeton, PAL 3427. Fig. 24, crossed-nicols, Hauterivian, Speeton, PAL 3410.

Plate 8.13, Fig. 25. *Corollithion rhombicum*. Bright field, Albian, Mundays Hill, Bedfordshire.

Plate 8.13, Figs. 26–28. *Stradnerlithus comptus*. Fig. 26, bright field, Hauterivian, Speeton, PAL 3403. Fig. 27, crossed-nicols, same specimen. Fig. 28, crossed-nicols, Hauterivian, Moorberg, PAL 2909.

Plate 8.13, Figs. 29, 30. *Cylindralithus laffittei*. Fig. 29, bright field, Hauterivian, Moorberg, PAL 2909. Fig. 30, bright field, same specimen.

Plate 8.13, Figs. 31, 32. *Conusphaera rothii*. Fig. 31, bright field, Hauterivian?, Otto Gott, PAL 2846. Fig. 32, crossed-nicols, same specimen.

Plate 8.13, Fig. 33. *Pseudolithraphidites parallelus*. Bright field, Hauterivian, Otto Gott, PAL 2852.

Plate 8.13, Figs. 34, 35. *Lithraphidites carniolensis*. Fig. 34, bright field, Hauterivian, Otto Gott, PAL 2852. Fig. 35, crossed-nicols, same specimen.

Pl. 8.13] **Biostratigraphy and palaeogeographical applications**

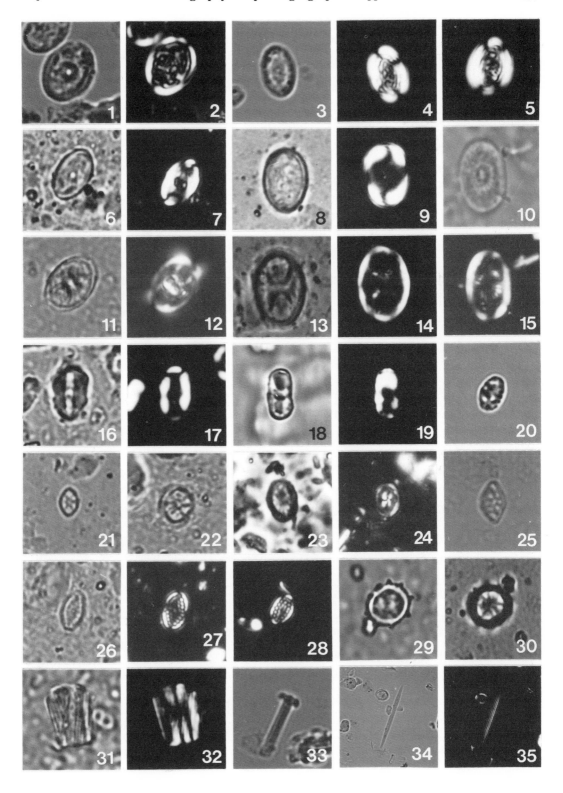

or a climatic change at the end of the Early Hauterivian, caused a decline in the nannofloras.

(e) The pattern of nannofloral development through the Late Hauterivian suggests a period of strong Boreal–Arctic influence (common *T. septentrionalis*) in the early Late Hauterivian followed by increasing Tethyan influence (three species of *Nannoconus*). The Boreal–Arctic influence probably occurred during the early part of a major transgression, while the Tethyan influence occurred in the later part of the same transgression. Further evidence for the latest Hauterivian transgression, which continued into the Early Barremian, comes from the onlap of sediments onto Sklinnabanken at this time.

(f) Three nannoconid species present in the uppermost Hauterivian and lowest Barremian are replaced by the endemic species *N. abundans* (and *N. borealis* (Mutterlose and Harding 1987a)) in the Lower Barremian. This is thought to reflect the closing of seaways to Tethys, possibly by a regression. An interval of poor circulation followed, during which the blätterton organic-rich laminated clays were deposited. This deposition may have occurred during a period of transgression as *R. asper*, a species found to be common during the earlier transgressions of the Early Cretaceous, is also common in these beds. The 'Hauptblätterton' deposition was brought to an end by an influx of northern waters which brought with them the return of *T. septentrionalis*. This influx was probably the

PLATE 8.14
All magnifications × 2430, except Fig. 15 which is × 1900.

Plate 8.14, Figs. 1, 2. *Micrantholithus obtusus*. Fig. 1, bright field, Hauterivian, Moorberg, PAL 2902. Fig. 2, crossed-nicols, same specimen.

Plate 8.14, Figs. 3, 4. *Assipetra infracretacea*. Fig. 3, bright field, Barremian, Otto Gott, PAL 2843. Fig. 4, crossed-nicols, same specimen.

Plate 8.14, Fig. 5. *Micrantholithus speetonensis*. Crossed-nicols, Valanginian, Speeton, PAL 743.

Plate 8.14, Figs. 6, 7. *Nannoconus* sp. (discs). Fig. 6, bright field, Ryazanian, Speeton, PAL 3428. Fig. 7, crossed-nicols, same specimen.

Plate 8.14, Figs. 8, 12, 17. *Nannoconus* sp. 2. Fig. 8, bright field, Hauterivian, Otto Gott, PAL 2850. Fig. 12, bright field, Hauterivian, Otto Gott, PAL 2850. Fig. 17, bright field, Hauterivian, Otto Gott, PAL 2850.

Plate 8.14, Figs. 9–1. *Nannoconus abundans*. Fig. 9, bright field, Barremian, Otto Gott, PAL 2839. Fig. 10, crossed-nicols, same specimen. Fig. 11, bright field, Barremian, Otto Gott, PAL 2889.

Plate 8.14, Figs. 13, 18, 23. *Eprolithus antiquus*. Fig. 13, bright field, Hauterivian, Speeton, PAL 3408. Fig. 18, bright field, Hauterivian, Moorberg, PAL 2917. Fig. 23, crossed-nicols, same specimen.

Plate 8.14, Fig. 14. 'Calcite rosette'. Crossed-nicols, Valanginian, Münchenhagen, PAL 2976.

Plate 8.14, Fig. 15. *Nannoconus steinmannii*. Bright field, Hauterivian?, Otto Gott, PAL 2847.

Plate 8.14, Fig. 16. *Nannoconus* sp. 1. Bright field, Hauterivian?, Otto Gott, PAL 2847.

Plate 8.14, Figs. 19–22, 24–27. *Tegulalithus septentrionalis*. Fig. 19, bright field, Barremian, Otto Gott, PAL 2886. Fig. 20, crossed-nicols, same specimen. Fig. 24, bright field, same specimen. Fig. 25, crossed-nicols, Barremian, Otto Gott, PAL 2890. Fig. 21, bright field, Hauterivian, Moorberg, PAL 2922. Fig. 26, crossed-nicols, same specimen. Fig. 22, bright field, Barremian, Otto Gott, PAL 2889. Fig. 27, crossed-nicols, Barremian, Otto Gott, PAL 2889.

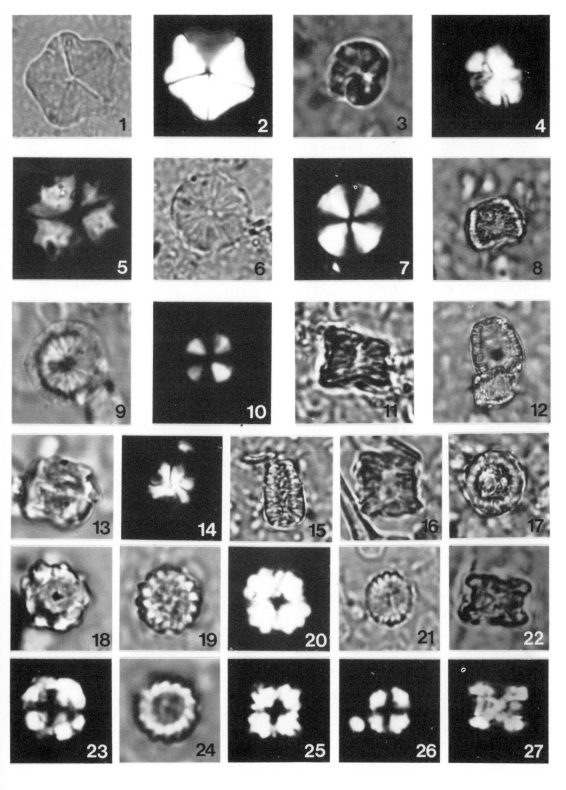

result of a seaway opening to the north, possibly as the result of the transgression.

(g) A decrease in the abundance and the diversity of the nannoflora in the Upper Barremian and the return of *N. abundans* probably indicates a period of regression.

These patterns of regressions and transgressions in the Early Cretaceous are in general agreement with those proposed by Haq *et al.* (1987). This similarity is to be expected, as many of the data used to create their chronology of fluctuating sea levels came from the north-western European area.

8.8 ACKNOWLEDGEMENTS

I would like to thank British Petroleum Plc. for permission to publish this paper. In particular I would like to thank C.P. Summerhayes for reading the manuscript and M.A. Partington who helped to collect the samples. I would also like to thank J. Mutterlose and P.F. Rawson for their help in sample collection. E. Erba and J. Mutterlose reviewed the manuscript and made many useful suggestions. Finally I thank J. Verdenius and IKU, Norway, for supplying me with prepared slides of the Sklinnabanken Core 7B.

8.9 REFERENCES

Aarhus, N., Verdenius, J. and Birkelund, T. 1986. Biostratigraphy of a Lower Cretaceous section from Sklinnabanken, Norway, with some comments on the Andøya exposure. *Norsk Geologisk Tidsskr., 66*, 17–43.

Black, M. 1968. Taxonomic problems in the study of coccoliths. *Palaeontology, 11*, 793–813.

Black, M. 1971a. The systematics of coccoliths in relation to the palaeontological record. In B. M. Funnell and W. R. Riedel (Eds.), *The Micropalaeontology of the Oceans*, Cambridge University Press, Cambridge, 611–624.

Black, M. 1971b. Coccoliths of the Speeton Clay and Sutterby Marl. *Proc. Yorks. Geol. Soc., 38*, 381–424.

Covington, J. M. and Wise, S. W. 1987. Calcareous nannofossil biostratigraphy of a Lower Cretaceous deep-sea fan complex: Deep Sea Drilling Project Leg 93 Site 603, lower continental rise off Cape Hatteras. *Init. Rep. DSDP, 92*, 617–660.

Crux, J. A. 1986. *Tegulalithus* a new genus of Early Cretaceous calcareous nannofossils. *INA Newsletter, 8*, 88–90.

Crux, J. A. 1987. Six new species of calcareous nannofossils from the Lower Cretaceous strata of England and Germany. *INA Newsletter, 9*, 30–35.

Erba, E. 1987. Mid-Cretaceous cyclic pelagic facies from the Umbrian–Marchean Basin: what do calcareous nannofossils suggest? *INA Newsletter, 9*, 52–53.

Fletcher, B. N. 1969. A lithological subdivision of the Speeton Clay C Beds (Hauterivian), East Yorkshire. *Proc. Yorks. Geol. Soc., 37*, 323–327.

Gartner, S. 1968. Coccoliths and related calcareous nannofossils from Upper Cretaceous deposits of Texas and Arkansas. *Univ. Kansas Paleontol. Contrib., 48*, 1–56.

Hallam, A. and Bradshaw, M. J. 1979. Bituminous shales and oolitic ironstones as indicators of transgressions. *J. Geol. Soc. Lond., 136*, 157–164.

Haq, B. U., Hardenbol, J. and Vail, P. R. 1987. Chronology of fluctuating sea levels since the Triassic. *Science, 235*, 1156–1166.

Jakubowski, M. 1986. New calcareous nannofossil taxa from the Lower Cretaceous of the North Sea. *INA Newsletter, 8*, 35–42.

Jakubowski, M. 1987. A proposed Lower Cretaceous calcareous nannofossil zonation scheme for the Moray Firth area of the North Sea. *Abh. Geol. B.-A., 39*, 99–119.

Kaye, P. 1964. Observations on the Speeton Clay (Lower Cretaceous). *Geol. Mag., 101*, 340–356.

Kemper, E., Rawson, P. F. and Thieuloy, J. P. 1981. Ammonites of Tethyan ancestry in the early Lower Cretaceous of north-west Europe. *Palaeontology, 24*, 251–311.

Kemper, E., Mutterlose, J. and Wiedenroth, K. 1987. Die Grenze Unter/Ober Hauterive in Nordwestdeutschland, Beispiel eines stratigraphisch zu nutzenden Klima-Unschwungs. *Geol. Jb., A96*, 209–218.

Keupp, H. 1976. Kalkiges Nannoplankton aus den Solnhofener Schichten (Unter-Tithon, Sudliche Frankenalb). *N. Jb. Geol. Paläont., Mh., 1976*, 361–381.

Kok, C. P. 1985. An Early Cretaceous nannofossil from the Central North Sea. *INA Newsletter, 7*, 38.

Köthe, A. 1981. Kalkiges Nannoplankton aus dem Unter-Hauterivum bis Unter-Barremium der Tongrube Moorberg/Sarstedt (Unter-Kreide, NW-Deutschland). *Mitt. Geol. Inst. Univ. Hannover, 21*, 1–95.

Lamplugh, G. W. 1889. On the subdivisions of the Speeton Clay. *Q. J. Geol. Soc. Lond., 45*, 575–618.

Mutterlose, J. 1984. Die Unterkreide-Aufschlusse (Valangin–Alb) in Raum Hannover–Braunschweig. *Mitt. Geol. Inst. Univ. Hannover, 24*, 1–61.

Mutterlose, J. and Harding, I. 1987. Phytoplankton from the anoxic sediments of the Barremian (Lower Cretaceous) of the North-West Germany. *Abh. Geol. B.-A., 39*, 177–215.

Mutterlose, J., Schmid, F. and Spaeth, C. 1983. Zur Paläobiogeographie von Belemniten der Unter-Kreide in NW-Europa. *Zitteliana, 10*, 293–307.

Neale, J. W. 1960. The subdivision of the Upper D Beds of the Speeton Clay of Speeton, East Yorkshire. *Geol. Mag., 97*, 353–362.

Neale, J. W. 1962. Ammonoidea from the Lower D Beds (Berriasian) of the Speeton Clay. *Palaeontology, 5*, 272–296.

Noël, D. 1965. *Sur les Coccolithes du Jurassique Européen et d'Afrique du Nord*, Editions du CNRS, Paris.

Perch-Nielsen, K. 1979. Calcareous nannofossils from the Cretaceous between the North Sea and the Mediterranean. In *Aspekte der Kreide Europas. IUGS Series A*, **6**, 223–272.

Perch-Nielsen, K. 1985. Mesozoic calcareous nannofossils. In H. M. Bolli, J. B. Saunders and K. Perch-Nielsen (Eds.), *Plankton Stratigraphy*, Cambridge University Press, Cambridge, 329–426.

Rawson, P. F. 1971. The Hauterivian (Lower Cretaceous) biostratigraphy of the Speeton Clay of Yorkshire, England. *Newsl. Stratigr.*, **1**, 61–75.

Rawson, P. F. 1973. Lower Cretaceous (Ryazanian–Barremian) marine connections and cephalopod migrations between the Tethyan and Boreal Realms. In R. Casey and P. F. Rawson (Eds.), *The Boreal Early Cretaceous*, Seel House Press, Liverpool, 131–144.

Rawson, P. F. and Mutterlose, J. 1983. Stratigraphy of the Lower B and basal Cement Beds (Barremian) of the Speeton Clay, Yorkshire, England. *Proc. Geol. Ass.*, **94**, 133–146.

Rawson, P. F. and Riley, L. A. 1982. Latest Jurassic–Early Cretaceous events and the 'Late Cimmerian unconformity' in North Sea Area. *Bull. Am. Assoc. Pet. Geol.*, **66**, 2628–2648.

Rawson, P. F. Curry, D., Dilley, F. C., Hancock, J. M., Kennedy, W. J., Neale, J. W., Wood, C. J. and Worssam, B. C. 1978. A correlation of Cretaceous rocks in the British Isles. *Geol. Soc. Lond., Special Report*, **9**, 1–70.

Stradner, H., Adamiker, D. and Maresch, O. 1968. Electron microscope studies on Albian calcareous nannoplankton from the Delft 2 and Leidschendeam 1 Deep Wells, Holland. *Verh. Kon. Ned. Akad. Wetensch. Afd. Natuurk.*, **24**, 1–107.

Taylor, R. J. 1978. The distribution of calcareous nannofossils in the Speeton Clay of Yorkshire and their biostratigraphical significance. *Proc. Yorks. Geol. Soc.*, **42**, 195–209.

Taylor, R. J. 1982. Lower Cretaceous (Ryazanian to Albian) calcareous nannofossils. In A. R. Lord (Ed.), *A Stratigraphical Index of Calcareous Nannofossils*, Ellis Horwood, Chichester, 40–80.

Taylor, R. J. and Hamilton, G. B. 1982. Techniques. In A. R. Lord (Ed.), *A Stratigraphical Index of Calcareous Nannofossils*, Ellis Horwood, Chichester, 11–15.

Thierstein, H.R. 1973. Lower Cretaceous calcareous nannoplankton biostratigraphy. *Abh. Geol. B.-A.*, **29**, 1–52.

Tyson, R. V. and Funnell, B. M. 1987. European Cretaceous shorelines, stage by stage. *Palaeogeog., Palaeoclimatol., Palaeoecol.*, **59**, 69–91.

Verdenius, J. G. 1978. A Valanginian calcareous nannofossil association from Kong Karls Land, Eastern Svalbard. *Norsk Polarinstitutt, Notiser*, 350-352.

Watkins, D. K. and Bowdler, J. L. 1984. Cretaceous calcareous nannofossils from the Deep Sea Drilling Project Leg 77, southeast Gulf of Mexico. *Init. Rep. DSDP.*, **77**, 221–251.

Wind, F.H. and Čepek, P. 1979. Lower Cretaceous calcareous nannoplankton from DSDP Hole 397A (Northwest African Margin). *Init. Rep. DSDP.*, **47**, 221–255.

References cited in the taxonomic section and not included here may be found in Perch-Nielsen 1985

9

Lower Cretaceous calcareous nannofossils from continental margin drill sites off North Carolina (DSDP Leg 93) and Portugal (ODP Leg 103): a comparison

Joseph L. Applegate, James A. Bergen, J. Mitchener Covington and Sherwood W. Wise, Jr.

Extensive Lower Cretaceous sequences drilled off the continental margins on opposite sides of the North Atlantic during DSDP–ODP Legs 93 and 103 provide an opportunity to study the introduction of neritic calcareous nannofossil taxa into the deep-sea environment, a phenomenon that appears to have been widespread within the circum-North Atlantic during Neocomian times.

DSDP Leg 93 Site 603 on the lower continental rise off North Carolina penetrated a siliciclastic deep-sea fan complex characterised by sharp vertical facies changes and alterations of lithology. Well-preserved assemblages in dark, carbonaceous claystones were probably displaced from the oxygen minimum zone along the upper slope or outer shelf; neritic taxa include the holococcolith *Zebrashapka vanhintei*, *Lithraphidites alatus* ssp. *magnus*, *Pickelhaube furtiva*, and a host of nannoconids and micrantholiths, all of which are few or absent in the interbedded pelagic, bioturbated carbonates.

ODP Leg 103 Sites 638 to 641 were drilled along the Galicia Bank off Portugal and penetrated pre-rift, syn-rift, and post-rift sediments often characterised by siliciclastic turbidites and calcareous microturbidites. *Micrantholithus*, *Nannoconus*, and *Lithraphidites* again dominate the displaced nannofossil assemblages. *Pickelhaube* is common in a Barremian shell lag where the excellent preservation of the entire assemblage demonstrates the minimal damage that nannofossils experience during transportation by turbidites. Holococcoliths, however, are absent, and a different lithraphiditid assemblage is present compared with Site 603. A modification of the Sissingh zonation proved most useful in subdividing the section.

A qualitative comparison of these northern Tethyan drill sites with sequences reported from the Boreal North Sea indicates major assemblage differences, even though the latitudinal distance between the two regions is not great. These differences are attributed to the presence of geographical land barriers that tended to isolate the North Sea from Tethys during Late Valanginian to Late Barremian times.

9.1 INTRODUCTION

Extensive Lower Cretaceous sequences drilled on opposite sides of the North Atlantic during Deep Sea Drilling Project Leg 93 off North Carolina (van Hinte *et al.* 1987) and Ocean Drilling Project Leg 103 off the Iberian Peninsula (Boillet *et al.* 1987) provide an opportunity to compare deep sea nannofossil assemblages of Tethyan affinity from continental margin

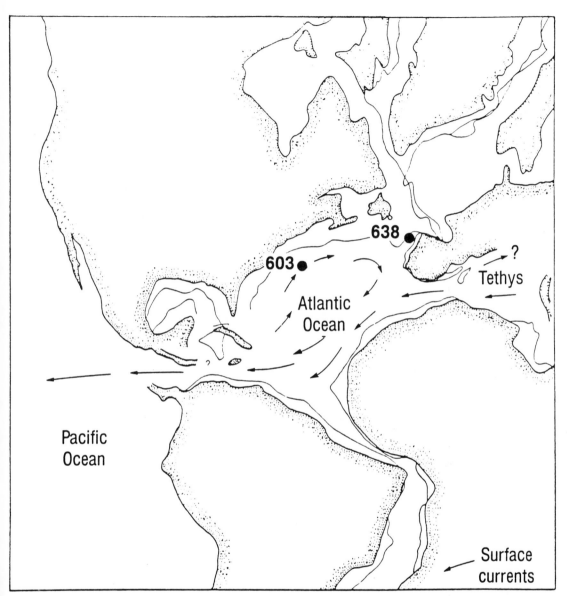

Fig. 9.1 — Location of DSDP Site 603 and ODP Site 638 on a palaeoreconstruction map of the North Atlantic in the Lower Cretaceous. Modified from Ogg *et al.* (1987).

settings. A reconstruction of the palaeo-geography of the Lower Cretaceous illustrates the close proximity of the principal drill holes, DSDP Hole 603B and ODP Holes 638B/C (Fig. 9.1). Both sites were drilled along passive continental margins where sedimentation progressed in a similar fashion during the opening of the North Atlantic Ocean, although the timing of rifting and turbidite activity was different at each locality.

As summarised by Wise and van Hinte (1987), Site 603 was drilled on the Lower Continental Rise in the central North Atlantic over oceanic crust of presumed Oxfordian age. Final rifting between the US margin and North Africa occurred during the Bathonian (Sheridan *et al.* 1983). Pelagic limestone deposition was interrupted during the Late Valanginian by mud and sand turbidite deposition, which then peaked during the Early Barremian, and again during the Late Barremian–earliest Aptian during a sharp sea level drop. These turbidites reached Site 603 some 320 km seaward of the Cretaceous shelf edge and contributed to the extensive Cape Hatteras Deep Sea Fan Complex.

In contrast, rifting in the northernmost Atlantic between Newfoundland and Iberia in the region of Site 638 began during the Late Jurassic and continued over an extended period of time until final rifting during the Late Aptian (ODP Leg 103, Part A, Boillet *et al.* 1987). Stretching and extension of the continental crust was accompanied by turbidite activity off northern Iberia, which peaked at Site 638 during the Valanginian. Tectonic stresses related to rifting of Iberia–Newfoundland may have been translated along transform fault zones across the North Atlantic and into the northern Appalachian region. This could have caused rejuvenation of source areas and the progradation of massive 'Wealden-type' deltas across the continental shelves of eastern North America. Upon reaching the shelf edge, the rivers fed the clastic turbidity currents that reached Site 603 (Wise and van Hinte 1987).

By Albian times, stretching and rifting of the continental crust off Iberia had left a series of tilted fault blocks, which underlie the present day Galicia Bank. The ODP Leg 103 sites were drilled into these blocks of continental crust.

9.2 LITHOSTRATIGRAPHY
The lithostratigraphy of the Lower Cretaceous at both sites consists of pelagic limestones interrupted by sand, silt and mud turbidites.

These sharply alternating sequences are composed of (1) interbedded sandstones, claystones, and nannofossil marlstone–claystone couplets, (2) alternating bioturbated limestones and marlstones, and (3) laminated nannofossil claystones–marlstone couplets. At Site 603, the sandstone beds occur in the Upper Valanginian to Lower Aptian; at Site 638 the sandstones are essentially confined to the Valanginian. Although the occurrence of the sand turbidites varied in response to local tectonic factors, the deposition of these turbidites corresponds to the emplacement of vast deltaic and sandstone turbidite deposits around the North Atlantic margin during the Neocomian times, including the Wealden deposits of England (see Emery and Uchupi 1984, Sarti and von Rad 1987, Wise and van Hinte 1987, for further discussion). The downslope processes associated with the deposition of these 'Wealden' siliciclastics may have strongly affected the composition and character of Lower Cretaceous nannofossil assemblages within this circum-North Atlantic realm, and this factor should be taken into account in interpreting these assemblages.

9.3 BIOSTRATIGRAPHY
At Site 603, Covington and Wise (1987) attempted to use the zonal compilation of Roth (1978, 1983), which had been developed for the North American Basin (western North Atlantic) from DSDP sites to the south in the Blake Bahama Basin. They were only partially successful in subdividing the section, however, and cited the following difficulties in applying some of the traditional index taxa: selective preservation or diagenetic alteration dependent on lithology (*Cyclagelosphaera deflandrei*), total ecological exclusion (*Lithraphidites bollii*); vague taxonomic concept (*Chiastozygus litterarius*); mimicking by previously undescribed taxa (*Zeugrabdotus? pseudoangustus* for *Rhagodiscus angustus; Rucinolithus terebrodentarius* for *Hayesites irregularis*); possible reworking by turbidites (*Nannoconus colomii/steinmannii, Micrantholithus hoschulzii,* and *Conusphaera*

mexicana). A number of the taxonomic problems have been dealt with by Covington and Wise (1987), Crux (1987) and Applegate and Bergen (in press).

Of particular concern was the potential for redeposition of older nannofossils into younger assemblages, as discussed above under Lithostratigraphy. Therefore, it seemed preferable to use first occurrence datums where downslope processes were prevalent. Applegate and Bergen (in press), therefore, modified the zonation of Sissingh (1977) in order to utilise first occurrence datums where possible. Their modified scheme of 'CC' zones (after Sissingh 1977) and subzones is given in Fig. 9.2 together with approximate correlations with Roth's 'NC' zones. The section at ODP Site 638 encompasses the modified Sissingh zones CC2 through CC6, which are Valanginian through Late Barremian in age.

For the present study, we have re-examined the section at DSDP Site 603, applying the modified Sissingh zonation of Applegate and Bergen (in press). This has resulted in a more direct correlation of the sections at Sites 603 and 638. Our results are given in Fig. 9.3. Total ranges of key taxa are given separately in Fig. 9.4.

It is immediately evident from Fig. 9.3 that the sedimentation rate in the Valanginian at Site 638 was higher than at Site 603. Conversely, the sedimentation rates in the Hauterivian to Barremian were much greater at Site 603. If these differences are taken into account, correlation of the relative ages of the sediment with respect to the nannofossil biostratigraphy is excellent.

The lowermost nannofossil datum observed at both localities is the first occurrence of *Calcicalathina oblongata*, which marks the base of Zone CC3. In Hole 638C it occurs within the sandstone turbidite sequence. It is questionable whether or not this is a true first occurrence in Hole 638C. Hole 639A, drilled updip of Holes 638B/C, is believed to represent a section that lies directly beneath the turbidite sequence. A Lower Valanginian nannofossil marlstone was recovered in 639A where *C. oblongata* was present and therefore this datum may lie below the base of Hole 638C.

The first occurrence of *C. oblongata* in Hole 603B is within a nannofossil limestone. The base of Hole 603B is dated as Berriasian by palynology (Habib and Drugg 1987), but this can be neither confirmed nor denied by nannofossils owing to the lack of suitable markers at this boundary. However, Bralower (1987) places this stage boundary only a short distance below the FOD of *C. oblongata*.

The first occurrence of *Eiffellithus windii*, a new species described by Applegate and Bergen (in press), marks the base of the Subzone CC3b.

STAGE	AGE (Van Hinte, 1976)	ZONES (Roth, 1983)	(Sissingh, 1977; modified)		SUBZONES		DATUMS
Aptian		NC7	CC7	Chiastozygus litterarius	CC7b	Rhagodiscus angustus	FAD Eprolithus floralis
	—115	NC6			CC7a	Hayesites irregularis	FAD Hayesites irregularis
Barremian		NC5b	CC6	Micrantholithus hoschulzii			LAD Calcicalathina oblongata
	—121	NC5a	CC5	Lithraphidites bollii			LAD Speetonia colligata
Hauterivian		NC4b	CC4	Cretarhabdus loriei	CC4b	Speetonia colligata	FAD Lithraphidites bollii
	—126	NC4a			CC4a	Eiffellithus striatus	FAD Eiffellithus striatus
Valanginian		NC3	CC3	Calcicalathina oblongata	CC3b	Eiffellithus windii	FAD Eiffellithus windii
	—131				CC3a	Tubodiscus verenae	FAD Calcicalathina oblongata
Berriasian		NC2	CC2	Retecapsa angustiforata			FAD Retecapsa angustiforata
		NC1	CC1	Nannoconus steinmannii			

Fig. 9.2 — Zonation used for the Lower Cretaceous in this study (based on Sissingh 1977, as modified by Applegate and Bergen in press). Note that all but two datums are first appearance datums.

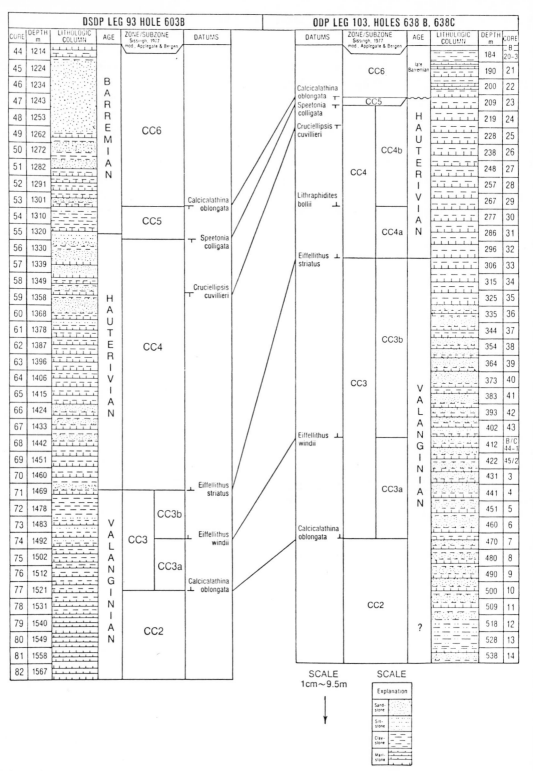

Fig. 9.3 — Comparison of lithologies, ages, nannofossil zones, and nannofossil datums of DSDP Hole 603B and ODP Hole 638B/C.

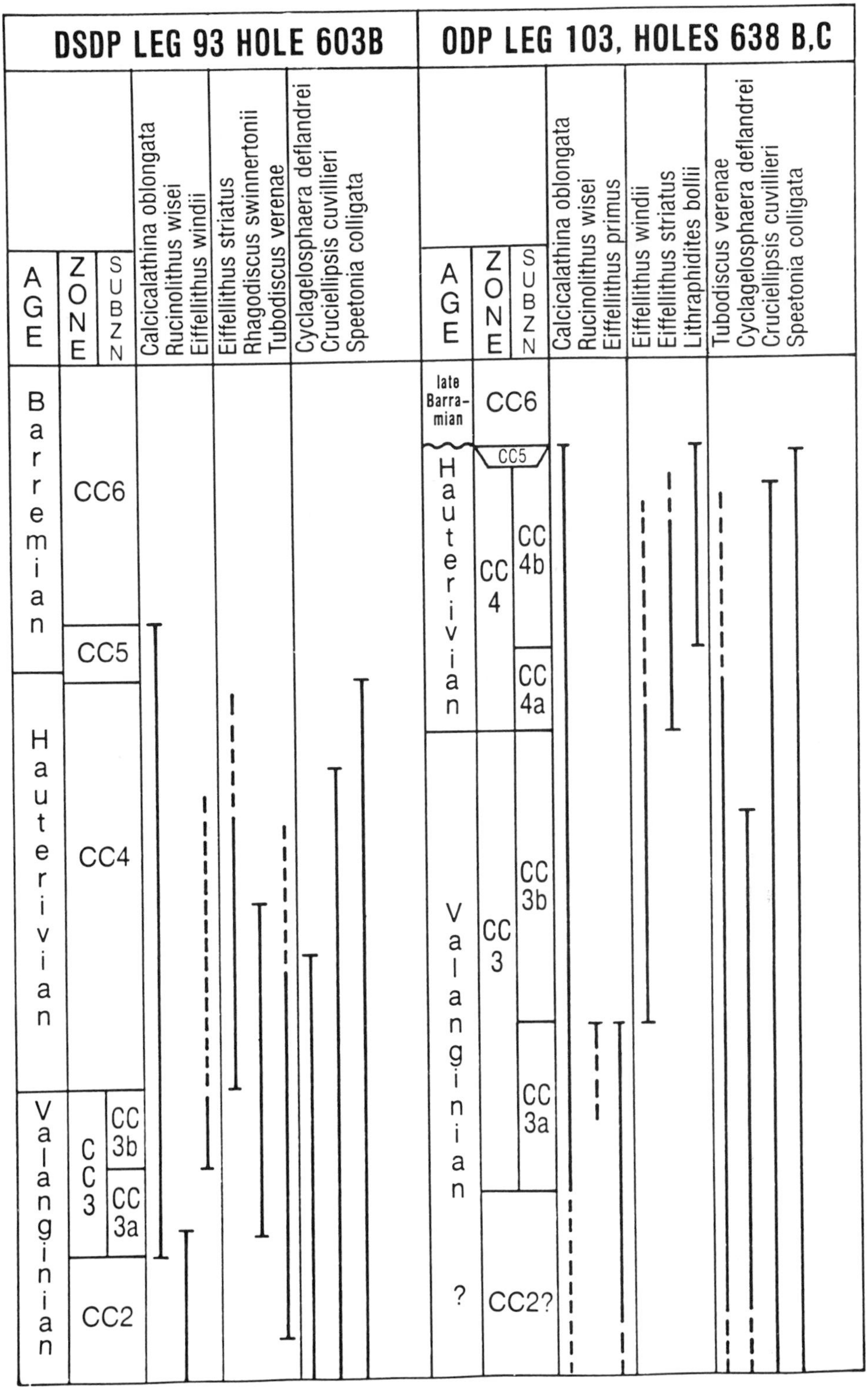

Fig. 9.4 — Comparison of the zonal markers and ranges of other important nannofossil species in DSDP Hole 603B and ODP Hole 638B/C.

This mid-Valanginian datum overlaps the range of *Rucinolithus wisei*. The Valanginian–Hauterivian boundary is approximated by the first occurrence of *Eiffellithus striatus* (= *Chiastozygus striatus*), which marks the base of Zone CC4. This species was identified as *Eiffellithus* sp. 1 by Covington and Wise (1987) and was later transferred to the genus *Eiffellithus* in Applegate and Bergen (in press). A transition between *Eiffellithus windii* and *Eiffellithus striatus* is observed in both sections and is discussed in detail in Applegate and Bergen (in press). Zone CC3 spans at least 160 m in Holes 638B/C, whereas in Hole 603B this zone encompasses only about 55 m. This reflects the much higher sedimentation rate in Holes 638B/C in the Valanginian.

The Lower Hauterivian is subdivided in Holes 638B/C by the first occurrence of *Lithraphidites bollii*, which marks the base of Zone CC4b. This species is not present in Hole 603B where the Lower Hauterivian could not be subdivided. A possible alternate datum that should be given further study is the last occurrence of *Rhagodiscus swinnertonii*, which in Hole 603B has its last occurrence in the mid-Hauterivian. The highest occurrence of *Speetonia colligata* marks the base of Zone CC5, slightly below the Hauterivian–Barremian boundary; thus the Hauterivian (most of Zone CC4) is approximately 95 m thick in Hole 638B and it is approximately 150 m in Hole 603B.

The last occurrence of *Calcicalathina oblongata* marks the base of Zone CC6. A hiatus is indicated in Hole 638B (Fig. 9.3) where part or all of the uppermost Hauterivian–Lower Barremian is missing. There, the LOD of *C. oblongata* occurs one core-section above the LOD of *S. colligata*. Hole 640A, located just west of 638B, recovered a similar sequence and again the Lower Barremian is absent. At Hole 603B, Zone CC5 spans only about 15 m, which also indicates either a condensed or a missing section.

The tops of the Lower Cretaceous sections in Holes 603B and 638B are assigned to Zone CC6, which according to Applegate and Bergen (in press) is Late Barremian in age. This interval is a gap zone that is defined from the last occurrence of *Calcicalathina oblongata* to the first occurrence of *Hayesites irregularis*. The first occurrence of *H. irregularis* has been correlated here with the base of the Aptian (Bralower 1987), and this taxon was not observed in either hole. However, this species was observed in Hole 641C (drilled up-section from Site 638) where its first occurrence is one core below the magnetic anomaly M-0 (Moullade *et al.* in press). These sediments in Hole 641C were dated by magnetostratigraphy as Late Barremian (Ogg *et al.* 1987) and are of similar composition to those in the top of the Lower Cretaceous sequence in Hole 638C. This finding suggests that *H. irregularis* should be an excellent marker for the base of the Aptian. Care should be taken, however, to distinguish it from the superficially similar *Rucinolithus terebrodentarius*.

By the above correlation, the top of the nannofossiliferous section at Site 603 is Barremian in age. This conflicts with the age assignment given Cores 603B–44 to 603B–46-4 by Habib and Drugg (1987). A similar discrepancy between the nannofossil and palynology placement of the Barremian–Aptian boundary is also noted at ODP Sites 638 and 641 (see Moullade *et al.* in press).

9.4 BIOGEOGRAPHY

(a) Northern Tethys

The nannofossil assemblages in Holes 603B and 638B/C are similar and can be characterised as Tethyan, where *Micrantholithus, Nannoconus* and *Lithraphidites* dominate the assemblages. The diversity of nannofossils is high in both sections, being slightly higher in Holes 638B/C. The occurrence of *Lithraphidites bollii* at Site 638 and its absence at Site 603 indicates that this species is restricted to the eastern North Atlantic. This taxon, therefore, can be used as a subzonal marker only in that region. Although both drill sequences were deposited in relatively deep water, they each contain redeposited nannofossils from shallower waters. Among these,

Lithraphidites alatus ssp. *magnus* and the holococcolith *Zebrashapka vanhintei* are recorded only at 603B, although the latter has been observed in sections from southern France (Bergen pers. commun.). Holococcoliths are dissolution susceptible and, unless redeposited, are found only in relatively shallow depositional environments. *Pickelhaube furtiva* is found at both sites and was rare except in a reworked Upper Barremian shell lag (Sample ODP 641C–6–3, 123 cm) where this rather delicate form was quite common and well preserved. Thus, it is clear that this is a 'neritic' taxon that was introduced into the deep-sea environment.

The redeposited forms are often accompanied by a host of nannoconids and micrantholiths, all of which are few or absent in the interbedded bioturbated, *in situ*, pelagic carbonates (see range chart of Covington and Wise 1987, Table 1). Instead these taxa are most abundant and best preserved in the mud turbidites, which were probably derived from the oxygen minimum zone of the upper slope (Covington and Wise 1987). Based on their Atlantic-wide quantitative studies, Roth and Krumbach (1986) also consider nannoconids and micrantholiths to be dissolution prone and/or 'neritic forms'.

The occurrence of species of *Lithraphidites* in deep-sea sediments is of particular interest because several of these have been used as index taxa. *Lithraphidites bollii* has long been used as a zonal or subzonal marker for the Upper Hauterivian–Lower Barremian, whereas *L. acutus* has been used in the Cenomanian. Other species recently described include *L. moray-firthensis* from the Upper Barremian–Aptian of the North Sea and *L. alatus* ssp. *magnus*. The latter occurs in the upper turbidite sequence at Site 603, which is here dated as Late Barremian.

Thierstein (1976) and Roth (1979) considered *Lithraphidites* to be a near-shore species common in the neritic environment. It appears that, during the Neocomian, this genus achieved a marked degree of provincialism in the near-shore basins of the circum-North Atlantic. The extent to which the various species may be encountered in deep-sea sediments of this region, however,

probably depends on two factors; (1) the given province one is in, and (2) the vigour of the downslope processes that would have transported specimens into the oceanic environment.

(b) Tethyan vs. Boreal Provinces

Having surveyed Neocomian nannofossil assemblages from two separate localities from the Northern Atlantic Tethys, we can compare these qualitatively with assemblages reported from the Boreal North Sea province. The differences are striking.

Abundance estimates for North Sea well sections are given by Jakubowski (1987 Fig. 5). There, the Valanginian through Lower Hauterivian is characterised by abundant to common *Crucibiscutum salebrosum*. This form also characterises Neocomian sediments from the high latitude Austral province (Falkland Plateau; see Wise 1983), but was absent from Holes 603B and 638B/C. Even some of the important zonal markers for the Valanginian-Hauterivian in the North Sea (*Micrantholithus speetonensis, Sollasites arcuatus* and *Lithastrinus septentrionalis*) are absent at our northern Tethyan drill sites.

Likewise, three zones and subzones used by Jakubowski (1987) to subdivide the Barremian to Upper Aptian of the North Sea are based on or named for taxa not recorded in the Tethyan province (his *Nannoconnus borealis* Zone, *Lithraphidites moray-firthensis* Zone, and *Eprolithus varolii* Subzone). Conversely, Jakubowski does not identify *Hayesites irregularis* in his material. Another Barremian zonal marker of Jakubowski, *Nannoconus abundans*, which is exceedingly common locally in northern Germany (Stradner and Grün 1973), was quite rare in the northern Tethys (Covington and Wise 1987, Plate 12, Fig. 8, noted only one poorly developed and perhaps questionable specimen). One species that does correlate well from Site 638 and the North Sea is *Tranolithus gabalus*. This species was very common in the lower Valanginian at Site 638 and was abundant in the North Sea in this interval. This range is

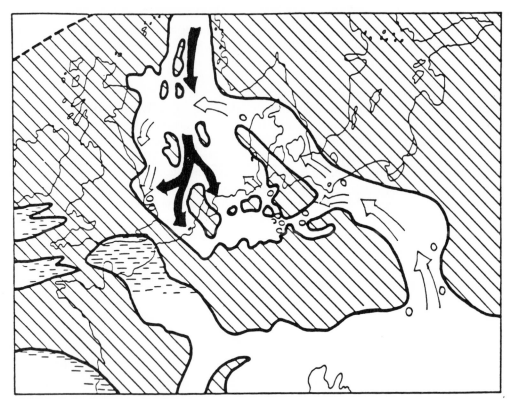

Fig. 9.5 — Palaeogeographical reconstruction for the Valanginian to Late Hauterivian of the North Sea–European Tethyan Region (from Michael 1978, Fig. 1C). Molluscan studies indicate Boreal conditions with some influx of Tethyan elements. Nannofossils also suggest predominantly boreal conditions (see text). Hatched areas = continental masses; dashed horizontal lines = lacustrian, brackish facies; open arrows = warm ocean currents from the Tethys; solid arrows = cold water currents from the Arctic Ocean.

lower than generally reported.

The Albian at DSDP Site 603 was devoid of nannofossils, and this interval was the only interval cored at ODP Site 641; therefore we have a less complete Tethyan record to compare with the North Sea. The North Sea Albian is characterised by abundant *Seribiscutum primitivum* with a peak of abundant *Repagulum parvidentatum* in the lower part (Jakubowski 1987, Fig. 5). These taxa are also common to abundant in the Albian of the high southern latitudes of the Falkland Plateau (Wise 1983). Neither of these taxa, however, was recorded by Applegate and Bergen (in press) or Covington and Wise (1987). The latter, however, is a small taxon that could easily be overlooked if not present in appreciable numbers.

It is apparent, therefore, that even a qualitative comparison of the fossil nannofloras shows strong provincial differences between the North Atlantic Tethys and the North Sea–northern Europe area, even though these localities are not greatly separated by latitude (Fig. 9.1). There must have been substantial geographic barriers between these two regions, barriers not shown on our very generalised Fig. 9.1. Such barriers have been hypothesised by molluscan workers, however, notably by Michael (1978), whose palaeogeographical maps for the Valanginian to Upper Barremian are reproduced in our Figs. 9.5 and 9.6. Michael found that, during the Barremian, cephalopod faunas became severely restricted in the North Sea region as a result of isolation and the severance of all central

Fig. 9.6 — Palaeogeographical reconstruction for the Late Hauterivian to Late Barremian of the North Sea – European Tethyan Region (from Michael 1978, Fig. 1C). Molluscan studies indicate strong Boreal conditions with no influx of Tethyan elements. The North Sea appears to have been totally isolated from Tethys. Nannofossils also suggest predominantly Boreal conditions (see text). Hatched areas = continental masses; dashed horizontal lines = lacustrian, brackish facies; open arrows = warm ocean currents from the Tethys; solid arrows = cold water currents from the Arctic Ocean.

European oceanic connections with the Tethys (Fig. 9.6). This was in strong contrast to Upper Valangianian–Lower Hauterivian and Aptian times when stenothermal benthic warm water cephalopod faunas from Tethys invaded the European Boreal sea basin, indicating the influx of warm Tethyan currents into that region.

9.5 SUMMARY

A comparison of extensive Lower Cretaceous sections of DSDP Site 603 and ODP Site 638 shows an excellent correlation of the nannofossil biostratigraphy, using Sissingh's zonal scheme (1977) modified to include mainly first occurrence datums. The occurrence of the shallow water forms *Micrantholithus*,

Nannoconus, *Lithraphidites*, *Pickelhaube* and *Zebrashapka* in sediments that were deposited in relatively deep water points to a prevalence of downslope processes along the continental margins of the circum-North Atlantic. Lower Cretaceous sedimentation along these passive continental margins proceeded as rifting and subsidence on the margins, as well as probable uplift on the continents. This led to the deposition of shallow water sediments to the deep sea via deep-sea canyons, where sediments were shed from the outer shelf and slope to abyssal depths in the form of turbidites.

A striking difference in nannofossil assemblages is observed between the North Atlantic–Tethyan and the North Sea Boreal regions. Although not separated by any great latitudinal distances, these regions must have been separated by some type of geographical barrier during much of the Valanginian–Barremian interval. Although our comparisons are only qualitative, a substantial data base is being accumulated from the extensive drilling activity in the North Sea. This should allow a future quantitative study that would define the differences and similarities between the two regions more sharply.

9.6 REFERENCES

Applegate, J. L. and Bergen, J. A. in press. Lower Cretaceous nannofossil biostratigraphy of sediments recovered from the Galicia Bank, ODP Leg 103. In G. Boillet, E. L. Winterer, A. W. Meyer *et al.*, *Proc., Init. Rep. (Part B)*, *ODP*, **103**.

Boillet, G., Winterer, E. L., Meyer, A. W., *et al.* 1987. *Proc., Init Rep. (Part A)*, *ODP*, **103**.

Bralower, T. 1987. Valanginian to Aptian calcareous nannofossil stratigraphy and correlation with the Upper M-sequence magnetic anomalies. *Mar. Micropaleont.*, **11**, 293–310.

Covington, M. and Wise, S. W. 1987. Calcareous nannofossil biostratigraphy of a Lower Cretaceous deep sea fan complex: DSDP Leg 93 Site 603, lower continental rise off Cape Hatteras, U.S.A. *Init. Rep. DSDP*, **93**, 617–660.

Crux, J. A. 1987. Six new species of calcareous nannofossils from the Lower Cretaceous strata of England and Germany. *INA Newsletter*, **9**, 30–35.

Emery and Uchupi 1984. *The Geology of the Atlantic Ocean*, Springer-Verlag, New York.

Habib, D. and Drugg, W. S. 1987. Palynology of Sites 603 and 605 Leg 93, Deep Sea Drilling Project. *Init. Rep. DSDP*, **93**, 751–776.

Hinte, J. E. van 1976. A Jurassic time scale. *Bull. Am. Assoc. Pet. Geol.*, **60**, 489–497.

Hinte, J. E. van and Wise, S. W., *et al.* 1987. *Proc., Init. Rep. DSDP*, **93**

Jakubowski, M. 1987. A proposed Lower Cretaceous calcareous nannofossil zonation scheme for the Moray

Firth area of the North Sea. *Abh. Geol. B.-A.*, **39**, 99–119.

Michael, E. 1978. Mediterrane Fauneneinflüsse in den borealen Unterkreide-Becken Europas, besonders Nordwestdeutschlands. In *Aspekte der Kreide Europas. IUGS Series A*, **6**, 305–321.

Moullade, M., *et al.* (in press). Ocean Drilling Program Leg. 103: biostratigraphic synthesis. *Proc., Init. Rep. (Part B) ODP*, **103**.

Ogg, J. G., Haggerty, J., Sarti, M. and Rad, U. v. 1987. Lower Cretaceous pelagic sediments of Deep Sea Drilling Project Site 603, Western North Atlantic: a synthesis. *Init. Rep. DSDP.*, **93**, 1305–1331.

Roth, P. H. 1978. Cretaceous nannoplankton biostratigraphy and oceanography of the north western Atlantic Ocean. *Init. Rep. DSDP*, **44**, 731–759.

Roth, P. H. 1979. Cretaceous calcareous nannoplankton diversity and paleoceanography. *IV Int. Palynol. Conf. Lucknow (1976–1977)*, **2**, 22–23.

Roth, P. H. 1983. Jurassic and Lower Cretaceous calcareous nannofossils in the western North Atlantic (Site 534): biostratigraphy, preservation, and some observations on biogeography and paleoceanography. *Init. Rep. DSDP*, **76**, 587–621.

Roth, P. H. and Krumbach, R. 1986. Middle Cretaceous calcareous nannofossil biogeography and preservation in the Atlantic and Indian Oceans: implications for paleoceanography. *Mar. Micropaleont.*, **10**, 235–236.

Sarti, M. and von Rad, U. 1987. Early Cretaceous turbidite sedimentation at Deep Sea Drilling Project Site 603, off Cape Hatteras (Leg 93). *Init. Rep. DSDP*, **93**, 891–940.

Sheridan, R. E., Gradstein, F. M., *et al.* 1983. *Init. Rep. DSDP*, **76**.

Sissingh, W. 1977. Biostratigraphy of Cretaceous calcareous nannoplankton. *Geol. Mijnbouw.*, **56**, 37–65.

Stradner, H. and Grün, W. 1973. On *Nannoconus abundans* nov. spec. and on laminated calcite growth in Lower Cretaceous nannofossils. *Verh. Geol. B.-A.*, **2**, 267–283.

Thierstein, H. R. 1976. Mesozoic calcareous nannoplankton biostratigraphy of marine sediments. *Mar. Micropaleont.*, **1**, 325–362.

Wise, S. W. 1983. Mesozoic and Cenozoic calcareous nannofossils recovered by Deep Sea Drilling Project Leg 71 in the Falkland Plateau Region, Southwest Atlantic Ocean. *Init. Rep. DSDP*, **71**, 481–550.

Wise, S. W. and Hinte, J. E. van 1987. Mesozoic-Cenozoic depositional environments revealed by deep sea drilling on the continental rise off the Eastern United States: cruise summary. *Init. Rep. DSDP*, **93**, 1367–1423.

10

Nannofossil provincialism in the Late Jurassic–Early Cretaceous (Kimmeridgian to Valanginian) Period

M. Kevin E. Cooper

During the Late Jurassic–Early Cretaceous period many macro- and microfossil groups show evidence of provincialism and calcareous nannofossils are no exception. A number of localities were studied: the Norwegian shelf, England, Germany, France, North Africa and three Deep Sea Drilling Project Sites. At these localities, it is possible to recognise two distinct calcareous nannofossil 'Realms': a low latitude Tethyan Realm and a northern high latitude Boreal Realm. Over 75% of all species found occurred in both realms and therefore the occurrence of individual species has only a limited use in the recognition of these realms. However, quantitative methods have revealed that there were differences in the dominance of the nannofossil families and important genera between these two realms.

10.1 INTRODUCTION

In the Mesozoic, nannofossils are considered to be of more use for biostratigraphical rather than palaeo-oceanographical studies. Palaeo-oceanographical studies using calcareous nannoplankton date back to Huxley (1858). However, these studies have generally been biological rather than geological and in the Mesozoic, and particularly the Jurassic, these palaeo-oceanographical studies are rare. Many macro- and microfossil groups showed evidence of provincialism during the Late Jurassic–Early Cretaceous. The present study suggests that calcareous nannofossils are no exception. When using calcareous nannofossils for palaeo-oceanographical studies, some broad guidelines are suggested.

(1) Calcareous nannofossils are planktonic rather than free swimming and their distribution is thus dependent upon ocean currents.

(2) Modern calcareous nannoplankton appear to show marked temperature dependence; a higher diversity is found in the tropics and they are rare or absent in the higher polar latitudes. Because of this temperature dependence, it can be assumed that calcareous nannofossils should show marked latitudinal changes but very little longitudinal variation.

(3) In modern nannoplankton assemblages, higher densities occur in nutrient-rich waters, e.g. at the edge of the continental shelf and in areas of oceanic upwelling.

(4) It is possible to see an environmental effect in the nannofossil assemblage between onshore localities where shallow marginal marine settings occur and DSDP sites. These effects are seen to influence the *incertae sedis* groups, usually by an increase in density and diversity. This is most noticable in the genus *Nannoconus*.

223

(5) Finally, all the previously mentioned guidelines can be altered by preservational effects, particularly dissolution of nannofossils in the rock or the difficulty in extracting nannofossils from certain lithologies, e.g. limestones.

The possibility that the features produced are lithological or environmental was considered unlikely. The percentage patterns produced for the Tithonian of south-eastern France and DSDP Site 547B are similar, but in south-eastern France the samples were primarily taken from limestones or hard marls, whilst at DSDP Site 547B they were from clays and silts.

In my study of samples from Djebel Zaghouan in Tunisia, a high abundance of *Nannoconus* occurs, possibly because of the shallow water setting of this locality. However, the other features associated with low latitude localities are still present. In this study of calcareous nannofossil provincialism, three criteria were applied:

(A)   species distribution
(B)   nannofossil species diversity
(C)   the distribution of nannofossils, primarily at a family level

For this study, at least six localities for each stratigraphic interval were sampled and examined personally. In the case of DSDP Sites 534A and 547B, these were examined before publication of the reports. However, only two sites in south-eastern France and DSDP Site 534A provided a continuous succession.

For the Kimmeridgian/Early Kimmeridgian, seven localities were examined (see Fig. 10.1): southern England (Ely), Volga Basin (Gorodishche), southern Germany (Bavaria), northern France (Boulogne), south-eastern France (Vocontian region), and DSDP Sites 534A and 547B off the coasts of Florida and Morrocco respectively.

For the Tithonian/Volgian/Upper Kimmeridgian to Portlandian, nine localities were examined (see Fig. 10.1): southern England (Kimmeridge Bay), Volga Basin (Gorodishche and Kasphir), southern Germany (Bavaria),

northern France (Boulogne), south-eastern France (Vocontian region), Tunisia (Djebel Oust) DSDP Site 261 (in the Timor Sea) and Sites 534A and 547B.

For the Berriasian/Ryazanian, seven localities were examined (see Fig. 10.2): north-eastern England (Nettleton and Speeton), the Norwegian continental shelf (Haltenbanken), south-eastern France (Vocontian region), Tunisia (Djebel Zaghouan), Morocco (north of Agadir) and DSDP Sites 261 and 534A.

For the Valanginian, there were six localities (see Fig. 10.2): north-eastern England (Nettleton), south-eastern France (Vocontian region), Tunisia (Djebel Zaghuoan) and DSDP Sites 261, 534A and 547B.

10.2 SPECIES DISTRIBUTION

Most nannofossil species are distributed over very large areas. In the localities studied, often as many as 75% of all species present were common to all localities for the corresponding time intervals. The major difference in species distribution between the two realms is in the age ranges of species. For example, *Stephanolithion bigotii* has a last occurrence level in south-eastern France within the Kimmeridgian, whilst at Gorodishche in the USSR it ranges up to the base of the Middle Volgian. A similar pattern could be seen with the first occurrence level. *Nannoconus steinmannii* is first found occurring in the Late Tithonian of DSDP Site 534A and in south-eastern France. Deres and Achéritéguy (1980) record it from the Late Berriasian in a well from offshore Ireland, while it is also present in the Late Ryazanian of a well on the Norwegian shelf (see Aarhus *et al.* 1986); it does not, however, occur until the Late Hauterivian at Speeton.

One species which seems to be confined to the Tethyan Realm throughout the period of time studied is *Cyclagelosphaera deflandrei*. In the Late Jurassic, it is recorded from DSDP Sites 261, 534A and 547B, south-eastern France, Tunisia and southern Germany, but it is absent from southern England, northern France and the Volga Basin. A similar pattern is seen in the Early

Cretaceous, with it being absent from the Norwegian shelf and north-eastern England, but present in all low latitude localities (Tethyan). The southern limit is based on other authors' work, e.g. DSDP Site 330 on the Falkland Plateau by Wise and Wind (1977), and on confidential well data from Africa and Asia. Other possible Tethyan restricted species include *Conusphaera mexicana*, *Hexalithus noeliae* and *Polycostella senaria*.

There are also two species which are restricted to high latitudes (Boreal Realm). The first is *Stephanolithion atmetros*, which is found in the Middle Volgian of Gorodishche and Kasphir, the Late Kimmeridgian of Dorset and the 'Portlandian' of Boulogne, but there is no record of this species in any low latitude site. In the Late Ryazanian, another Boreal restricted species is *Sollasites arcuatus* which occurs in both high latitude localities but was absent from the low latitude localities. One problem in recognising Boreal restricted species is the absence of a complete Boreal succession.

10.3 SPECIES DIVERSITY

The advantage of using species diversity is that the results of other authors' work can be incorporated into the study (i.e. the number of species they record from their sections). There is a disadvantage in that species diversity can vary through time as well as space.

In the Kimmeridgian (Fig. 10.3), a marked area of low diversity is seen in North Africa where only 16 species were recorded at DSDP Site 547B, while Moshkovitz and Ehrlich (1980) recorded only nine from Sinai. There is an increase in species number northward with 19 in south-eastern France, 22 in southern Germany, 23 in northern France and 39 at Gorodishche in the USSR. Southwards at DSDP Site 534A, 25 species were found, while on the Falkland Plateau only 20 species were found. Thus, in the Kimmeridgian, there is very little evidence for a polar decrease in species numbers. What was found was evidence for a zone of low diversity between approximately 30° N and 45° N

palaeolatitude. In the Tithonian (Fig. 10.4), the zone of low diversity is a more noticable feature as more localities have been examined. Species diversity again increases northwards and southwards of this zone with the highest diversity (51 species) being recorded at DSDP Site 534A, but again there is no clear evidence of a decrease in species diversity polewards.

In the Berriasian (Fig. 10.5), there is a change in the pattern from that seen in the Late Jurassic. The zone of low diversity has disappeared although there is still no decrease in species diversity polewards. It is in the Valanginian (Fig. 10.6) that a decrease in species diversity towards the pole is seen. The highest species diversities (52 and 50) are recorded from DSDP sites off Florida. Numbers decrease to around 40 species in the area of North Africa and southern Europe while, in high latitudes, 20 and 11 species respectively are recorded from north-eastern England and the Norwegian shelf.

10.4 NANNOFOSSIL ASSEMBLAGE DISTRIBUTION

This quantitative study was based on a count in one × 1000 field of view per slide and the percentage of each family was then calculated.

(a) Total Nannofossil Assemblages

The first quantitative study concerns the nannofossil assemblages. These are generally composed of three main components during this period.

(1) The Ellipsagelosphaeraceae coccolithophorid family dominates throughout the period.

(2) 'Other' coccolithophorid families: Biscutaceae, Retecapsaceae, Eiffellithaceae, Stephanolithiaceae and Podorhabdaceae.

(3) 'Rest'—a category including such groups as the Braarudosphaeraceae, Microrhabdulaceae, *Nannoconus* and *Conusphaera*.

In the Kimmeridgian (see Fig. 10.7), only the Ellipsagelosphaeraceae and 'other' coccolithophorids are present in the nannofossil assemblage.

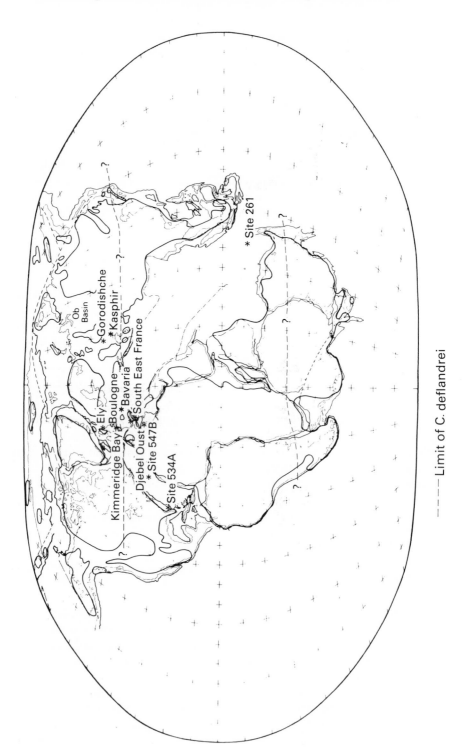

------ Limit of C. deflandrei

Fig. 10.1 — Late Jurassic localities.

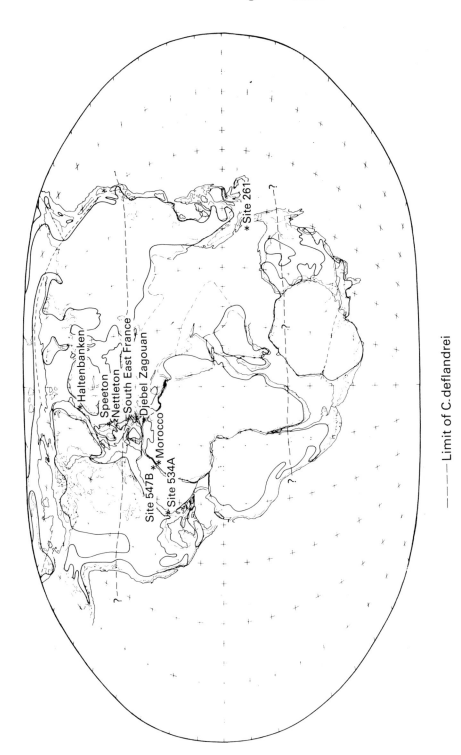

----- Limit of C. deflandrei

Fig. 10.2 — Early Cretaceous localities.

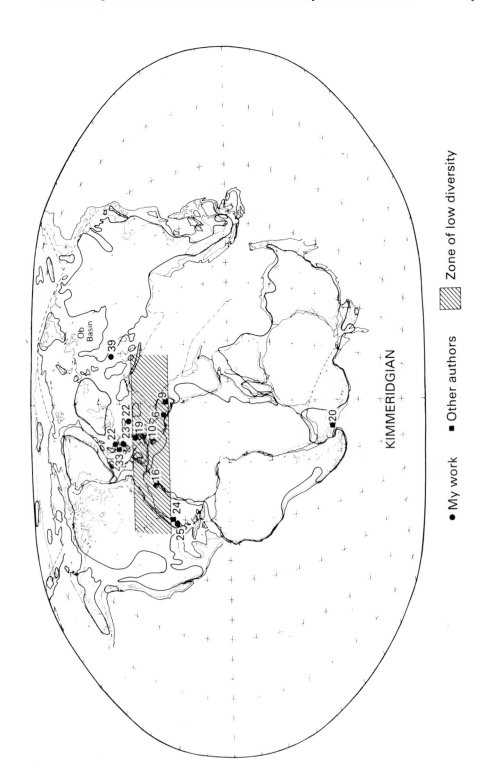

Fig. 10.3 — Species diversity for the **Kimmeridgian**.

Nannofossil assemblage distribution

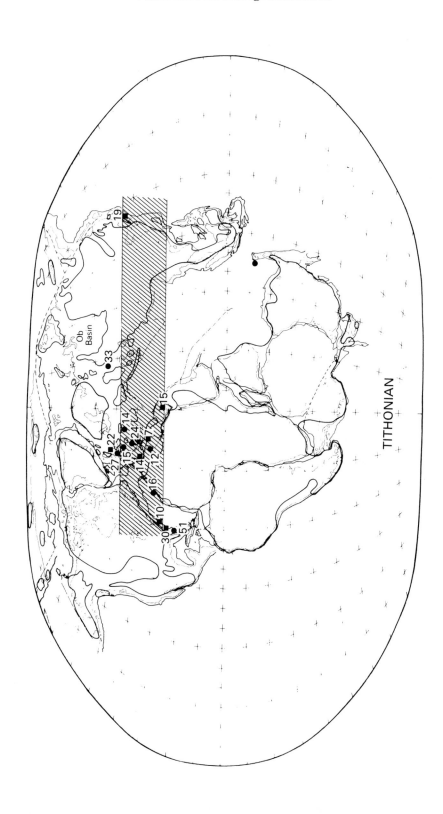

Fig. 10.4 — Species diversity for the Tithonian.

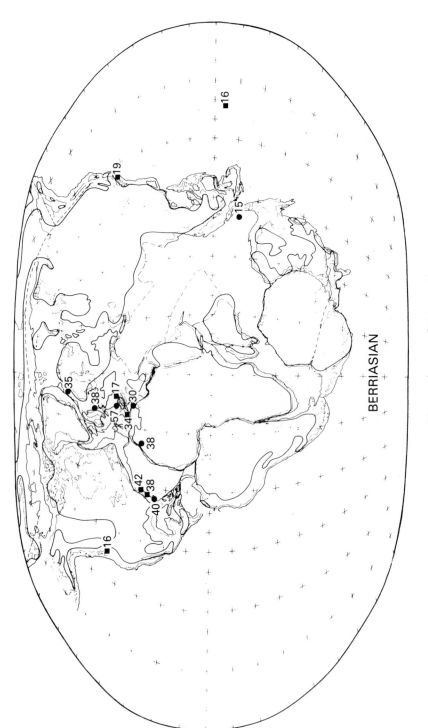

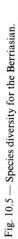

Fig. 10.5 — Species diversity for the Berriasian.

● My work   ■ Other authors

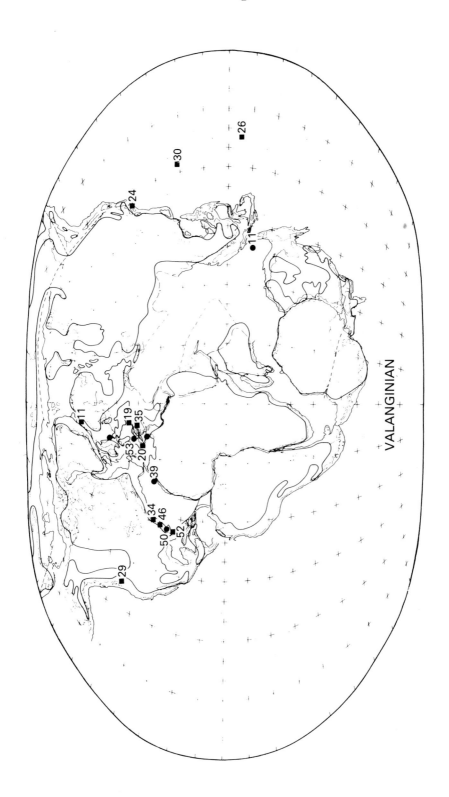

Fig. 10.6 — Species diversity for the Valanginian.

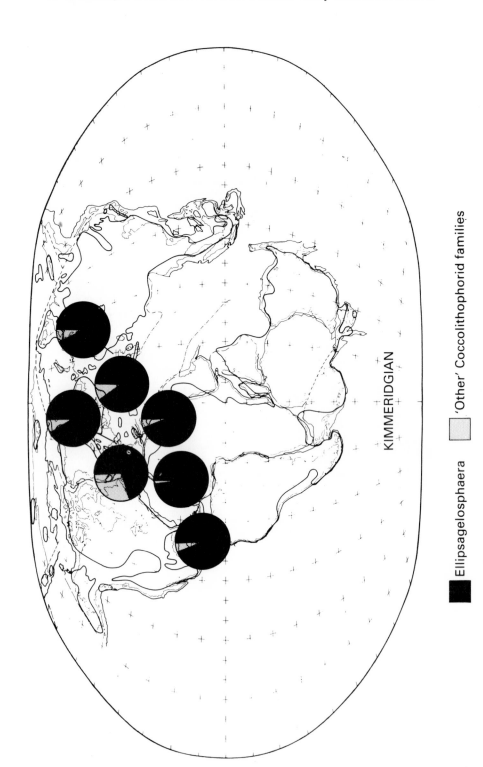

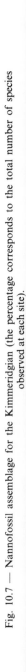

Fig. 10.7 — Nannofossil assemblage for the Kimmeridgian (the percentage corresponds to the total number of species observed at each site).

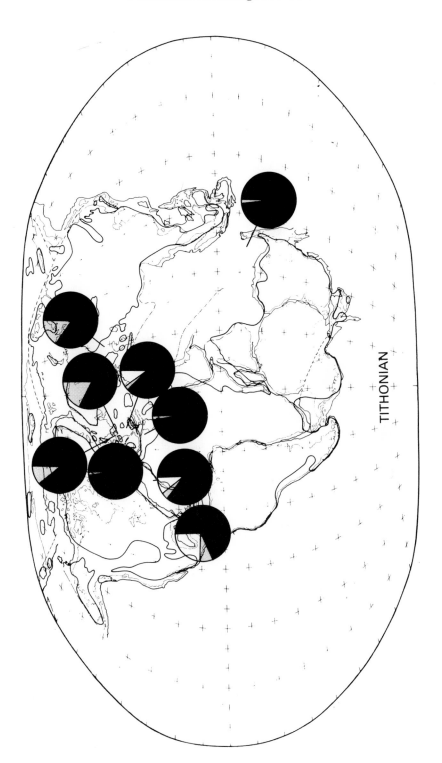

Fig. 10.8 — Nannofossil assemblage for the Tithonian (the percentage corresponds to the total number of species observed at each site).

TITHONIAN

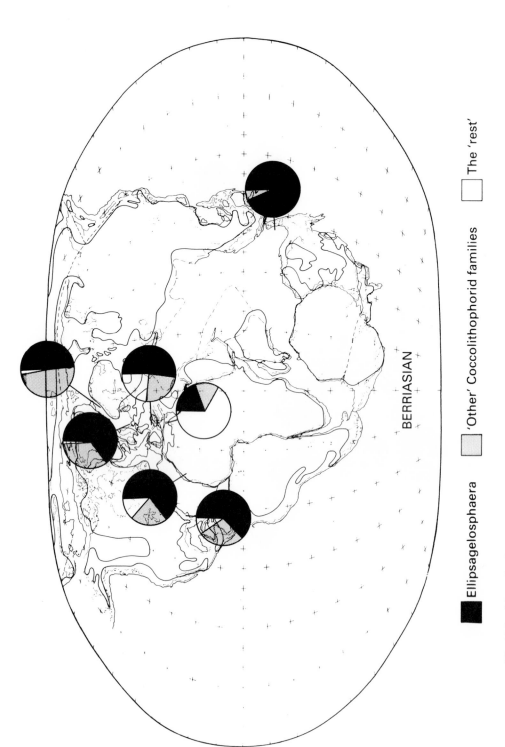

Fig. 10.9 — Nannofossil assemblage for the Berriasian (the percentage corresponds to the total number of species observed at each site).

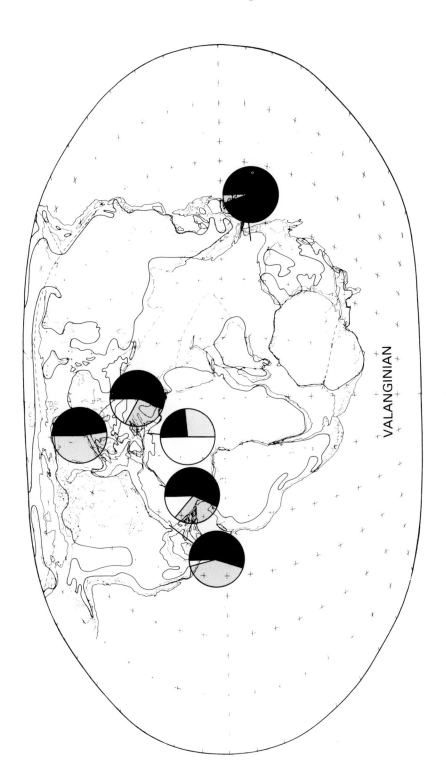

VALANGINIAN

Fig. 10.10 — Nannofossil assemblage for the Valanginian (the percentage corresponds to the total number of species observed at each site).

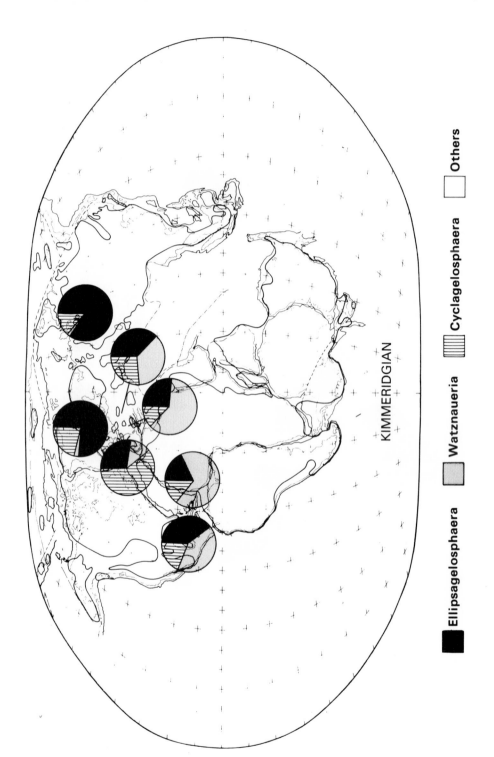

Fig. 10.11 — Distribution of Ellipsagelosphaeraceae in the Kimmeridgian.

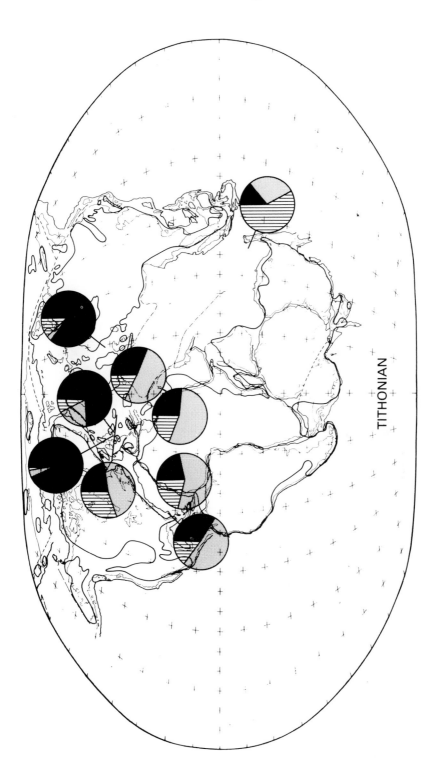

TITHONIAN

Fig. 10.12 — Distribution of Ellipsagelosphaeraceae in the Tithonian.

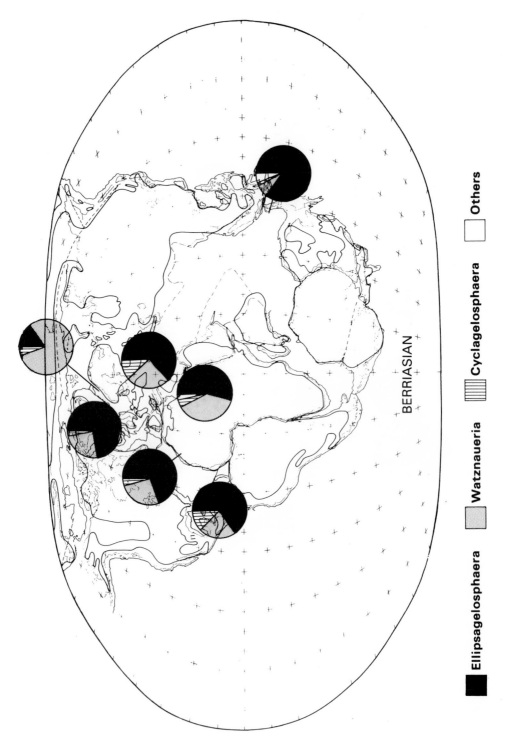

Fig. 10.13 — Distribution of Ellipsagelosphaeraceae in the Berriasian

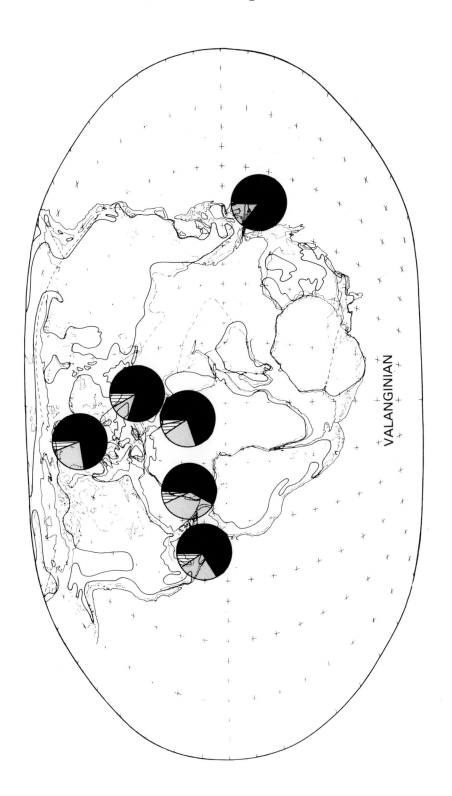

Fig. 10.14 — Distribution of Ellipsagelosphaeraceae in the Valanginian.

VALANGINIAN

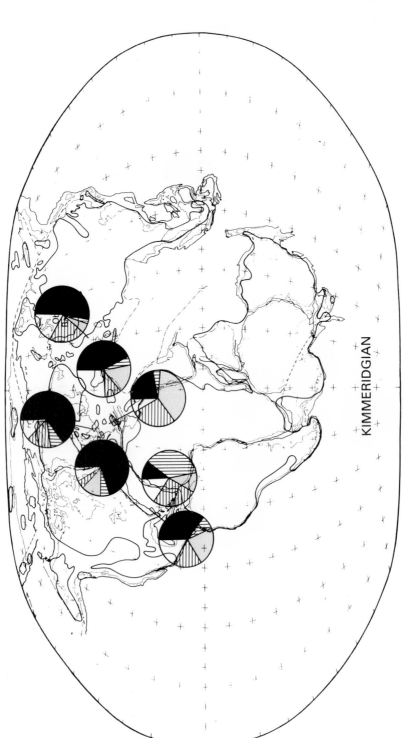

Fig. 10.15 — Distribution of Coccolithophorids in the Kimmeridgian.

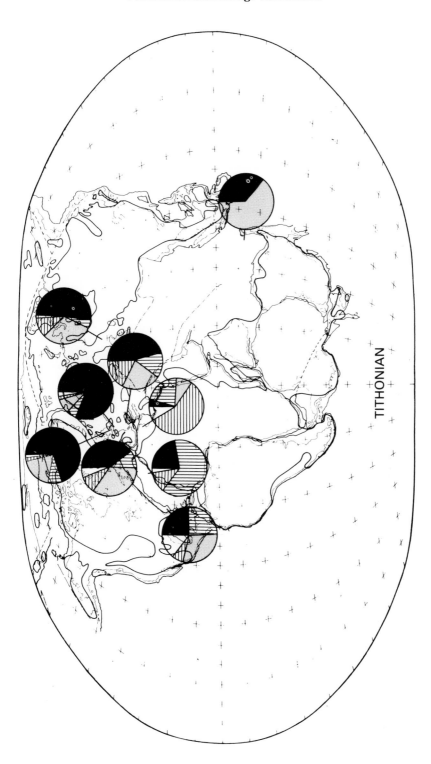

Fig. 10.16 — Distribution of Coccolithophorids in the Tithonian.

TITHONIAN

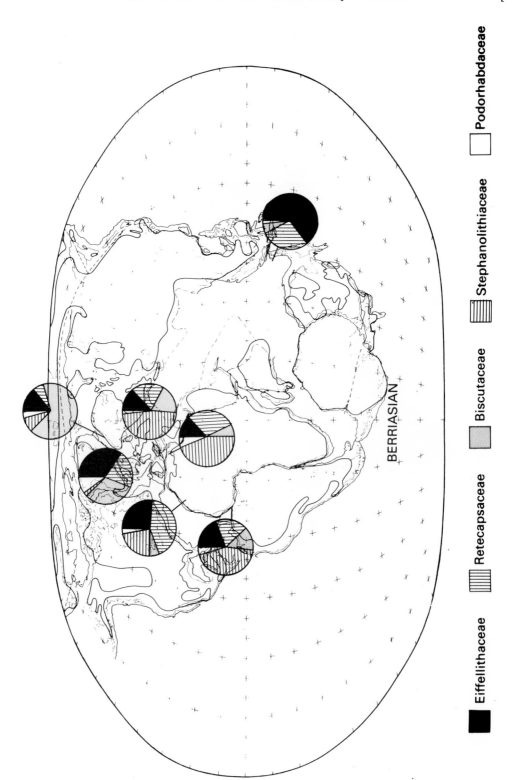

Fig. 10.17 — Distribution of Coccolithophorids in the Berriasian.

BERRIASIAN

Eiffellithaceae   Retecapsaceae   Biscutaceae   Stephanolithiaceae   Podorhabdaceae

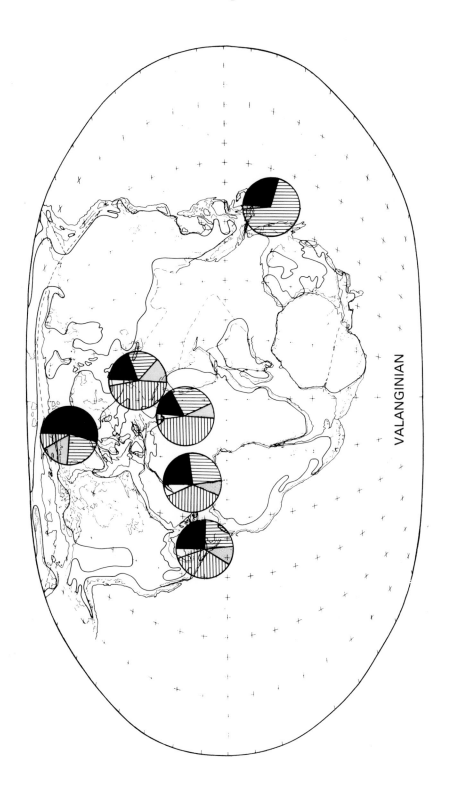

Fig. 10.18 — Distribution of Coccolithophorids in the Valanginian.

VALANGINIAN

The 'other' coccolithophorids seem to be more abundant in high latitude localities, e.g. Boulogne, Ely and Gorodishche. They are also found in abundance in southern Germany (Bavaria), the most northern of the low latitude localities (e.g. *C. deflandrei* is present in the assemblage).

In the Tithonian (see Fig. 10.8), this pattern is again seen with the coccolithophorids being more prevalent in high latitude assemblages. The southern German locality is now similar to other low latitude sites. The only low latitude site with a high proportion of coccolithophorids is DSDP Site 534A. In the three most westerly low latitude sites, the 'rest' grouping now appears. The low percentages of coccolithophorids found in low latitude localities may be related to the low species diversity, which has been shown to occur in most of these sites during the Late Jurassic. In the Berriasian (see Fig. 10.9), it can be noted that the 'other' coccolithophorids are more abundant in the high latitude localities. The 'rest' group has a higher percentage in the low latitude localities than in the high latitude ones, but this grouping now occurs at all localities except the most easterly, DSDP Site 261. In the Valanginian (see Fig. 10.10), this pattern appears to break down. High coccolithophorid percentages are seen in north-eastern England and also in low latitude localities such as DSDP Site 534A. The 'rest' are restricted to low latitude localities during this period. They have high percentages in Tunisia and also fairly high percentages in south-eastern France, whilst the low latitude DSDP Site 534A has a fairly low percentage.

(b) Ellipsagelosphaeraceae Assemblage

The Ellipsagelosphaeraceae are the most important family of the nannofossil assemblage. Three individual genera have been distinguished: *Ellipsagelosphaera*, *Watznaueria* and *Cyclagelosphaera*. A final grouping includes all the other genera in this family, e.g. *Haqius*. In the Kimmeridgian (see Fig. 10.11), there are two notable features. Firstly, in all the high latitude localities except Boulogne, *Ellipsagelosphaera* is

the dominant genus. This genus also occurs in fairly high percentages in southern Germany (the most northern of all the low latitude sites). It also has a high percentage at DSDP Site 534A. Secondly, in the low latitude sites, it is *Watznaueria* that is dominant. In both southern Germany and DSDP Site 534A, *Ellipsagelosphaera* and *Watznaueria* occur in equal percentages. At Boulogne, *Cyclagelosphaera* is the dominant genus.

This pattern is repeated in the Tithonian (see Fig. 10.12). Here, all Boreal sites show the dominance of *Ellipsagelosphaera*, whilst all the low latitude sites show the dominance of *Watznaueria*, even in southern Germany and DSDP Site 534A. The only exception is DSDP Site 261 where *Cyclagelosphaera* dominates (although at this site it is *C. deflandrei* and not *C. margerelii*, as in the case of the Kimmeridgian at Boulogne).

The distribution pattern seen in the Ellipsagelosphaeraceae in the Late Jurassic changes in the Early Cretaceous. In the Berriasian (see Fig. 10.13), *Ellipsagelosphaera* is the dominant genus in all locations and they are probably now more abundant in the low latitude sites. The only exception to this pattern is on the Norwegian shelf where *Watznaueria* is more abundant. *Cyclagelosphaera* is now present in consistently small proportions. The 'other' grouping is restricted to low latitude localities in the Berriasian.

The Valanginian (see Fig. 10.14) repeats all the features seen in the Berriasian, with *Ellipsagelosphaera* dominating all the localities and the other groups restricted to the low latitude sites.

(c) Other Coccolithophorid Assemblages

The final quantitative study concerns the 'other' coccolithophorid families: the Eiffellithaceae, Retecapsaceae, Biscutaceae, Stephanolithiaceae and Podorhabdaceae. In the Kimmeridgian, the family Eiffellithaceae dominates all high latitude localities (see Fig. 10.15) and southern Germany (although it is not as abundant as in the high latitude localities). In the other low latitude localities, there is no one family which dominates the coccolithophorid assemblage. This pattern is repeated in the Tithonian (see Fig. 10.16) with the Eiffellithaceae dominating all the high latitude localities. It is also the dominant family in southern Germany and south-eastern France, although it is not represented in as high a proportion as in the high latitude localities, where it usually constitutes more than 50% of the assemblage.

In the Berriasian (see Fig. 10.17), there is a change in the other coccolithophorid families. The Eiffellithaceae still dominate the assemblages in north-eastern England, although in the other high latitude locality (the Norwegian shelf site) it represents a smaller factor, with the Biscutaceae dominant. In the low latitude localities, except Morocco and DSDP Site 261, the family Stephanolithiaceae is dominant, whilst it is rare in the Boreal localities.

In the Valanginian (see Fig. 10.18), this pattern becomes clearer, with all low latitude sites showing the dominance of the Stephano-lithiaceae, while in the high latitude localities the Eiffellithaceae are dominant. Throughout this study, DSDP Site 261 seems to have had its own particular assemblage, not truly fitting in with either the high latitude or the low latitude assemblages seen here. DSDP Site 534A, even though it is a low latitude locality, is the most southern low latitude locality. There are periods when the assemblage shows this. In the Kimmeridgian, however, both the Ellipsage-losphaeraceae and the other coccolithophorid families appear more typical of a high latitude locality site. The third member of the total nannofossil assemblage, the 'rest', seems to have little use as its composition was generally consistent throughout sites in both high and low latitude localities. It is characterised by a very high dominance of the genus *Nannoconus*.

10.5 CONCLUSIONS

(1) Low and high latitude realm nannofossil assemblages can be seen from these results. The boundary may be defined on the distribution of *C. deflandrei*. This is possibly a simplistic view as there are also other differences in the nannofossil assemblages, e.g. the abundances of families within the two realms.

(2) The species *C. deflandrei* was restricted to low latitudes during the period studied. *S. atmetros* is a Late Jurassic, high latitude, restricted species and *S. arcuatus* is an Early Cretaceous, high latitude restricted species.

(3) The use of species diversity is limited as it has failed to show any polar decrease in this study, although it did recognise the apparent zone of low diversity between approximately 30° N and 45° N palaeolatitude.

(4) In the total nannofossil assemblage, the 'other' coccolithophorid families are more prevalent in high latitude localities.

(5) Amongst the Ellipsagelosphaeraceae during the Late Jurassic, the genus *Ellipsage-losphaera* dominates high latitude localities, whilst *Watznaueria* dominates low latitude localities.

(6) Amongst the other coccolithophorid families present during this period, the Eiffellithaceae are more abundant in high latitude localities, while in the Early Cretaceous the Stephano-lithiaceae are more abundant in low latitude localities.

(7) The most likely cause of this provincialism is temperature. However, this is probably a simplistic view which requires more detailed work, particularly statistical studies of localities from the Southern Hemisphere.

10.6 ACKNOWLEDGEMENTS
I would like to thank Dr. Alan Lord (University College London) for assistance and advice given whilst undertaking the research project from which some of the results are taken from. The financial support of NERC is gratefully acknowledged, as is the provision of samples by the DSDP and Dr. J. Verdenius. I would also like to acknowledge help from Stratigraphic Services International, in particular Dr. R. Bate and Dr. L. Sheppard.

10.7 APPENDIX. LIST OF AUTHORS AND LOCALITIES USED FOR FIGS. 7–10

Kimmeridgian

Sinai	Moshkovitz and Ehrlich 1980
DSDP Sites 4 and 5A	Worsley 1971
DSDP Site 330	Wise and Wind 1977
Southern England	Gallois and Medd 1979

Tithonian

Northern England	Gallois and Medd 1979
Sinai	Moshkovitz and Ehrlich 1980
Southern Spain	Grün and Allemann 1975
DSDP Site 105	Thierstein 1975
DSDP Sites 4 and 5A	Worsley 1971
Japan	Aita and Okada 1986

Berriasian

Japan	Aita and Okada 1986
Southern Spain	Grün and Allemann 1975
Switzerland	Thierstein 1973
Northern California	Worsley 1979
DSDP Site 105	Thierstein 1975
DSDP Sites 4 and 5A	Worsley 1971
DSDP Site 167	Roth 1973

Valanginian

Japan	Aita and Okada 1986
Southern Spain	Grün and Allemann 1975
Sicily	De Wever et al. 1986
Northern California	Worsley 1979
DSDP Site 105	Thierstein 1975
DSDP Sites 4 and 5A	Worsley 1971
DSDP Site 167	Roth 1973
Norway	Aarhus et al. 1986

10.8 REFERENCES

Aarhus, N., Verdenius, J. and Birkelund, T. 1986. Biostratigraphy of a Lower Cretaceous section from Sklinnabanken, Norway, with some comments on the Andøya exposure. *Norsk Geologisk Tidsskr.*, **66**, 17–43.

Aita, Y. and Okada, H. 1986. Radiolarians and calcareous nannofossils from the uppermost Jurassic and Lower Cretaceous strata of Japan and Tethyan regions. *Micropaleontology*, **32**, 197–228.

Deres, F. and Achériteguy, J. 1980. Biostratigraphie des Nannoconides. *Bull. Cent. Rech. SNEA*, **4**, 1–53.

De Wever, P., Geyssant, J. R., Azema, J., Devos, I., Duee, G., Manivit, H. and Vrielynck, B. 1986. La coupe de Santa Anna (zone de Sciacca, Sicile): une synthèse biostratigraphique des apports macro- micro- et nannofossiles de Jurassique supérieur et Crétacé inférieur. *Rev. Micropaléontol.*, **29**, 141–186.

Gallois, R. W. and Medd, A. W. 1979. Coccolith-rich marker bands in the English Kimmeridge Clay. *Geol. Mag.*, **116**, 247–260.

Grün, W. and Allemann, F. 1975. The Lower Cretaceous of Caravaca (Spain) Berriasian calcareous nannoplankton of the Miravetes section (Subbetic Zone, Prov of Murcia). *Eclogae. Geol. Helv.*, **68**, 147–211.

Huxley, T. H. 1858. Appendix A in J. Dayman, *Deepsea Sounding in the North Atlantic Ocean Between Ireland and Newfoundland and made in H.M.S. Cyclops*, Lords Commissioners of the Admiralty, London, 63–68.

Moshkovitz, S. and Ehrlich, A. 1980. Late Jurassic Calcareous nannofossils in Israel's offshore and onland areas. *Geol. Surv. Israel, Current Research 1980*, 65–72.

Roth, P. H. 1973. Calcareous nannofossils Leg 17. DSDP. *Init. Rep. DSDP*, **17**, 695–793.

Thierstein, H. R. 1973. Lower Cretaceous Calcareous nannoplankton biostratigraphy. *Abh. Geol. B.-A.*, **29**, 1–52.

Thierstein, H. R. 1975. Calcareous nannoplankton biostratigraphy at the Jurassic Cretaceous boundary. *Mém. Bur. Rech. Géol. Minières*, **86**, 84–94.

Wise, S. W. and Wind, F. H. 1977. Mesozoic and Cenozoic calcareous nannofossils recovered by DSDP Leg 36 drilling in the Falkland plateau Region, Southwest Atlantic Ocean. *Init. Rep. DSDP*, **36**, 269–392.

Worsley, T. R. 1971. Calcareous nannofossil zonation of Jurassic and Lower Cretaceous sediments from the Western Atlantic. In A. Farinacci (Ed.), *Proc. II Plankt. Conf. Roma 1970*, **2**, Tecnoscienza, Rome, 1301–1322.

Worsley, T. R. 1979. Lower Cretaceous calcareous nannofossils from the Great Valley Sequence of Northern California. A preliminary survey. *Newsl. Stratigr.*, **8**, 124–132.

Part III

Biostratigraphical Applications

Despite the little taxonomic work done and the difficulties involved, present data indicate the probability that the discoasters, especially, may aid in world-wide correlation of certain Tertiary time horizons.

Bramlette and Riedel 1954

11

Calcareous nannoflora and planktonic foraminifera in the Tortonian–Messinian boundary interval of East Atlantic DSDP sites and their relation to Spanish and Moroccan sections

José-Abel Flores and Francisco-Javier Sierro

A quantitative and qualitative study was made of the calcareous nannofossil associations from the Upper Miocene of four northeastern Atlantic DSDP sites: 410, 334, 544A and 397, together with an analysis of samples obtained in the Oued Akrech section (Morocco). The nannoplankton and foraminifera were analysed for a scheme of biostratigraphical events proposed for the Guadalquivir basin (Spain) by Sierro (1984), Flores (1985) and Flores and Sierro (1987). From bottom to top the events recognised are as follows.

Calcareous nannoplankton:
(1) Earliest regular record of *Eu-discoaster berggrenii.*
(2) Marked increase of the 'small placoliths' (reticulofenestrids) relative to the *Reticulofenestra haqii/minutula* group.
(3) Earliest record of *Amaurolithus* (ceratolith).
(4) Earliest record of *Amaurolithus delicatus.*
(5) Reduction in the proportion of *Eu-discoaster berggrenii*

Foraminifera:
(1) Reduction of group 1 of '*Globorotalia menardii*' (sinistral coiling).

(2) Abrupt appearance of group 2 of '*Globorotalia menardii*' (dextral coiling) after an interval in which the keeled Globorotalids are practically absent.
(3) Replacement of '*Globorotalia menardii*' group 2 by the *Globorotalia miotumida* group.
(4) Change in coiling direction from sinistral to dextral of the *Neogloboquadrina acostaensis* group.
(5) Reduction of the frequency of the *Globorotalia miotumida* group.
(6) Earliest frequent record of *Globorotalia margaritae* s.s.

According to the findings, the stratigraphical interval studied is of Late Tortonian–Messinian age. The changes in the association from bottom to top are interpreted as a response to a relative cooling in surface water masses, with intermittent pulses of warmer waters of different intensities at the different sites.

11.1 INTRODUCTION AND GEOGRAPHICAL SITUATION
The main purpose of the present work was to test the succession of events in the calcareous

249

nannoflora and microfauna of the northeastern Atlantic, previously established in the Guadalquivir basin (Spain) by Sierro (1984), Flores (1985) and Flores and Sierro (1987). To do so, material from four DSDP sites in the northeastern Atlantic was selected: 410, 334, 544A, and 397. Previous biostratigraphic studies allowed the authors to select the Upper Tortonian–Messinian intervals (Aumento *et al.* 1977, Howe 1977, Miles 1977, Poore 1978, Luyendyk *et al.* 1978, Steinmetz 1978, Bukry 1977, 1978, Rad *et al.* 1979, Čepek and Wind 1979, Salvatorini and Cita 1979, Mazzei *et al.* 1979, Cita and Ryan 1979, Hinz *et al.* 1984). A section obtained in the Bou Regreg basin (Morocco) was also studied ('Oued Akrech'). According to earlier work (Feinberg and Lorentz 1970, Bossio *et al.* 1976, Wernli 1971, Sierro *et al.* in press) this section can be correlated with some sections in the Guadalquivir basin. Fig. 11.1 shows the location of the cores and land based sections.

11.2 TECHNIQUES AND COUNTING METHOD
(a) Calcareous nannoplankton
The methodology and techniques employed in the present study were the same as those used in the Guadalquivir basin as described in Flores (1985, 1987a, 1987b) and Flores and Sierro (1987). In the processing of the samples for observation with a petrographical microscope (× 1250), a series of parameters were maintained constant in order to make the results comparable, i.e. the initial mass of sediment, the volume of encompassing fluid, the volume of suspension extracted (keeping a constant time for decantation) and the surface of the slide covered.

Two different kinds of analysis were performed: a medium-resolution analysis, considering the specimens whose proportions in the samples were greater than 0.1–0.5% (for which between 200 and 600 nannoliths per sample were studied), and a high-resolution analysis to detect the specimens appearing in a

smaller proportion (greater than 0.005), for which between 15 000 and 20 000 nannoliths were observed. The results of both types of analysis are shown in separate figures. In the case of the DSDP samples, the specimens with no clear biostratigraphical significance were ignored. Data concerning the total abundance of nannoliths and their preservational status are included, following the guidelines proposed by Flores (1985) which are based on the work of Bukry (1973) and Roth and Thierstein (1972). For the legend of the distribution chart see Figure 11.2.

(b) Planktonic foraminifera
For the study of planktonic foraminifera all samples were wet-sieved at 61, 149 and 500 μm but only the 149–500 μm residue was analysed. Successive representative fractions of the residue were analysed on the tray. About 100 *Globoratalia* specimens were picked from each sample.

11.3 SECTIONS AND DSDP SITES
(a) Site 410
As can be seen in Fig. 11.1, this is the northernmost site of those studied, situated at 45°, 30.51′ N, 29° 28.56′ W. The interval studied comprises Cores 34 to 18 (Figs. 11.3 and 11.4).

The proportion of nannoliths in the samples is relatively high (in general more than 50 specimens per field of view, at × 1250), although in some samples they are appreciably less common. Such samples seem to be distributed at random. The preservational status of the specimens is usually good, but in some samples a certain degree of dissolution has occurred, and sometimes overgrowth (Fig. 11.3)

Until Section 27-1, forms assignable to the taxon *Eu-discoaster berggrenii* are rarely found, while from this point upwards these forms are found with a certain regularity. This change marks the event that we called 'earliest regular record of *E. berggrenii*'. Above Section 26-1 (Fig. 11.4) a drastic and continuous increase is seen in

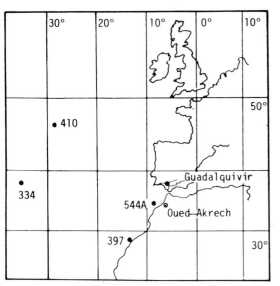

Fig. 11.1 — Geographical location of the DSDP sites and sections studied.

the abundance of the small placoliths. There is an inverse relationship between the distribution of this group and the *Reticulofenestra haqii/minutula* group. Following a short low frequency interval, an increase in the proportion of *Dictyococcites antarcticus* is noted at the base of Core 25, while *Coccolithus pelagicus* shows the opposite distribution. Slightly below, specimens of *Geminilithella jafarii* reach a relatively high proportion with respect to the related species *Geminilithella rotula*. Other morphotypes and/or taxa show a less pronounced frequency pattern.

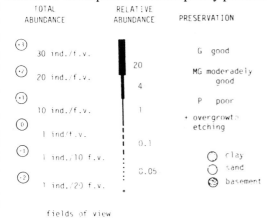

Fig. 11.2 — Legend for Figs. 11.3–11.12. Between 15 000–20 000 nannoliths of each sample were analysed.

The first representative of *Amaurolithus*, in this case *A. primus*, was seen in Section 24-3. In Section 23-5, the first characteristic *Amaurolithus delicatus* was recorded. Specimens of *Amaurolithus ninae* with a morphology intermediate between *A. delicatus* and *Amaurolithus tricorniculatus* first appear in Section 24-1.

The above-mentioned *E. berggrenii* is regularly present up to Core 21. In the top part of this Section 21-2 we have positioned the event called 'reduction in the proportion of *E. berggrenii*'. As may be seen from Fig. 11.3, this event is difficult to pinpoint because, in the samples above, sporadic occurrences were recorded. In turn, *E. quinqueramus* was regularly observed from about Core 26 up to Section 18-2.

The last sample studied (18-1, 12–24 cm) shows a characteristic Pliocene association with specimens of *Ceratolithus cristatus* and *Eudiscoaster asymmetricus* s.s. Hence one may assume that a hiatus exists between this sample and the one immediately below (18-2, 50–52 cm).

The events recorded in the planktonic foraminifera are shown in Fig. 11.4 (see Fig. 11.13 for a definition of the events). Event 1 almost coincides with the increase in the group of small placoliths relative to the *R. haqii/minutula* group. Event 2 is located slightly below the first recording of *Amaurolithus* whereas event 3 nearly coincides with the first occurrence of *A. delicatus*. Events 4, 5 and 6 span a short stratigraphical interval in the top part of the sequence.

(b) Site 334

This site is located at approximately the same latitude as the sections of the Guadalquivir basin (37° 02.13′ N, 34° 24.87′ W). A noteworthy feature is the richness in calcareous nannoflora (always more than 50 specimens per field of view). The specimens are in a good state of preservation (Fig. 11.5). Cores 8 to 2 have been studied.

Below Section 7-3 representatives of *E. berggrenii* occur sporadically. We have located the earliest regular record of this taxon at about this level. The increase in the small placoliths

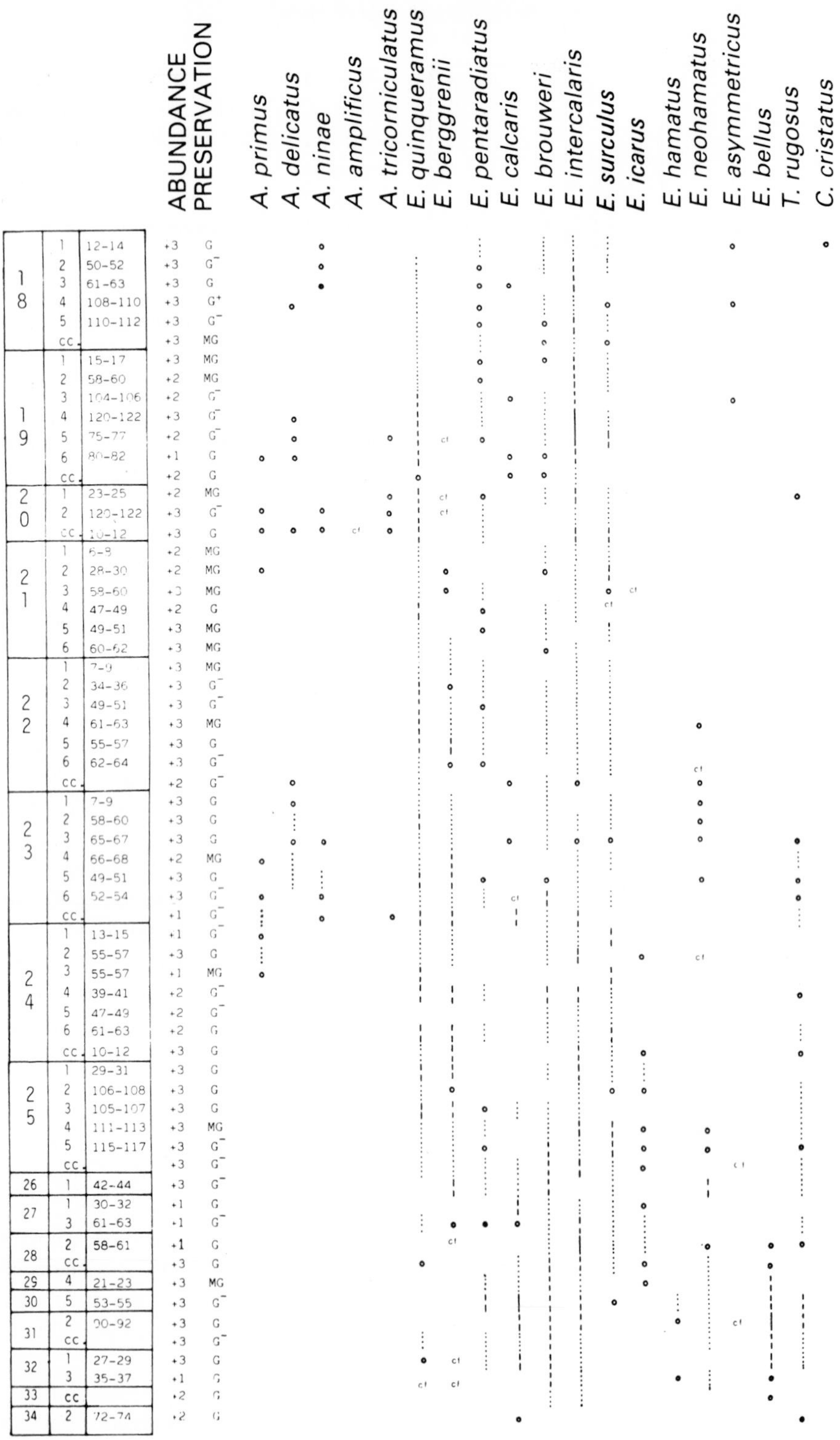

Fig. 11.3 — Total abundance, nannoplankton preservation and relative abundance of selected species in Site 410. Legend in Fig. 11.2.

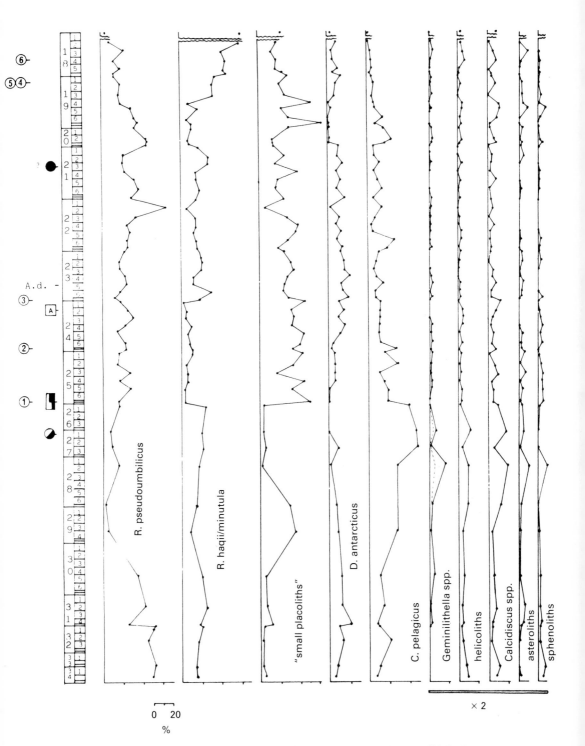

Fig. 11.4 — Proportion of some taxa, morphotypes or groups of taxa in Site 410, For legend of biohorizons see Fig. 11.13.

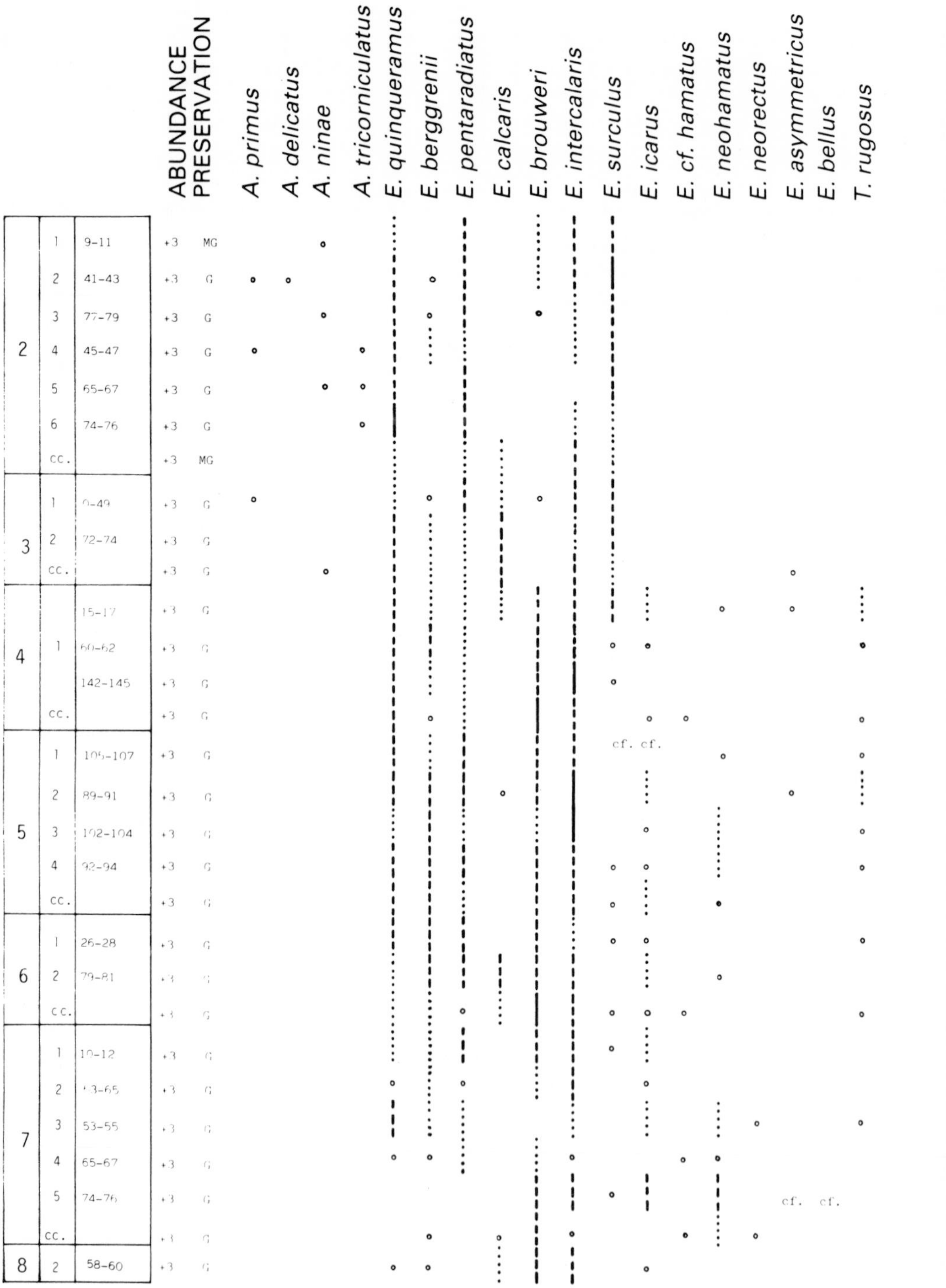

Fig. 11.5 — Total abundance, nannoplankton preservation and relative abundance of selected species in Site 334. Legend in Fig. 11.2.

relative to the *R. haqii/minutula* group occurs around Section 5-1 (Fig. 11.6). The frequency of the small placoliths in Sample 4-CC is very low. Nevertheless, the starting position of the changes in other groups, such as the reduction in *G. jafarii* compared with *G. rotula* (at about 5-4), support our positioning of the small placolith event. The inversion of *C. pelagicus* and *D. antarcticus* is especially clear in the middle part of our core sequence. There is a hiatus between Cores 4 and 3 (Fig. 11.6). This interpretation is based on the abrupt disappearance of siliceous organisms and on the changes in the associations of Globorotaliids. The previous recording of forms such as *A. tricorniculatus* versus *A. primus* and *A. ninae* is in agreement with our succession of events (see for example Bukry 1973). One could assume that the first recording of *Amaurolithus* and the first occurence of *A. delicatus* correspond

to the hiatus. The event marked by the significant reduction in *E. berggrenii* can be located in Section 3-1 because of the existence of a broad interval above in which this taxon is absent.

Event 1 of the planktonic foraminifera can be observed a few centimetres below the level of the drastic increase of small placoliths. Events 2 and 3 characterise the presence of the hiatus referred to above. Events 4, 5 and 6 are situated very close to one another and coincide approximately with an important reduction in the abundance of *E. berggrenii*, and they are somewhat higher than the first recording of *A. delicatus* (Figs. 11.5 and 11.6).

(c) Site 544A

This site is situated to the west of Casablanca (33° 46.0′ N, 09° 24.3′ W). Cores 7 to 5 were studied. Noteworthy is the abundance of the nannoliths,

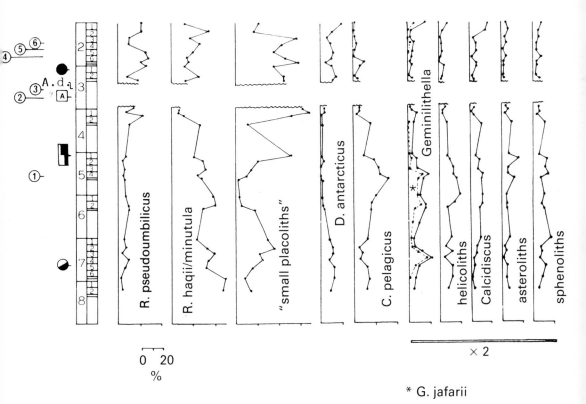

0   20
%

× 2

* G. jafarii

Fig. 11.6 — Proportion of some taxa, morphotypes or groups of taxa in Site 334. For legend of biohorizons see Fig. 11.13.

despite the fact that Cores 6 and 7 exhibit a certain degree of dissolution (Fig. 11.7). Selective dissolution may have influenced the relative abundance of some groups, such as the small placoliths (Fig. 11.8). Therefore the event called 'marked increase of the small placoliths relative to the *R. haqii/minutula* group' is undetectable, if present at all. The distribution of other taxa such as *Geminilithella, D. antarcticus* and *C. pelagicus* suggests that the horizon is located below Core 7. The first *Amaurolithus* recorded is *A. ninae*, suggesting that one is not dealing with the first true record of *Amaurolithus*; this is probably situated below Core 7. In turn *A. delicatus* is observed in the core catcher of Core 6. *Eu-discoaster berggrenii* seems to be recorded with a certain regularity up to the top of Core 6.

Regarding the planktonic foraminifera, event 4 is located slightly above the first occurrence of *A. delicatus*. Events 5 and 6 nearly coincide, and they are situated slightly above the reduction in *E. berggrenii*.

(d) Site 397
This is the southernmost site (26° 50.7′ N, 15° 10.8′ W), situated off Cape Bojador; Cores 60 to 43 were studied. The number of nannoliths recorded in the samples is high (more than 50 specimens per field of view). The nannofossils are in a good state of preservation. Occasionally certain features of dissolution can be observed although never to a pronounced extent (Fig. 11.9).

Eu-discoaster berggrenii is recorded regularly

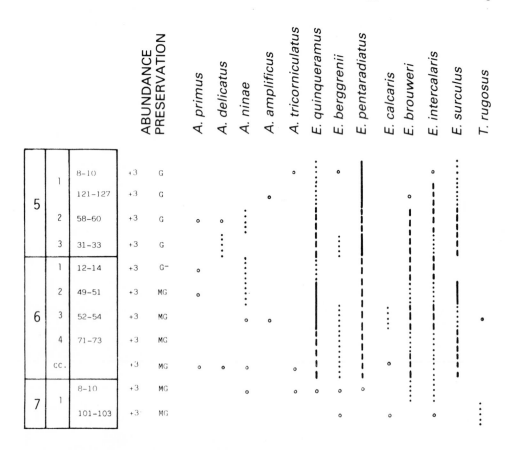

Fig. 11.7 — Total abundance, nannoplankton preservation and relative abundance of selected species in Site 544A. Legend in Fig. 11.2.

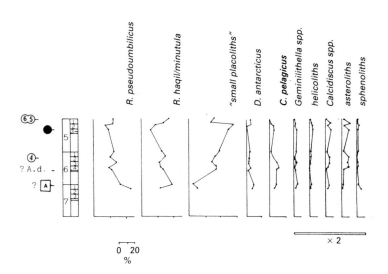

Fig. 11.8 — Proportion of some taxa, morphotypes or groups of taxa in Site 544A. For legend of biohorizons see Fig. 11.13.

from Section 56-2 and upwards, following an interval with relatively common *Eu-discoaster calcaris*, *Eu-discoaster neohamatus* and *Eu-discoaster bellus*, and with occasional *Eu-discoaster neorectus* and *Eu-discoaster hamatus* (Fig. 11.9).

The small placoliths increase in relative proportion to the rest of the nannoflora in the Section 53-1, while *G. jafarii* is reduced in abundance with respect to other coccoliths such as *G. rotula*. *Dictyococcites antarcticus* is present only in very low numbers throughout the cores. Other taxa such as *C. pelagicus*, *Calcidiscus* and the asteroliths, sphenoliths, and helicoliths seem to show a progressive decrease from bottom to top (Fig. 11.10).

The first *Amaurolithus (A. primus)* is seen in the core catcher of Core 51 whereas *A. delicatus* first appears in Section 50-1.

It is rather difficult to indicate the top of the interval with regular *E. berggrenii*; it may occur in Section 44-2.

Event 1 of the planktonic foraminifera is difficult to locate since the disappearance of keeled Globorotaliids of group 1 of '*G. menardii*' coincides with the appearance of forms of group 2. We interpret this to be due to a persistence of group 1 as compared with its occurrence in other areas. The general pattern of our section is that event 2 is close to the horizon defined by the lowest level of predominance of the small placoliths over the *R. haqii/minutula* group. In Site 397 event 2 occurs a few metres above, while event 3 coincides with, or is close to, the earliest record of *Amaurolithus*. Events 4 and 5 are situated above the earliest record of *A. delicatus* and below but close to the highest level of regular *E. berggrenii*. The later event coincides at this site approximately with event 6 of the foraminifera.

(e) The Oued Akrech section
This section is located in the Bou Regreg basin in Morocco in the neighbourhood of Rabat. The abundance of nannoliths varies considerably, samples from the basal half being poorer and also showing a certain degree of dissolution and

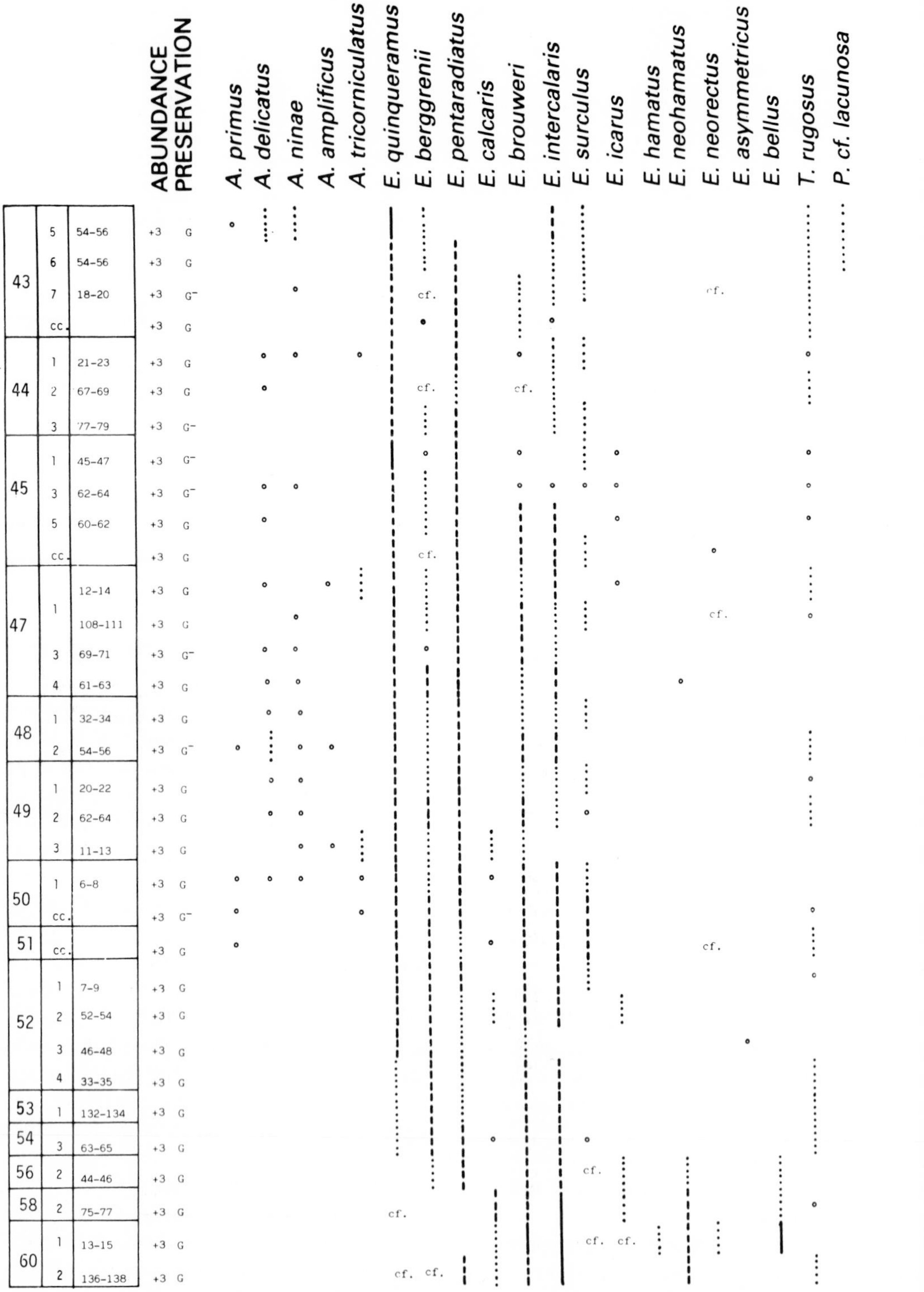

		Interval (cm)	Abundance	Preservation
43	5	54–56	+3	G
	6	54–56	+3	G
	7	18–20	+3	G⁻
	cc		+3	G
44	1	21–23	+3	G
	2	67–69	+3	G
	3	77–79	+3	G⁻
45	1	45–47	+3	G⁻
	3	62–64	+3	G⁻
	5	60–62	+3	G
	cc		+3	G
47	1	12–14	+3	G
		108–111	+3	G
	3	69–71	+3	G⁻
	4	61–63	+3	G
48	1	32–34	+3	G
	2	54–56	+3	G⁻
49	1	20–22	+3	G
	2	62–64	+3	G
	3	11–13	+3	G
50	1	6–8	+3	G
	cc		+3	G⁻
51	cc		+3	G
52	1	7–9	+3	G
	2	52–54	+3	G
	3	46–48	+3	G
	4	33–35	+3	G
53	1	132–134	+3	G
54	3	63–65	+3	G
56	2	44–46	+3	G
58	2	75–77	+3	G
60	1	13–15	+3	G
	2	136–138	+3	G

Species columns (relative abundance shown as dots / cf.): *A. primus*, *A. delicatus*, *A. ninae*, *A. amplificus*, *A. tricorniculatus*, *E. quinqueramus*, *E. berggrenii*, *E. pentaradiatus*, *E. calcaris*, *E. brouweri*, *E. intercalaris*, *E. surculus*, *E. icarus*, *E. hamatus*, *E. neohamatus*, *E. neorectus*, *E. asymmetricus*, *E. bellus*, *T. rugosus*, *P.* cf. *lacunosa*

Fig. 11.9 — Total abundance, nannoplankton preservation and relative abundance of selected species in Site 397. Legend in Fig. 11.2.

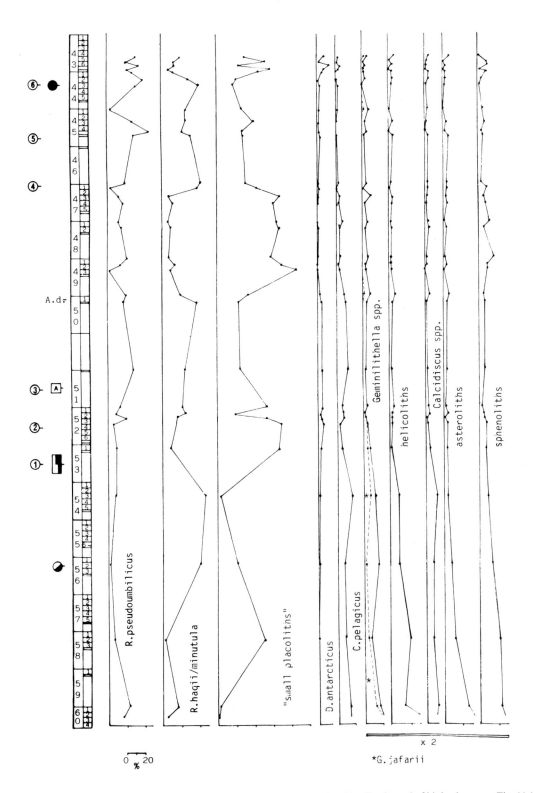

Fig. 11.10 — Proportion of some taxa, morphotypes or groups of taxa in Site 397. For legend of biohorizons see Fig. 11.13.

overgrowth. The material from the upper half is richer and the specimens are better preserved (Fig. 11.11).

From sample OA-5 upwards *E. berggrenii* was found discontinuously. In view of the preservation of lower samples it is uncertain whether this species was present below this level (note the rare occurrence of asteroliths in the underlying samples; Fig. 11.11). From sample OA-9 and upward there is an important increase in the proportion of small placoliths while the numbers of the *R. haqii/minutula* group are reduced. *Dictyococcites antarcticus* exhibits a noteworthy increase, which is contrasted to the trend of *C. pelagicus* (Fig. 11.12). *Amaurolithus primus* is first recorded in OA-16, while forms very similar to *A. delicatus* appear from OA-18 upwards. As can be seen quite clearly in Figs. 11.11 and 11.12, the asteroliths disappear in the upper samples. Therefore, the highest level of regular *E. berggrenii* cannot be detected.

In this section event 1 of the planktonic foraminifera occurs slightly above the earliest regular record of *E. berggrenii* and immediately below the increase in the proportion of small placoliths. The level of the lowermost *A. primus* is situated halfway between events 2 and 3 of the foraminifera, while the forms close to *A. delicatus* were recorded in the same sample in which event 3 is situated.

The succession of events with little differences is the same as in DSDP sites.

11.4 BIOSTRATIGRAPHICAL AND CHRONOSTRATIGRAPHICAL CONSIDERATIONS

The Late Tortonian–Messinian events recorded in the calcareous nannoflora and planktonic foraminifera, both in the DSDP sites and in the sections in the Guadalquivir basin are shown in Fig. 11.13. The definition of these events differs slightly from those described by Sierro (1984), Flores (1985) and Flores and Sierro (1987). We distinguished the event known as 'reduction in the proportion of *E berggrenii*' (not 'reduction of

E. berggrenii and *E. quinqueramus*'; Flores 1985, Flores and Sierro 1987), because *E. quinqueramus* has regularly been observed above.

Regarding the planktonic foraminifera events mentioned in Flores and Sierro (1987), it should be noted that the definition of some events has been changed as a result of recognising a new event 4, which is characterised by the reduction in the *G. miotumida* group. The previous events 4 and 5 now become events 5 and 6 respectively.

The succession of events is very similar in all the sections studied. It is possible to establish a detailed correlation on the basis of the events as shown in Fig. 11.14.

Chronostratigraphically, these sites and sections examined are situated in the Upper Tortonian or Messinian. The boundary between the stages coincides with event 3 of the foraminifera, immediately after or synchronous with the earliest record of *Amaurolithus* (see Sierro 1984, and Flores and Sierro 1987). The first representative of *A. delicatus* appears slightly above. In Site 397, the marked increase in the abundance of the small placoliths relative to the *R. haqii/minutula* group and events 1 and 2 of the foraminifera are probably situated in the upper part of the geomagnetic Chron 7 (Hamilton 1979). The strong reduction in the numbers of *E. berggrenii* coincident with event 6 of the foraminifera could be situated towards the base of the upper normal magnetic interval of Chron 5. According to the ideas of Mazzei *et al.* (1979), Salvatorini and Cita (1979) and Berggren *et al.* (1985), the Miocene–Pliocene boundary could be situated between magnetic Chron 5 and the Gilbert polarity Zone, above the studied interval. Zijderveld *et al.* (1987) situated this boundary in a section in South Italy immediately below the base of the Thuera Subchron of the Gilbert polarity Zone (about 4.83 m.y.).

11.5 PALAEOENVIRONMENTAL CONSIDERATIONS

As postulated in our study of material from the Guadalquivir basin (Flores and Sierro, 1987), the

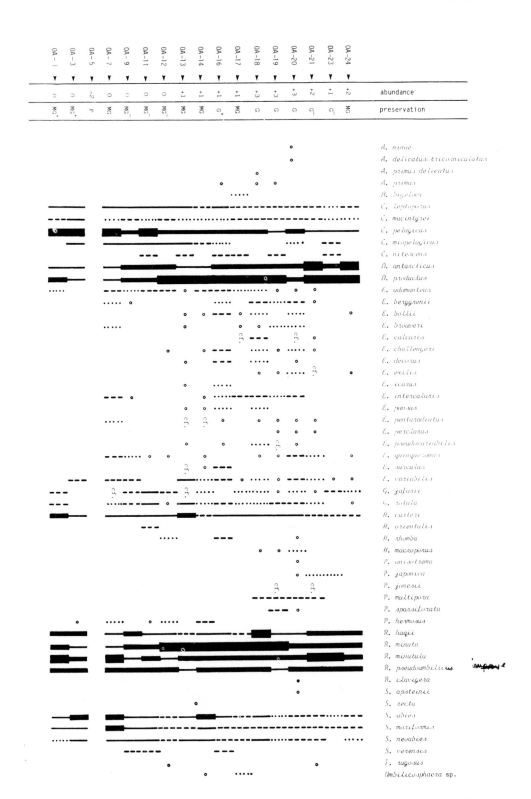

Fig. 11.11 — Total abundance, nannoplankton preservation and relative abundance of species in the Oued Akrech section.
Legend in Fig. 11.2.

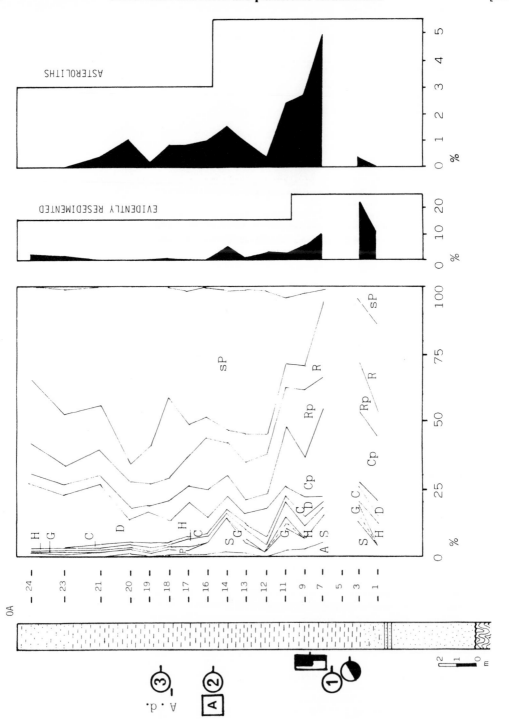

Fig. 11.12 — Proportion of some taxa, morphotypes or groups of taxa in the Oued Akrech section. A, asteroliths; S, sphenoliths; G, *Geminilithella*; H, helicoliths; C, *Calcidiscus*, D, *Dictyococcites antarcticus*; Cp, *Coccolithus pelagicus*; Rp, *Reticulofenestra pseudoumbilicus*; R, *Reticulofenestra haqii/minutula* group; sP, small placoliths.

NANNOPLANKTON

Earliest regular record of *Eu-discoaster berggrenii*

Marked increase of the "small placoliths"
    relative to the *R.haqii/minutula* group

 Earliest record of *Amaurolithus*

A.d.   Earliest record of *Amaurolithus delicatus*

 Reduction in the proportion of *Eu-discoaster berggrenii*

FORAMINIFERA

①   Reduction of group 1 of *"Globorotalia menardii"*
      (sinistral coiling)

②   Abrupt appearance of group 2 of *"Globorotalia menardii"*
      (dextral coiling)

③   Replacement of *"Globorotalia menardii"*group 2
      by *Globorotalia miotumida* group.

④   Change in coiling direction from sinistral to
      dextral of the *Neogloboquadrina acostaensis*
      group.

⑤   Reduction of the frequency of the *Globorotalia miotumida*
      group.

⑥   Earliest frequent record of *Globorotalia margaritae*
      s.s.

Fig. 11.13 — Legend for Calcareous nannoplankton and planktonic foraminifera events.

abundance increase in the small placoliths seems to be related to the presence of cool water masses in the North Atlantic towards the end of the Tortonian and during a large part of the Messinian. Similarly, the decrease in abundance of the same group observed in younger levels could be due to a relative warming of the water. However, as may be concluded from the data presented in Figs. 11.4, 11.6, 11.8 and 11.10, the frequency pattern of the different taxa and/or groups is not identical throughout the DSDP sites. The northernmost sites exhibit a more clear frequency pattern in forms such as *D. antarcticus* and *C. pelagicus* (which show an inverse distribution), while towards the south the patterns are not so pronounced. In contrast, towards low latitudes the progressive reduction in asteroliths, sphenoliths and helicoliths becomes more marked. This corroborates the idea of a general progressive cooling throughout the Late Tortonian–Messinian, on which several minor climatic fluctuations are superimposed.

In the Oued Akrech section, as well as in the Guadalquivir basin, the changes are much more pronounced. Perhaps this is due to the geographical position. Possibly, certain palaeo-oceanographical factors (cold currents) had a greater effect in these areas.

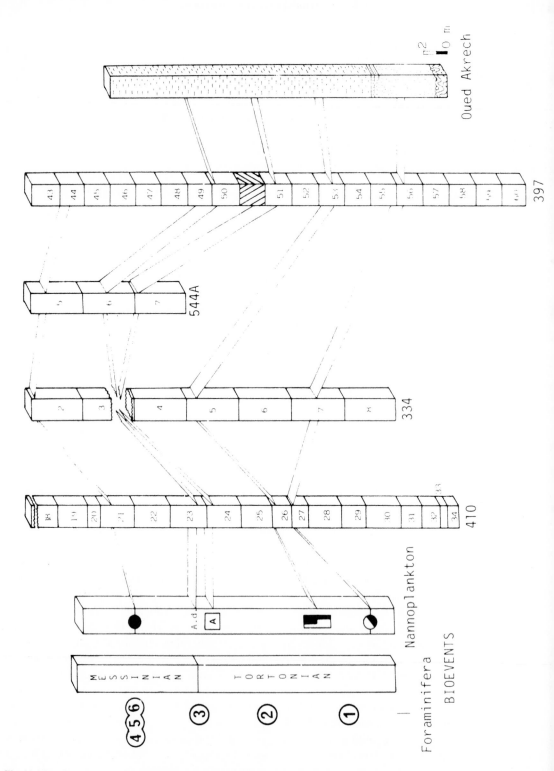

Fig. 11.14 — Correlation of the DSDP sites (410, 334, 544A and 397) and the Oued Akrech section, based on the events defined with calcareous nannoplankton and foraminifera.

11.6 TAXONOMY

The biostratigraphical importance of certain taxa requires a series of considerations concerning the systematic criteria that have been taken into account. The first group to be considered is of the reticulofenestrids. For the nomenclature of the taxa that occur in our samples we have followed Backman (1980) and Flores (1985). The small placoliths include *Reticulofenestra minuta* and *Dictyococcites 'productus'* (coccolith size smaller than 3 μm). Another group is the reticulofenestrid coccoliths, with a size between 3.5 μm and 4.5 μm. This group includes *Reticulofenestra haqii* and *Reticulofenestra minutula*. Our *Reticulofenestra pseudoumbilicus* includes all the forms with a more or less distally opened central area longer than 4.5 μm. Finally, we include in *Dictyococcites antarcticus* the reticulofenestrids with a maximum diameter greater than 4.5 μm and with a distally closed area. Flores (1985)

recorded in the sections from the Guadalquivir basin a morphological continuity between the taxa *R. pseudoumbilicus, R. minutula, R. minuta, R. haqii, D. antarcticus* and *D. productus* (Fig. 11.15). They are morphologically differentiated in the present study, and the species can be used in ecological and biostratigraphical studies.

Another controversial group is *E. quinqueramus*, and its relation to *E. berggrenii*. Intermediate forms are very common. We have included in *E. berggrenii* all the pentaradiate asteroliths with a pentagonal central area, and a ratio between the central area diameter and the radius length equal to, or greater than, one. When this ratio is less than one, they have been included in *E. quinqueramus*. In both taxa a stellate or polygonal knob is present and the length of the radii is constant.

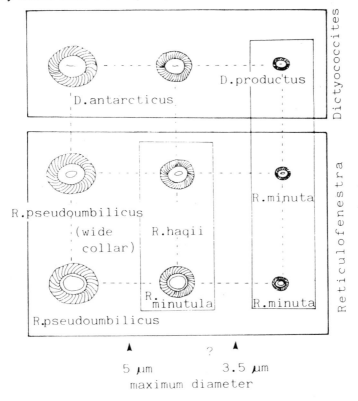

Fig. 11.15 — Variability in size and characteristics of the central area in some reticulofenestrids (modified by Flores 1985).

11.7 ACKNOWLEDGEMENTS

This work was supported by a grant for Project PB85-0315-CO2-00 of the CAICYT (Spain), and included in the International Program for the Study of the Origin of the Mediterranean (IPOM). The material was supplied by the DSDP Organization which is thanked by the authors.

The help of Professor J. Civis in critically reading the manuscript and for his invaluable comments and the moral support of the team at the Area of Palaeontology of the University of Salamanca are also gratefully acknowledged.

11.8 REFERENCES

Aumento, F., *et al.* (Shipboard Scientific Party). 1977. Site 334. *Init. Rep. DSDP*, **47**(1), 239–287.

Backman, J. 1980. Miocene–Pliocene nannofossils and sedimentation rates in the Hatton–Rockall Basin, NE Atlantic Ocean. *Stockholm Contr. Geol.*, **36**, 1–91.

Berggren, W. A., Kent, D. V. and Van Couvering, J. A. 1985. The Neogene: part 2, Neogene geochronology and chronostratigraphy. In N. J. Snelling (Ed.), *The Chronology of the Geological Record. The Geol. Soc. (London)*, *Mem.*, **10**, 211–250.

Bossio, A., El-Bied Rakich, K., Gianelli, L., Mazzei, R., Russo, A. and Salvatorini, G. 1976. Corrélations de quelques sections stratigraphiques du Mio-Pliocène de la zone Atlantique du Maroc avec les stratotypes du Bassin Mediterranéen sur la base des Foraminiferes Planctoniques, Nannoplancton calcaire et Ostracodes. *Att. Soc. Toscana Sci. Nat. Mem. Ser. A*, **83**, 121–126.

Bukry, D. 1973. Coccolith stratigraphy. Eastern Equatorial Pacific, Leg 16 Deep Sea Drilling Project. *Init. Rep. DSDP*, **16**, 653–611.

Bukry, D. 1977. Coccolith and silicoflagellate stratigraphy, Central North Atlantic Ocean, Deep Sea Drilling Project Leg 37. *Init. Rep. DSDP*, **37**, 917–927.

Bukry, D. 1978. Coccolith and silicoflagellate stratigraphy, Northern mid-Atlantic Ridge and Reykjanes Ridge, Deep Sea Drilling Project Leg 49. *Init. Rep. DSDP*, **49**, 551–581.

Čepek, P. and Wind, F. M. 1979. Neogene and Quaternary calcareous nannoplankton from DSDP Site 397 (Northwest African Margin). *Init. Rep. DSDP*, **47**(1), 289–315.

Cita, M. B. and Ryan, W. B. F. 1979. Late Neogene environmental evolution. *Init. Rep. DSDP*, **47**(1), 447–459.

Feinberg, M. and Lorentz, M. G. 1970. Nouvelles données stratigraphiques sur le Miocène supérieur et le Pliocene de Maroc-occidental. *Not. Serv. Géol. Maroc*, **30**(225), 21–26.

Flores, J. A. 1985. *Nanoplancton calcáreo en el Neógeno del borde noroccidental de la Cuenca del Guadalquivir (SO de España)*. Tesis Doctoral Universidad de Salamanca, 1–715 (unedited). Abstract, Universidad de Salamanca, 1–37.

Flores, J. A. 1987a. El nanoplancton calcáreo en la formación 'Arcillas de Gibraleón': síntesis bio-estratigráfica y paleoecológica. In J. Civis (Ed.), *Paleontología del Neógeno de Huelva*, Universidad de Salamanca, Salamanca, 65–68.

Flores, J. A. 1987b. Las asociaciones de nanoplancton calcáreo en algunas series del Mioceno superior-Plioceno inferior en el Oeste de la Cuenca del Guadalquivir (España). In J. Civis (Ed.), *Paleontología del Neógeno de Huelva*, Universidad de Salamanca, Salamanca, 69–88.

Flores, J. A. and Sierro, F. J. 1987. Calcareous plankton in the Tortonian/Messinian Transition Series of the Northwestern Edge of the Guadalquivir Basin. *Abh. Geol. B.-A.*, **39**, 67–84.

Hamilton, N. 1979. A paleomagnetic study of sediments from Site 397 Northwest African Continental margin. *Init. Rep. DSDP*, **47**(1), 463–477.

Hinz, K., *et al.* (Shipboard Scientific Party). 1984. Site 544. *Init. Rep. DSDP*, **79**, 25–80.

Howe, P.R. 1977. Calcareous nannofossils Leg 37, Deep Drilling Project. *Init. Rep. DSDP*, **37**, 909–916.

Luyendyk, B. P. *et al.* (Shipboard Scientific Party). 1978. Site 410. *Init. Rep. DSDP* **49**, 227–313.

Mazzei, R., Raffi, I., Rio, D., Hamilton, N. and Cita, M.B. 1979. Calibration of Late Neogene Calcareous plankton with the paleomagnetic record of Site 397, and correlation with Moroccan and Mediterranean sections. *Init. Rep. DSDP*, **47**(1), 375–389.

Miles, G. A. 1977. Planktonic foraminifera from Leg 37 of the Deep Sea Drilling Project. *Init. Rep. DSDP* **37**, 927–965.

Poore, R. Z. 1978. Oligocene through Quaternary foraminiferal biostratigraphy of the North Atlantic: DSDP Leg 49. *Init. Rep. DSDP*, **49** 447–476.

Rad. V. von, *et al.* (Shipboard Scientific Party). 1979. Site 397. *Init. Rep. DSDP*, **47**(1), 17–217.

Roth, P. H. and Thierstein, H. 1972. Calcareous nannoplankton: Leg 14 of the Deep Sea Drilling Project. *Init. Rep. DSDP*. **14**, 421–485.

Salvatorini, G. and Cita, M. B. 1979. Miocene foraminiferal stratigraphy DSDP Site 397 (Cape Bojador, North Atlantic). *Init. Rep. DSDP*, **47**(1), 317–373.

Sierro, F. J. 1984. *Foraminíferos planctónicos y bioestratigrafía del Mioceno superior-Plioceno del borde occidental de la Cuenca del Guadalquivir (SO de España)*. Tesis Doctoral Universidad de Salamanca, 1–391 (unedited). Abstract, Universidad de Salamanca, 1–34.

Sierro, F. J., Flores, J. A., Civis, J. and González Delgado, J. A. in press. Variations in the assemblages of keeled Globorotaliids of the NE Atlantic and Mediterranean during the Upper Miocene (submitted to *Mar. Micropaleont.*).

Steinmetz, J. C. 1978. Calcareous nannofossils from the North Atlantic Ocean, Leg 49, Deep Sea Drilling Project. *Init. Rep. DSDP*, **49**, 519–531.

Wernli, R. 1971. Les foraminifères planctoniques de la limite Mio-Pliocène des environs de Rabat (Maroc). *Eclogae Geol. Helv.*, **70**, 143–191.

Zijderveld, J. D. A., Zachariasse, J. W., Verhallen, P. J. J. M. and Hilgen, F. J., 1987. The age of the Miocene–Pliocene boundary. *Newsl. Stratigr.*, **16**, 169–181.

12

Palaeocene calcareous nannofossil biostratigraphy

Osman Varol

Palaeocene zonation schemes are proposed for use in the oil industry on a global scale (low to mid latitude) and for the North Sea area in particular. The proposed schemes are particularly suited to the oil industry's requirements since zonal boundaries are defined upon last occurrences (extinctions) of species and their acmes. First occurrences (evolutionary appearances) of selected species and their acmes have been utilised to mark subzonal boundaries applicable when working with sidewall core, core or field sample material.

In the proposed low to mid latitude zonation scheme, 20 zones and 24 subzones are introduced and for practical purposes each of these is prefixed NTp (nannofossil, Tertiary–Palaeocene). For the North Sea area, 10 zones and 17 subzones are proposed. Each zone and subzone is prefixed NNTp (North Sea, nannofossil, Tertiary–Palaeocene) for practical use. The criteria utilised in the latter zonation scheme are only applicable to the North Sea area.

A correlation of the proposed zonation schemes with previously published schemes is presented and briefly discussed.

Two new genera, *Futyania* and *Neobiscutum*, are described as well as a further eight new species: *Biantholithus hughesii*, *Biscutum harrisonii*, *Fasciculithus chowii*, *Futyania attewellii*, *Heliolithus aktasii*, *Lanternithus jawzii*, *Multipartis ponticus* and *Munarinus emrei*. Four new combinations, *Futyania petalosa*, *Neobiscutum parvulum*, *Neobiscutum romeinii*

and *Neocrepidolithus rimosus*, are also introduced.

12.1 INTRODUCTION

Following the biotic crisis at the end of the Cretaceous, calcareous nannofossils exhibited rapid evolutionary growth rates during the Paleocene. According to Perch-Nielsen (1985) 'one should be able to do better, on a worldwide basis, than the nine-fold subdivision suggested by Martini (1971) or the ten-fold one by Okada and Bukry (1980)'.

The application of standard zonation schemes is undoubtedly of great benefit in integrating the work of individual biostratigraphers and in helping geologists and geophysicists to assimilate biostratigraphical information. Deviations from standard zonation schemes are, however, often necessary to accommodate local variations.

Since the introduction of the first standard zonation scheme by Martini in 1971, a great deal of knowledge and experience has been accumulated by calcareous nannofossil workers worldwide. These more recent studies suggest that a number of the original boundaries were poorly defined and that certain of the marker taxa are particularly susceptible to environmental and preservational constraints. Examination of a greater number of sections has revealed that the ranges of certain species differ from those originally proposed and, in addition, the species concept of some marker taxa has changed completely.

For these reasons it has become necessary to revise drastically the standard zonation schemes. Moreover, neither the standard zonation scheme of Martini (1971) nor any of the previously published zonation schemes for the Palaeocene could be readily applied within the oil industry because they are based on first occurrences which are difficult to determine in well sections where only ditch cuttings samples are available.

During extensive examination of Palaeocene material, mainly from well sections, in various parts of the world it became evident that a zonation scheme based primarily on last occurrences would be practical, reliable and easily applicable. Incorporating last occurrences of taxa into the standard zonation scheme of Martini (1971) proved more difficult than establishing an entirely new zonation scheme.

In the present study, an attempt has been made to establish zonation schemes for use in the oil industry in low to mid latitude areas (Fig. 12.1) and in the North Sea area (fully discussed in Section 12.5). One objective of this study is to stimulate interest in achieving finer zonation schemes than those currently in use. The currently proposed zonation schemes are likely to be improved with information gained from future studies.

12.2 PREVIOUS STUDIES

A comparison of Palaeocene zonation schemes is presented in Fig. 12.2.

The first formal calcareous nannofossil zones for the Palaeocene were proposed and named by Bramlette and Sullivan (1961). The zones were not defined but a nannofloral range chart was given. They identified the *Heliolithus riedelii* Zone and the overlying *Discoaster multiradiatus* Zone in the lower part of the Lodo Formation in Lodo Gulch, California. An interval of about 100 ft was barren of calcareous nannofossils and separated the proposed zones.

Hay (1964) introduced three zones in the Palaeocene sediments of the Schlierenflysch of Switzerland.

The most significant contribution to the nannofloral zonations of the Palaeocene was made by Mohler and Hay in Hay *et al.* (1967). Seven zones were introduced spanning the Palaeocene and every nannopalaeontologist including Martini (1971) has adopted them with some modifications.

Edwards (1971) studied the Palaeocene of New Zealand and adopted the zonation scheme of Mohler and Hay but modified it substantially. The *Cruciplacolithus tenuis* Zone was replaced by the *Chiasmolithus danicus* Zone and *Prinsius martinii* Zone using the first occurrence of *Prinsius martinii*. Edwards (1971) did not study the interval below the first occurrence of *Chiasmolithus danicus*.

In the standard zonation scheme of Martini (1971) a nine-fold zonation of the Palaeocene was introduced prefixed NP. He subdivided the *Cruciplacolithus tenuis* Zone of Mohler and Hay (in Hay *et al.* 1967) into the *C. tenuis*, *Chiasmolithus danicus* and *Ellipsolithus macellus* Zones using the first occurrences of *C. danicus* and *Ellipsolithus macellus*. All the remaining zones were defined exactly as in Mohler and Hay (in Hay *et al.* 1967).

In 1979, Romein identified eight zones in the Palaeocene of Spain and Israel. He subdivided the zonal interval NP1 to NP3 of Martini (1971) differently using the first occurrences of *Cruciplacolithus primus*, *Prinsius dimorphosus* and *Cruciplacolithus tenuis*.

Okada and Bukry (1980) introduced a ten-fold zonation for the Palaeocene, prefixed CP, as part of their Cenozoic zonation scheme. The zonation was similar to those of Mohler and Hay (in Hay *et al.* 1967) and Martini (1971) although they subdivided Zone NP9 using the evolutionary appearance of *Campylosphaera eodela*.

Perch-Nielsen (1981b, 1981c), working on material from Tunisia, subdivided the zonal interval NP1 to NP2 in a different manner to any of the previously published schemes. She obtained a five-fold division using the first occurrences of *Neobiscutum romeinii*, *Neobiscutum parvulum*, *C. primus* and *Futyania petalosa*. She re-defined the NP2–NP3 boundary

AGE		ZONES/SUBZONES			
PALAEOCENE	LATE	NTp20	*Fasciculithus involutus*		
PALAEOCENE	LATE	NTp19	*Fasciculithus hayi*		
PALAEOCENE	LATE	NT 18	*Fasciculithus lillianiae*		
PALAEOCENE	LATE	NTp17	*Placozygus sigmoides*		
PALAEOCENE	LATE	NTp16	*Heliolithus cantabriae*	NTp16B	*Discoaster lenticularis*
PALAEOCENE	LATE			NTp16A	*Heliolithus aktasii*
PALAEOCENE	LATE	NTp15	*Hornibrookina australis*		
PALAEOCENE	LATE	NTp14	*Munarinus emrei*		
PALAEOCENE	LATE	NTp13	*Zygodiscus clausus*		
PALAEOCENE	LATE	NTp12	*Discoaster drieveri*		
PALAEOCENE	LATE	NTp11	*Fasciculithus billii*	NTp11B	*Sphenolithus anarrhopus*
PALAEOCENE	LATE			NTp11A	*Zygodiscus herlynii*
PALAEOCENE	LATE	NTp10	*Fasciculithus pileatus*	NTp10C	*Chiasmolithus bidens*
PALAEOCENE	LATE			NTp10B	*Cruciplacolithus latipons*
PALAEOCENE	LATE			NTp10A	*Multipartis ponticus*
PALAEOCENE	LATE	NTp9	*Neochiastozygus saepes*		
PALAEOCENE	EARLY	NTp8	*Cruciplacolithus subrotundus*	NTp8C	*Fasciculithus janii*
PALAEOCENE	EARLY			NTp8B	*Fasciculithus ulii*
PALAEOCENE	EARLY			NTp8A	*Sphenolithus primus*
PALAEOCENE	EARLY	NTp7	*Fasciculithus chowii*	NTp7B	*Chiasmolithus edentulus*
PALAEOCENE	EARLY			NTp7A	*Ellipsolithus distichus*
PALAEOCENE	EARLY	NTp6	*Neochiastozygus eosaepes*		
PALAEOCENE	EARLY	NTp5	*Neocrepidolithus cruciatus*	NTp5C	*Neochiastozygus imbriei*
PALAEOCENE	EARLY			NTp5B	*Prinsius martinii*
PALAEOCENE	EARLY			NTp5A	*Prinsius tenuiculus*
PALAEOCENE	EARLY	NTp4	*Hornibrookina edwardsii*		
PALAEOCENE	EARLY	NTp3	*Futyania attewellii*	NTp3C	*Coccolithus subpertusus*
PALAEOCENE	EARLY			NTp3B	*Chiasmolithus danicus*
PALAEOCENE	EARLY			NTp3A	*Prinsius dimorphosus*
PALAEOCENE	EARLY	NTp2	*Neocrepidolithus fossus*	NTp2B	*Coccolithus pelagicus*
PALAEOCENE	EARLY			NTp2A	*Cruciplacolithus intermedius*
PALAEOCENE	EARLY	NTp1	*Neobiscutum parvulum*	NTp1D	*Biscutum harrisonii*
PALAEOCENE	EARLY			NTp1C	*Micrantholithus fornicatus*
PALAEOCENE	EARLY			NTp1B	*Cruciplacolithus primus*
PALAEOCENE	EARLY			NTp1A	*Cyclagelosphaera alta*

Fig. 12.1 — Proposed low to mid latitude Palaeocene nannofossil zonation.

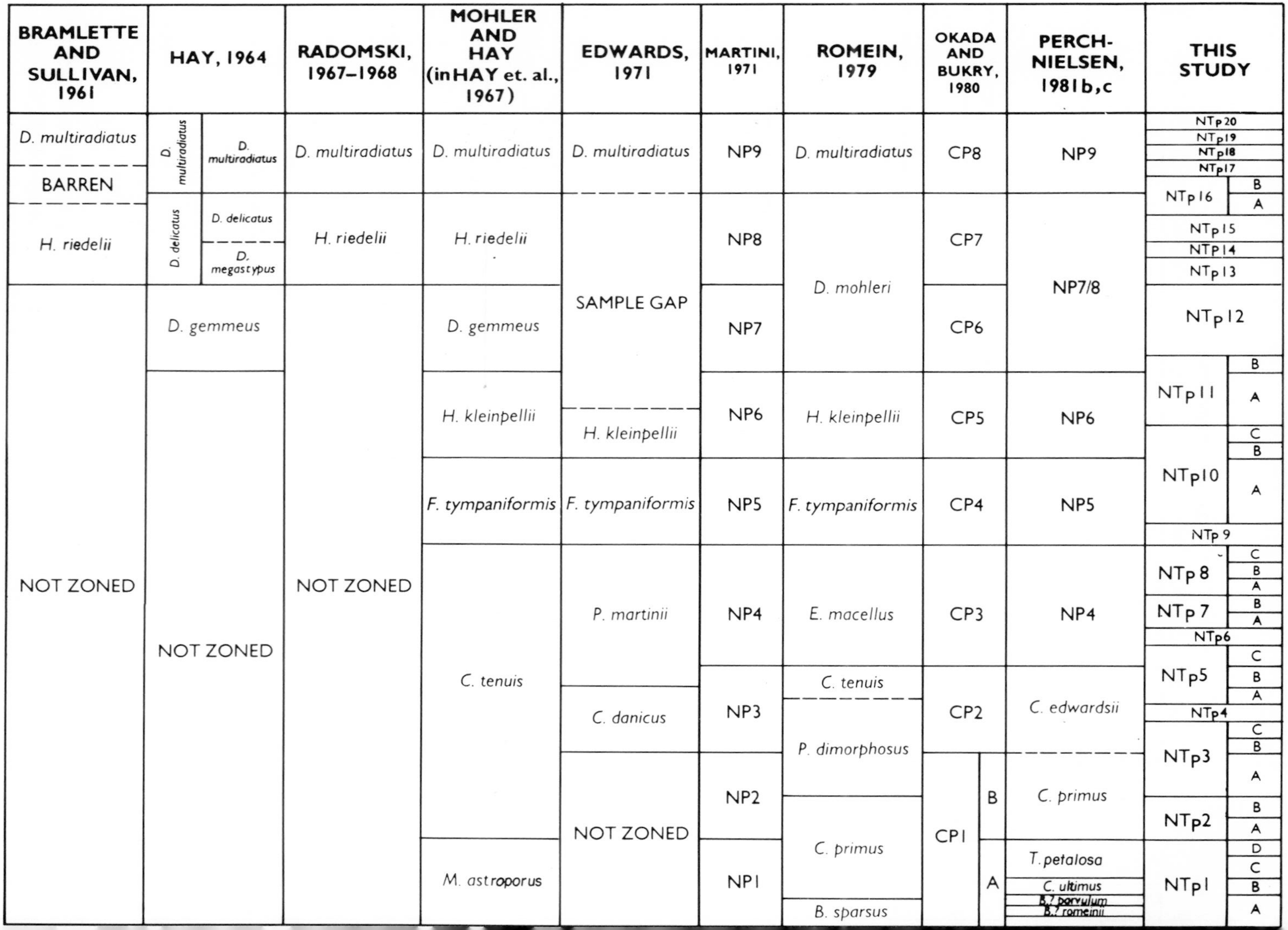

A comparison table of Palaeocene nannofossil zonation schemes, aligning the following columns: **THIS STUDY**, **PERCH-NIELSEN, 1981 b,c**, **OKADA AND BUKRY, 1980**, **ROMEIN, 1979**, **MARTINI, 1971**, **EDWARDS, 1971**, **MOHLER AND HAY (in HAY et al., 1967)**, **RADOMSKI, 1967–1968**, **HAY, 1964**, and **BRAMLETTE AND SULLIVAN, 1961**.

THIS STUDY	PERCH-NIELSEN, 1981 b,c	OKADA AND BUKRY, 1980	ROMEIN, 1979	MARTINI, 1971	EDWARDS, 1971	MOHLER AND HAY (in HAY et al., 1967)	RADOMSKI, 1967–1968	HAY, 1964	BRAMLETTE AND SULLIVAN, 1961
NTp20, NTp19, NTp18, NTp17 (B, A)	NP9	CP8	D. multiradiatus	NP9	D. multiradiatus	D. multiradiatus	D. multiradiatus	D. multiradiatus / D. delicatus (D. delicatus, D. megastypus)	D. multiradiatus
NTp16, NTp15, NTp14, NTp13		CP7	D. mohleri	NP8	SAMPLE GAP	H. riedelii	H. riedelii		BARREN / H. riedelii
NTp12 (B, A)	NP7/8	CP6		NP7		D. gemmeus		D. gemmeus	
NTp11 (C, B, A)	NP6	CP5	H. kleinpellii	NP6	H. kleinpellii	H. kleinpellii			NOT ZONED
NTp10 (C, B, A)	NP5	CP4	F. tympaniformis	NP5	F. tympaniformis	F. tympaniformis	NOT ZONED		
NTp9	NP4	CP3	E. macellus	NP4	P. martinii	C. tenuis		NOT ZONED	
NTp8 (C, B, A), NTp7 (B, A), NTp6			C. tenuis	NP3	C. danicus				
NTp5 (C, B, A), NTp4	C. edwardsii	CP2	P. dimorphosus	NP2	NOT ZONED				
NTp3 (C, B, A)	C. primus	CP1 (B)	C. primus	NP1		M. astroporus			
NTp2 (B, A)	T. petalosa	CP1 (A)	B. sparsus						
NTp1 (D, C, B, A)	C. ultimus, B.? parvulum, B.? romeinii								

Fig. 12.2 — A comparison of Palaeocene nannofossil zonation schemes.

using *Cruciplacolithus edwardsii* instead of *Chiasmolithus danicus*.

As part of her contribution to Plankton Stratigraphy (Bolli *et al.* 1985) Perch-Nielsen integrated and summarised all the major studies of the Cenozoic in which calcareous nannofossils had been used. Species ranges were presented and individual species illustrated with photomicrographs. Her work represents the most up to date and complete 'catalogue' of the Cenozoic nannofossils.

12.3 STUDIED MATERIAL

This study is mainly based on the examination of ditch-cutting samples obtained from well sections. Field samples, sidewall cores and core samples which have been examined by the author during the last few years were also re-examined for this study and these results are also incorporated in the zonation schemes.

The material examined was obtained from the Solomon Islands, New Zealand, Australia, Papua New Guinea, Indonesia (mainly Irian Jaya), the Philippines, Malaysia, Brunei, Burma, Sri Lanka, India, East Africa (including Sudan), Middle Eastern countries, Libya, Turkey, West Africa, Northwest Europe and the Caribbean area.

The Kokaksu Section of Turkey has been intensively studied by the author and compared with the results obtained from the new well section and field samples here considered.

Kokaksu Section: The results of calcareous nannofossil analysis of the Kokaksu Section, Zonguldak, northern Turkey, are presented in Fig. 12.3. The preliminary results of the nannofloral study of this section were published by Varol (1983). Since then this section has been intensively investigated by the author because, in some respects, the lower part of this section displays characteristics common to both low and high latitude assemblages. The currently proposed low to mid latitude zonation scheme is applied to the Kokaksu Section and Zones NTp1 to NTp16 or Zones NP1 to NP8 of Martini (1971) were identified. The lowermost part of

Zone NTp1 was not represented in the samples. However, since a sample gap of about 2 m occurs between Subzone NTp1C and the top of the Cretaceous, this section may contain one of the most complete sequences known across the Cretaceous–Tertiary boundary.

12.4 PROPOSED LOW TO MID LATITUDE ZONATION SCHEME

The zonal markers used for the definition of the proposed zones have well-defined stratigraphic tops, are easily recognisable under the light microscope, are widely distributed and are present in relatively frequent numbers. Such criteria are essential in order to establish a practical and applicable zonation scheme.

20 zones based upon last occurrences (LO) or acmes of species and 24 subzones based upon first occurrences (FO) or acmes of species are introduced (Fig. 12.4). The currently proposed low to mid latitude zonation scheme is not applicable to high latitude areas.

The ranges of the stratigraphically important species and generalised abundance of selected species are shown in Fig. 12.5. The names of authors are presented in square brackets in the definition of zonal boundaries.

The following relative abundances are used in the present study:

Rare	less than 1 individual in each 10 fields of view
Common	more than 1 individual in each 10 fields of view
Abundant	more than 1 individual in each 3 fields of view
Major influx	individuals dominate the assemblage (usually constitute more than 50% of total assemblage).

Neobiscutum parvulum Zone (NTp1)

Definition: interval from the FO of *Biantholithus sparsus* [Romein 1979] and/or *Cyclagelosphaera alta* [this study] to the LO of *N. parvulum* and/or *N. romeinii* [this study] and/or the first

This is a biostratigraphic range chart. The left-hand columns give the stratigraphic framework and the right-hand columns show the ranges of individual nannofossil species.

AGE	MARTINI, 1971	THIS STUDY		SAMPLE NO.	LITHOLOGY
LATE PALAEOCENE	NP8	NTp16	A	1	
		NTp15		2	
		NTp14		4	
				5	
				6	
		NTp13		7	
				8	
	NP7	NTp12		9	
				10	
	NP6	NTp11	B	11	
			A	12	
		NTp10	C	13	
				14	
	NP5		B	15	
				16	
				17	
			A	18	
				19	
				20	
		NTP9		21	
				22	
EARLY PALAEOCENE	NP4	NTp8	C	23	
			B	24/25	
			A	26/27	
				28/29	
				30/31	
				32/33	
				34	
				35	
				36	
				37/38	
				39	
				40/41	
				42/43	
				44	
				45	
				46	
				47	
		NTp7	A	48/49	
				50	
				51/52	
				53/54	
				55/56	
				57/59	
		NTp6		60	
				61	
				62	
				63	
	NP3	NTp5	C	64	
			B	65	
				66	
				67	
			A	68	
				69	
				70	
		NTp4			
	NP2	NTp3	C	71	
			B	72	
			A		
		NTp2	B	73	
			A		
	NP1	NTp1	B	74	
			A	75	
		NOT SAMPLED			
MAASTRICHTIAN					

Species columns (left to right): *B. harrisonii*, *C. bidens*, *C. consuetus*, *C. danicus*, *C. edentulus*, *C. frequens*, *C. intermedius*, *C. latipons*, *C. tenuis*, *C. cavus*, *C. pelagicus*, *C. subpertusus*, *D. mohleri*, *E. macellus*, *E. distichus*, *F. involutus*, *F. tympaniformis*, *N. distentus*, *N. modestus*, *N. perfectus*, *Neochiastozygus* spp., *N. rimosus*, *P. sigmoides*, *P. martinii*, *S. anarrhopus*, *S. apertus*, *S. primus*, *T. pertusus*, *Z. herlynii*, *B. bigelowii*, *C. edwardsii*, *H. aktasii*, *H. kleinpellii*

Scale bar: 10 m

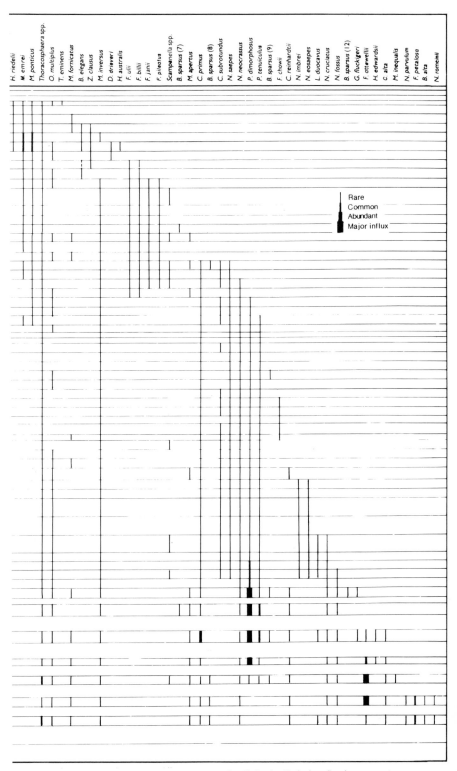

Fig. 12.3 — Distribution of calcareous nannofossils in the Kokaksu Section.

occurrence of *Cruciplacolithus intermedius* [Mohler and Hay in Hay *et al.* 1967].

Remarks: in ditch cuttings samples the last major influx of Cretaceous forms can be taken as the lower limit of this zone. The upper boundary is usually defined on the extinction of *N. parvulum*. However, the last occurrence of *N. romeinii* can also be used to mark this boundary although this latter species is extremely rare towards the upper limit of its range. The decline in the common occurrence of *Futyania petalosa* also corresponds to this boundary. Extremely rare occurrences of the latter species and some forms intermediate between *Futyania attewellii* and *F. petalosa* are recorded above the upper limit of Zone NTp1. The first occurrence of *C. intermedius* approximates to this boundary.

Zone NTp1 is subdivided into four subzones using the following criteria in ascending order: the first occurrence of *Cruciplacolithus primus*, the first occurrence of *F. petalosa* and the first major influx of *Futyania* spp.

Cyclagelosphaera alta Subzone (NTp1A)

Definition: interval from the FO of *B. sparsus* [Romein 1979] and/or *C. alta* [this study] to the FO of *C. primus* [Romein 1979].

Remarks: this subzonal interval was subdivided into three subzones by Perch-Nielsen (1981c) using the first occurrence of *N. romeinii* and *N. parvulum*. This subzone is extremely short (Perch-Nielsen 1981c) and has only rarely been sampled and examined.

Throughout this zone *Thoracosphaera* spp. and *Braarudosphaera bigelowii* are the most abundant forms represented, while in the upper part of the subzone *N. parvulum* is also recorded abundantly.

Cruciplacolithus primus Subzone (NTp1B)

Definition: interval from the FO of *C. primus* [Romein 1979] to the FO of *F. petalosa* [Perch-Nielsen 1981c].

Remarks: within this subzone the nannofloral assemblages are usually dominated by *N.*

parvulum, *Thoracosphaera* spp. and *B. bigelowii*, while *N. romeinii* becomes very rare.

Micrantholithus fornicatus Subzone (NTp1C)

Definition: interval from the FO of *F. petalosa* [Perch-Nielsen 1979] to the first major influx of *Futyania* spp. [this study].

Remarks: the nannofloral assemblages of this subzone are similar to those of the underlying Subzone NTp1B but are distinguished by the presence of *F. petalosa* and by a reduction in the number of *N. parvulum*, *M. fornicatus* persistently occurs in this subzone.

Biscutum harrisonii Subzone (NTp1D)

Definition: interval from the first major influx of *Futyania* spp. [this study] to the LO of *N. parvulum* and/or *N. romeinii* [this study] and/or FO of *C. intermedius* [Mohler and Hay in Hay *et al.* 1967].

Remarks: in this subzone *C. primus* becomes locally common to abundant. *Futyania* spp. dominates the nannofloral assemblages. *Futyania attewellii* is numerically superior to *F. petalosa*, the last common occurrence of which approximately corresponds to the top of this subzone. In some sections a slight overlap in the ranges of *C. intermedius* and *N. parvulum* was observed. For practical purposes, however, this can be ignored and both species are taken to mark the upper boundary of this subzone. *B. harrisonii* is usually common in this subzone.

Neocrepidolithus fossus Zone (NTp2)

Definition: interval from the LO of *N. parvulum* and/or *N. romeinii* [this study] and/or first occurrence of *C. intermedius* [Mohler and Hay in Hay *et al.* 1967] to the last major influx of *F. attewellii* [this study] and/or the FO of *Prinsius dimorphosus* [Perch-Nielsen 1979a].

Remarks: *Futyania* spp. dominate the nannofloral assemblages. Both *Thoracosphaera* spp. and *B. bigelowii* decrease gradually in abundance towards the top of this zone. *N. fossus* is frequently observed in this subzone.

Cruciplacolithus intermedius Subzone (NTp2A)

Definition: interval from the LO of *N. parvulum* and/or *N. romeinii* [this study] and/or the FO of *C. intermedius* [Mohler and Hay in Hay *et al.* 1967] to the FO of *Coccolithus pelagicus* [this study].
Remarks: *Thoracosphaera* spp. and *B. bigelowii* are still common to abundant in this subzone.

Coccolithus pelagicus Subzone (NTp2B)

Definition: interval from the first occurrence of *C. pelagicus* [this study] to the last major influx of *F. attewellii* [this study] and/or the FO of *Prinsius dimorphosus* [Perch-Nielsen 1979a].

Futyania attewellii Zone (NTp3)

Definition: interval from the last major influx of *F. attewellii* [this study] and/or the FO of *Prinsius dimorphosus* [Perch-Nielsen 1979a] to the LO of *F. attewellii* and/or *C. alta* [this study].
Remarks: shortly after making its first appearance *P. dimorphosus* dominates the calcareous nannofossil assemblages. The first major influx of this species can provide a useful datum for correlation when core, sidewall core or field samples are available. Forms similar to *Prinsius tenuiculus* are present in the lower part of this zone.

Prinsius dimorphosus Subzone (NTp3A)

Definition: interval from the last major influx of *F. attewellii* [this study] and/or FO of *P. dimorphosus* [Perch-Nielsen 1979a] to the FO of *Chiasmolithus danicus* [Martini 1970] and/or *Hornibrookina edwardsii* [this study].
Remarks: circular forms of *P. dimorphosus* greatly outnumber the elliptical and subelliptical forms.

Chiasmolithus danicus Subzone (NTp3B)

Definition: interval from the FO of *C. danicus* [Martini 1970] and/or *H. edwardsii* [this study] to the FO of *Coccolithus subpertusus* [this study].
Remarks: owing to presence of forms intermediate between *C. danicus* and *Cruciplacolithus edwardsii*, particularly in areas of low and mid latitude, consistent identification of *C. danicus* is not always possible. It is also possible that the species concept of *C. danicus* differs from specialist to specialist. The first occurrence of *H. edwardsii*, approximately corresponding to the first occurrence of *C. danicus*, can be recognised globally.

Coccolithus subpertusus Subzone (NTp3C)

Definition: interval from the FO of *C. subpertusus* [this study] to the LO of *F. attewellii* and/or *C. alta* [this study].

Hornibrookina edwardsii Zone (NTp4)

Definition: interval from the LO of *F. attewellii* and/or *C. alta* [this study] to the LO of *H. edwardsii* [this study].
Remarks: *P. dimorphosus* and *P. tenuiculus* are the most abundant species in this zone. *H. edwardsii* becomes locally common.

Neocrepidolithus cruciatus Zone (NTp5)

Definition: interval from the LO of *H. edwardsii* [this study] to the LO of *Neocrepidolithus cruciatus* [this study].
Remarks: *P. dimorphosus* dominates the nannofloral assemblages in this zone. Using the evolutionary appearance of *Prinsius martinii* and *Ellipsolithus macellus* this zone can be subdivided into three subzones. The last persistent occurrences of *Lanternithus duocavus* and *Lanternithus jawzii* were observed in this zone.

Prinsius tenuiculus Subzone (NTp5A)

Definition: interval from the LO of *H. edwardsii*

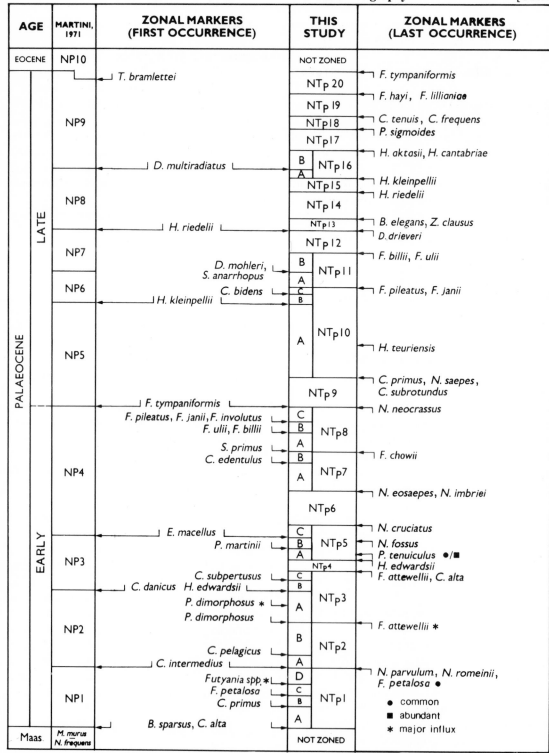

Fig. 12.4 — A correlation of standard and proposed zonation schemes.

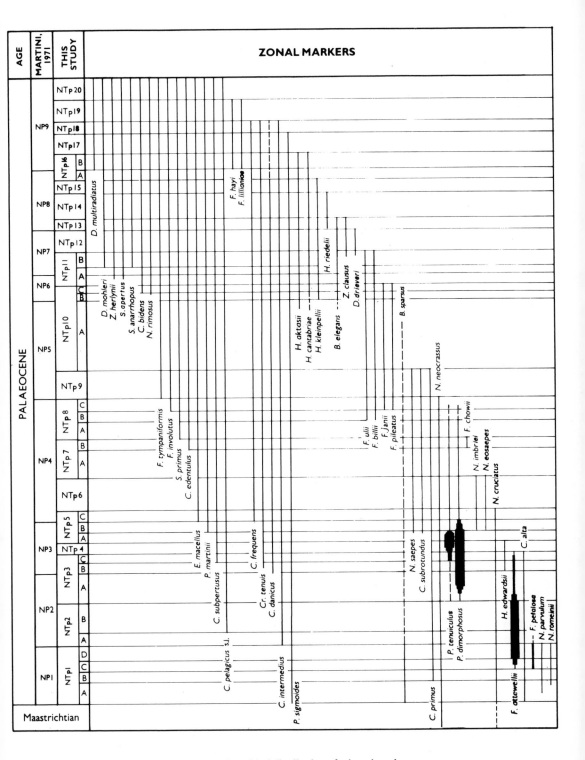

Fig. 12.5 — Stratigraphical distribution of selected markers.

[this study] to the FO of *P. martinii* [Edwards 1971].

Remarks: the last common to abundant occurrence of *P. tenuiculus* was observed in the lower part of this subzone. The last major influx of *P. dimorphosus* (circular forms) corresponds approximately to the top of this subzone. The elliptical to subelliptical forms of this species, similar to those found in the North Sea area, were observed very rarely in samples from low to mid latitudes.

Prinsius martinii Subzone (NTp5B)

Definition: interval from the FO of *P. martinii* [Edwards 1971] to the FO of *Ellipsolithus macellus* [Martini 1970].

Remarks: the first occurrences of *Neochiastozygus imbriei*, *Neochiastozygus eosaepes* and *Neochiastozygus saepes* are in close proximity to the base of this subzone. The last common to abundant occurrence of *P. dimorphosus* was observed in the lowermost part of this subzone where *P. martinii* is rare. In the upper part of the subzone *P. martinii* becomes common to abundant and dominates the assemblages. The last occurrence of *Neocrepidolithus fossus* was also observed in this subzone. Specimens very similar to *Neochiastozygus perfectus* were also recorded in this subzone in low and mid latitude.

Neochiastozygus imbriei Subzone (NTp5C)

Definition: interval from the FO of *E. macellus* [Martini, 1970] to the LO of *N. cruciatus* [this study].

Remarks: the occurrence of *E. macellus* is extremely rare in its lowermost range and consistent identification of this datum is always difficult. *N. imbriei* is frequently found in this subzone.

Neocrepidolithus eosaepes Zone (NTp6)

Definition: interval from the LO of *N. cruciatus* [this study] to the LO of *Neochiastozygus*

eosaepes and/or *Neochiastozygus imbriei* [this study].

Remarks: the first persistent occurrence of *Cruciplacolithus latipons* was observed in this zone. Extremely rare isolated specimens of this species were, however, observed in the underlying zone.

Fasciculithus chowii Zone (NTp7)

Definition: interval from the LO of *N. eosaepes* and/or *N. imbriei* [this study] to the LO of *Fasciculithus chowii* and/or the FO of *Sphenolithus primus* [this study].

Remarks: the last occurrence of *Fasciculithus magnicordis* and/or *Fasciculithus magnus* can be taken to mark the upper boundary of this zone in the absence of the nominate taxon since their last occurrence is very close to the upper limit of this zone.

Ellipsolithus distichus Subzone (NTp7A)

Definition: interval from the LO of *N. eosaepes* and/or *N. imbriei* [this study] to the FO of *Chiasmolithus edentulus* [Perch-Nielsen 1979a].

Remarks: *E. distichus* is consistently observed in this subzone.

Chiasmolithus edentulus Subzone (NTp7B)

Definition: interval from the FO of *C. edentulus* [Perch-Nielsen 1979a] to the LO of *F. chowii* and/or FO of *S. primus* [this study].

Cruciplacolithus subrotundus Zone (NTp8)

Definition: interval from the LO of *F. chowii* and/or FO of *S. primus* [this study] to the LO of *Neocrepidolithus neocrassus* [this study] and/or the FO of *Fasciculithus tympaniformis* [Mohler and Hay in Hay *et al.* 1967].

Remarks: in some of the studied sections a slight overlap in the ranges of *F. tympaniformis* and *N. neocrassus* was observed but it was so small as to be difficult to detect. For practical reasons both species can be taken to approximate the upper limit of this zone depending on the type of samples available (ditch cuttings as opposed to core, sidewall core or field sample material). *C. subrotundus* is frequently observed in this zone.

Sphenolithus primus Subzone (NTp8A)

Definition: interval from the LO of *F. chowii* and/ or the FO of *S. primus* [this study] to the FO of *Fasciculithus ulii* and/or *Fasciculithus billii* [this study].

Fasciculithus ulii Subzone (NTp8B)

Definition: interval from the FO of *F. ulii* and/or *F. billii* [this study] to the FO of *Fasciculithus janii* and/or *Fasciculithus pileatus* [this study].

Fasciculithus janii Subzone (NTp8C)

Definition: interval from the FO of *F. janii* and/or *F. pileatus* [this study] to the LO of *N. neocrassus* [this study] and/or the FO of *F. tympaniformis* [Mohler and Hay in Hay *et al*. 1967].

Neochiastozygus saepes Zone (NTp9)

Definition: interval from the LO of *N. neocrassus* [this study] and/or FO of *F. tympaniformis* [Mohler and Hay in Hay *et al*. 1969] to the last occurrence of *C. primus* and/or *N. saepes* [this study].

Remarks: the last occurrence of *N. saepes*, *C. primus* and *C. subrotundus* are very close to each other and any of them can be taken to mark the upper limit of this zone. *C. subrotundus* is, however, extremely rare towards the upper limit of its range.

Fasciculithus pileatus Zone (NTp10)

Definition: interval from the LO of *C. primus* and/or *N. saepes* [this study] to the LO of *F. pileatus* and/or *F. janii* [this study].

Remarks: the last occurrence of *Hornibrookina teuriensis* falls within this zone but this species does not have a sufficiently wide geographical distribution to be used as a zonal marker. Locally, however, it could have very useful correlatable value. This zone is subdivided into three subzones using the first occurrences of *Heliolithus kleinpellii* and *Chiasmolithus bidens*.

Multipartis ponticus Subzone (NTp10A)

Definition: interval from the LO of *C. primus* and/or *N. saepes* [this study] to the FO of *H. kleinpellii* [Mohler and Hay in Hay *et al*. 1967].

Remarks: at present this subzonal interval cannot be subdivided satisfactorily owing to the absence of reliable last occurrences of species. However, as mentioned above, the last occurrence of *H. teuriensis* can locally be used as a correlation point. *M. ponticus* is usually common in this subzone.

Cruciplacolithus latipons Subzone (NTp10B)

Definition: interval from the FO of *H. kleinpellii* [Mohler and Hay in Hay *et al*. 1967] to the FO of *C. bidens* [this study].

Remarks: extensive examination of outcrop samples in the present study confirmed that the ranges of *H. kleinpellii* and *F. pileatus* and/or *F. janii* overlap. The fact that this overlap had not been recorded in previous publications may possibly be due to the rare occurrence of *H. kleinpellii* towards the lower limit of its range. *C. latipons* persistently occurs in this subzone.

Chiasmolithus bidens Subzone (NTp10C)

Definition: interval from the FO of *C. bidens* [this study] to the LO of *F. pileatus* and/or *F. janii* [this study].

Remarks: the evolutionary appearance of *Neocrepidolithus rimosus* occurs in close proximity to the lower boundary of this subzone.

Fasciculithus billii Zone (NTp11)

Definition: interval from the LO of *F. pileatus* and/or *F. janii* [this study] to the LO of *F. billii* and/or *F. ulii* [this study].

Remarks: this zone can be subdivided into two subzones using the first occurrence of *Discoaster mohleri* and/or *Sphenolithus anarrhopus*. In the high latitude areas *F. billii* and *F. ulii* have their last occurrences between the first occurrence of *Discoaster multiradiatus* and the last occurrence of *F. tympaniformis*.

Zygodiscus herlynii Subzone (NTp11A)

Definition: interval from the LO of *F. pileatus* and/or *F. janii* [this study] to the FO of *Discoaster mohleri* [Hay 1964] and/or *Sphenolithus anarrhopus* [this study].
Remarks: the first persistent occurrence of *Z. herlynii* was observed towards the base of this subzone.

Sphenolithus anarrhopus Subzone (NTp11B)

Definition: interval from the FO of *D. mohleri* [Hay, 1964] and/or *S. anarrhopus* [this study] to the LO of *F. billii* and/or *F. ulii* [this study].
Remarks: the first persistent occurrences of *Scapholithus apertus* and *Zygodiscus clausus* were observed in this subzone.

Discoaster drieveri Zone (NTp12)

Definition: interval from the LO of *F. billii* and/or *F. ulii* [this study] to the LO of *D. drieveri* [this study] and/or the FO of *Heliolithus riedelii* [Bramlette and Sullivan 1961].
Remarks: a slight overlap in the ranges of *D. drieveri* and *H. riedelii* was observed in some of the studied sections. However, since the interval of overlap is so small and difficult to detect, for practical purposes both species may be taken to mark the upper boundary of this zone. The first occurrence of *Hornibrookina australis* was often observed in this zone.

Zygodiscus clausus Zone (NTp13)

Definition: interval from the LO of *D. drieveri* [this study] and/or the FO of *Heliolithus riedelii* [Bramlette and Sullivan 1961] to the LO of

Zygodiscus clausus and/or *Bomolithus elegans* [this study].
Remarks: consistent identification of *H. riedelii* is difficult unless samples are examined in mobile mounting where the presence of this species is expected. The variation in the range of *H. riedelii* presented in the literature is believed to be the result of inconsistent identification.

Munarinus emrei Zone (NTp14)

Definition: interval from the LO of *Z. clausus* and/or *B. elegans* [this study] to the last occurrence of *H. riedelii* [this study].
Remarks: the last occurrence of *M. emrei* and *Multipartis ponticus*, both of which have a wide geographical distribution, were observed in this zone. Extremely rare occurrences of *Octolithus multiplus* are also recorded in this zone.

Hornibrookina australis Zone (NTp15)

Definition: interval from the LO of *H. riedelii* [this study] to the LO of *H. kleinpellii* [this study].
Remarks: *H. australis* is frequently observed in this zone.

Heliolithus cantabriae Zone (NTp16)

Definition: interval from the LO of *H. kleinpellii* [this study] to the LO of *Heliolithus aktasii* and/or *Heliolithus cantabriae* [this study].
Remarks: *H. aktasii* is more widely distributed and more common than *H. cantabriae*. The first occurrence of *Discoaster multiradiatus* subdivides this zone into two subzones.

Heliolithus aktasii Subzone (NTp16A)

Definition: interval from the LO of *H. kleinpellii* [this study] to the FO of *D. multiradiatus* [Bramlette and Sullivan 1961].

Discoaster lenticularis Subzone (NTp16B)

Definition: interval from the FO of *D. multiradiatus* [Bramlette and Sullivan 1961] to the LO of *H. aktasii* and/or *H. cantabriae* [this study].

Remarks: the first occurrences of *Fasciculithus hayi*, *Fasciculithus lillianiae* and *Discoaster lenticularis* were found in this subzone.

Placozygus sigmoides Zone (NTp17)

Definition: interval from the LO of *H. aktasii* and/or *H. cantabriae* [this study] to the LO of *P. sigmoides* [this study].

Fasciculithus lillianiae Zone (NTp18)

Definition: interval from the LO of *P. sigmoides* [this study] to the LO of *Cruciplacolithus tenuis* and/or *Cruciplacolithus frequens* [this study].

Fasciculithus hayi Zone (NTp19)

Definition: interval from the LO of *C. tenuis* and/ or *C. frequens* [this study] to the LO of *F. hayi* [this study].

Remarks: the last occurrences of *Fasciculithus alanii* and *Fasciculithus tonii* are very close to top of this zone.

Fasciculithus involutus Zone (NTp20)

Definition: interval from the last occurrence of *F. hayi* [this study] to the LO of *F. tympaniformis* [this study].

Remarks: the last occurrence of *F. tympaniformis* is tentatively taken to approximate the Palaeocene–Eocene boundary. The first occurrence of *Tribrachiatus bramlettei* was observed in the uppermost part of this zone. According to Romein (1979), Figs. 18 and 19 in Plate 14 of Bramlette and Sullivan (1961) depict *T. bramlettei* rather than *Rhomboaster cuspis*. If this is true, *T. bramlettei* definitely ranges into the Palaeocene because the specimens illustrated by Bramlette and Sullivan (1961) were obtained from samples of confirmed Palaeocene age. The age of the type section of the Lodo Formation was discussed by Sullivan (1964, pp. 172–173) who, in addition, summarised all previous

studies based upon different fossil groups. The lower 150 ft at the type locality of the Lodo Formation (which includes units 1 and 2 of Bramlette and Sullivan 1961) is assigned to a Palaeocene age based upon molluscs and foraminifera in addition to other macrofossil species mentioned by Sullivan (1964).

12.5 NANNOFOSSIL BIOSTRATIGRAPHY OF THE NORTH SEA
(a) Introduction
The discovery of hydrocarbons in the Danian chalk of the North Sea area focused attention on the Palaeocene sediments of that area and generated the need for a more detailed biostratigraphical zonation than was previously available in the global schemes of Martini (1971) and Okada and Bukry (1980).

Perch-Nielsen (1979a) refined the zonation for the Danian to Selandian stages and established a series of subzones, prefixed with the letters D and S respectively, which were directly applicable to the North Sea area (Fig. 12.6). Similarly, a more refined zonation scheme for the central North Sea was introduced by van Heck and Prins (1987) who defined their zonal boundaries using first occurrences of species and variations in the abundance of selected species.

During routine examination of Palaeocene well sections in the North Sea area, because the great majority of the samples studied are ditch-cuttings, extensive caving of species means that none of the above-mentioned zonation schemes can be easily applied since they all rely heavily on first occurrences. It has, therefore, become necessary to formulate a scheme for use in the oil industry of the North Sea area in which zonal boundaries are defined upon last occurrences of species and their acmes. The criteria utilised in the proposed zonation scheme, while relevant in the North Sea area, are of little value elsewhere or have very limited use.

In the present study, significant evolutionary appearances, the majority of which have been incorporated into previous zonation schemes,

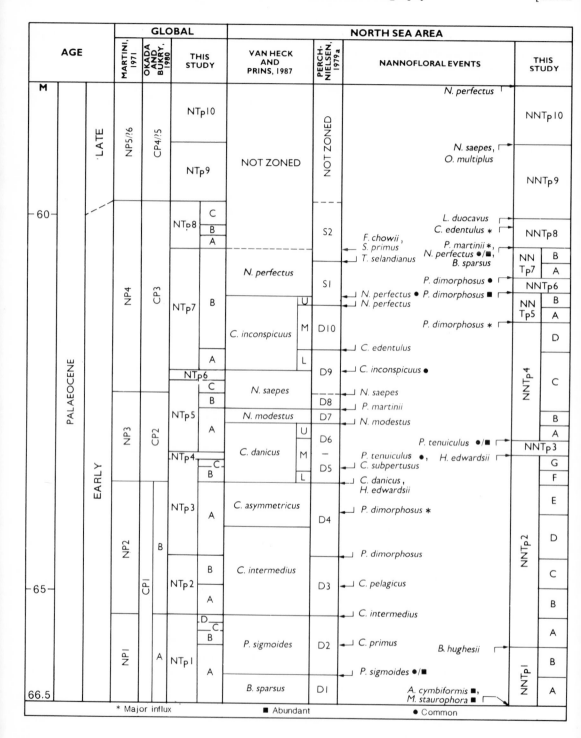

Fig. 12.6 — A correlation of the global and North Sea area zonation schemes (time in Myears after Haq *et al.* 1987).

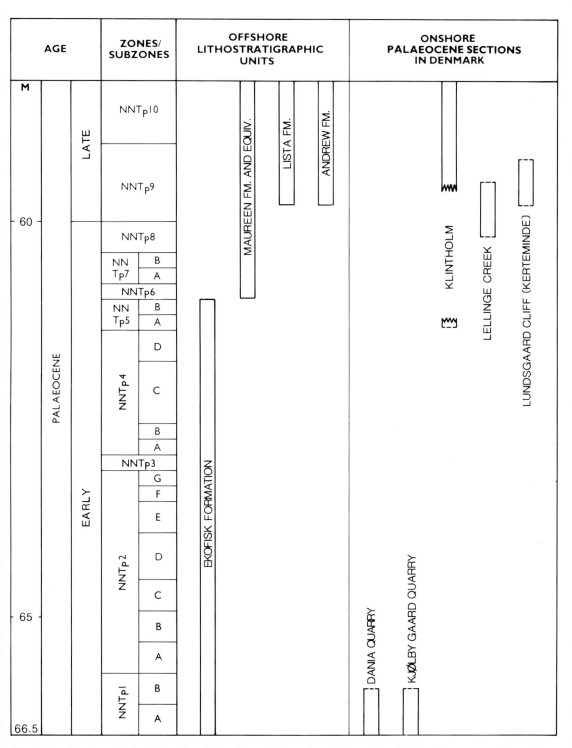

Fig. 12.7 — Stratigraphical relationships of the offshore lithostratigraphical units in the North Sea area and the onshore Palaeocene sections in Denmark (time in Myears after Haq *et al.* 1987).

have been chosen to mark subzonal boundaries applicable when working with sidewall core, core or field sample material. Each zone and subzone has been prefixed NNTp (North Sea, nannofossils, Tertiary–Palaeocene) for ease of use (Fig. 12.6).

The reliability of the proposed zones and subzones has been tested and confirmed using other biostratigraphical tools, in particular palynoflora in the Maureen Formation and foraminifera in the Ekofisk Formation.

(b) Palaeocene stratigraphy of the North Sea area
The present study concentrates on the calcareous nannofossils of the Ekofisk Formation, the Maureen Formation and its equivalents, and the lowermost parts of the Lista and Andrew Formations. The stratigraphical relationships of the formations, based upon the present calcareous nannofossil scheme, are given in Fig. 12.7.

The name Ekofisk Formation originates from the Ekofisk Field in the Norwegian block 2/4 (Deegan and Scull 1977). The formation consists of firm, white to very light grey, chalky limestones which are locally argillaceous. The Ekofisk Formation rests non-sequentially on the Tor Formation, the boundary being marked using nannofossils by the last abundant occurrence of *Arkhangelskiella cymbiformis* together with *Micula staurophora* and/or the first occurrence of *Biantholithus sparsus*. The Ekofisk formation is overlain by the clastic deposits of the Montrose or Rogaland Group. The boundary is often gradational from calcareous claystones to argillaceous limestones. The top of the Ekofisk Formation is taken on the top of the first clean limestone downhole, with the last abundant occurrence of *Prinsius dimorphosus* (top of Zone NNTp5).

The Montrose Group consists of the Maureen, Andrew, Lista and Forties Formations. In this study the calcareous nannofossils of the Maureen Formation and its equivalents and the lowermost parts of the Andrew and Lista Formations were examined. The studied intervals of the Andrew and Lista Formations

are laterally equivalent to the Maureen formation or its equivalents. The Maureen Formation is composed of claystones and sandstones with reworked Upper Cretaceous (partially upper Maastrichtian) and Lower Palaeocene limestones (Figs. 12.8–12.10). The formation takes its name from the Maureen Field in UK block 16/29 (Deegan and Scull 1977). In most places the Maureen Formation conformably overlies the Ekofisk Formation. The upper boundary with the overlying Andrew or Lista Formation is difficult to define and it is generally placed at the top of the reworked Cretaceous limestones (Fig. 12.8). The last occurrence of *Neochiastozygus perfectus* is recorded just below the base of these limestones.

(c) Studied onshore Palaeocene sections in Denmark
The offshore Palaeocene sections examined in the present study can be correlated biostratigraphically with onshore sequences of Denmark based upon the biostratigraphic results obtained in this study and in those of Perch-Nielsen (1969, 1979a, 1979b).

The Danskekalk of Denmark correlates with the Ekofisk Formation and is Danian in age. The Danish Selandian is represented by the Lellinge Greensand, the Kerteminde Marl and an overlying, grey, non-calcareous clay which together are equivalent to the Maureen Formation (Fig. 12.7).

A limited number of samples of Danian and Selandian age from Denmark were studied for their calcareous nannofossil content (Figs. 12.11 and 12.12).

(i) Danian sections
The Danian stage was established by Desor (1846). Stevns Klint and the Faxe quarry on Sjaelland Island are the classical outcrops and are regarded as the type locality. The calcareous nannofossils of the Stevns Klint section were studied by Perch-Nielsen (1969, 1979a) and Romein (1979). In the present study samples from the Kjølby Gaard and Dania Quarry were examined for their calcareous nannofossil

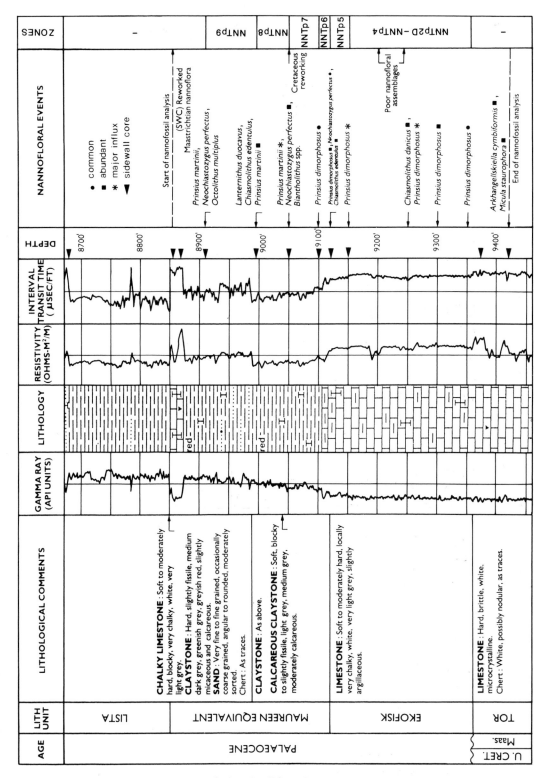

Fig. 12.8 — Summary log of selected well from Quadrant 29 (Palaeocene section).

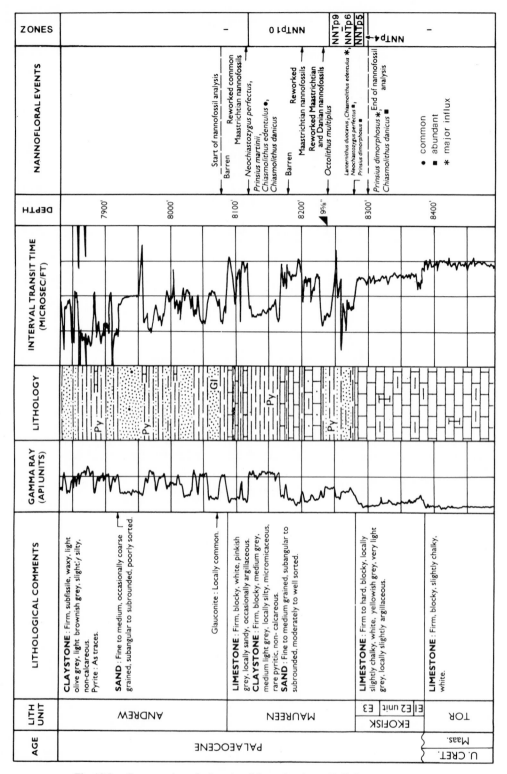

Fig. 12.9 — Summary log of selected well from Quadrant 21 (Palaeocene section).

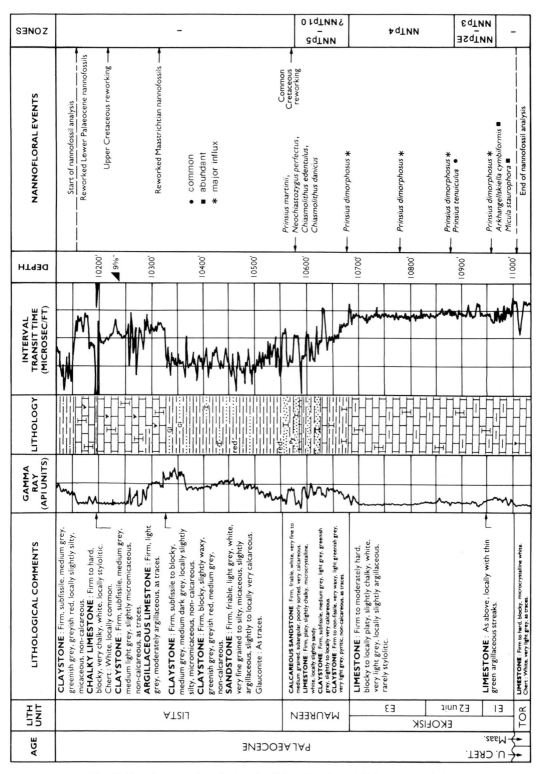

Fig. 12.10 — Summary log of selected well from Quadrant 1 (Palaeocene section).

content (Fig. 12.11). Both these sections were also examined by Perch-Nielsen (1968, 1969, 1979a) and Romein (1979).

In Kjølby Gaard, Danian sediments were identified by the presence of *Biantholithus* species, including *Biantholithus hughesii*, and the absence of abundant *Arkhangelskiella cymbiformis* and *Micula staurophora*. Danian sediments from this section are assigned to Zone NNTp1 and two subzones were identified by the marked increase of *Placozygus sigmoides*, following Perch-Nielsen (1979a).

Sediments from the Dania Quarry are also assigned to Zone NNTp1 and the two subzones were recognised using the above criterion (marked increase of *Placozygus sigmoides*).

(ii) Selandian sections

The Danish Palaeocene sediments above the Danian Limestone and below the clay with volcanic ash-beds were designated the type section for the Selandian stage by Rosenkrantz (1924). Detailed information about the Selandian coccoliths and stage was given by Perch-Nielsen (1979) and Perch-Nielsen and Hansen (1981).

In the present study, samples from Lellinge Creek, Klintholm and Lundsgaard Cliff (Kerteminde) were examined for their calcareous nannofossil content (Fig. 12.12). In the Lellinge Creek samples, Zones NNTp8 and NNTp9 were identified on the basis of the last occurrence of *Lanternithus duocavus*. The lowermost sample studied from the section in Klintholm was assigned to Subzone NNTp5a (Danian). Samples from the Kerteminde Marl are assigned to Zone NNTp9. The presence of an unconformity between the Danian and Selandian in the Klintholm section is also confirmed by the present study on the absence of Subzones NNTp5B to Zone NNTp8. Samples from the type locality of the Kerteminde Marl (Gry 1935) at Lundsgaard Cliff were also studied and were assigned to Zone NNTp9.

(d) Zonation scheme

Reworking is a major problem in the studied area

and, in order to solve the complex geological problems encountered, care must be taken to recognise the presence and significance of reworked blocks, repeated sequences and common reworking when applying any established zonation scheme (Figs. 12.8–12.10).

The zonation scheme proposed consists of ten zones (Fig. 12.13) based exclusively on last occurrences and acmes of selected species which in some cases also coincide with the first occurrence of certain species. The 17 subzones are defined on first occurrences or acmes of species. The majority of the subzonal markers have been used in previous schemes or have been suggested to have stratigraphical significance in the literature.

The ranges of stratigraphically important species and their generalised abundances are shown in Fig. 12.14. The first occurrences and abundances of the species in the North Sea area have been determined through the use of core and sidewall core samples.

A correlation between the proposed zonation scheme for the North Sea area and the zonation schemes of Perch-Nielsen (1979a) and van Heck and Prins (1987) is presented in Fig. 12.6. This diagram also shows the relationships with the global schemes.

In the present study the last occurrence of *Lanternithus duocavus* coincides with the Lower–Upper Palaeocene boundary and is consistent with palynological evidence.

Cyclagelosphaera alta Zone (NNTp1)

Definition: interval from the last abundant occurrence of *Arkhangelskiella cymbiformis*, together with *Micula staurophora* [this study], and/or the FO of *B. sparsus* [Romein 1979] to the LO of *Biantholithus hughesii* [this study] and/or the FO of *Cruciplacolithus primus* [Romein 1979].

Remarks: the first occurrence of *B. sparsus* is taken to mark the Cretaceous–Tertiary boundary by most nannofossil workers. This species, however, has been observed in Maastrichtian sediments in the North Sea area

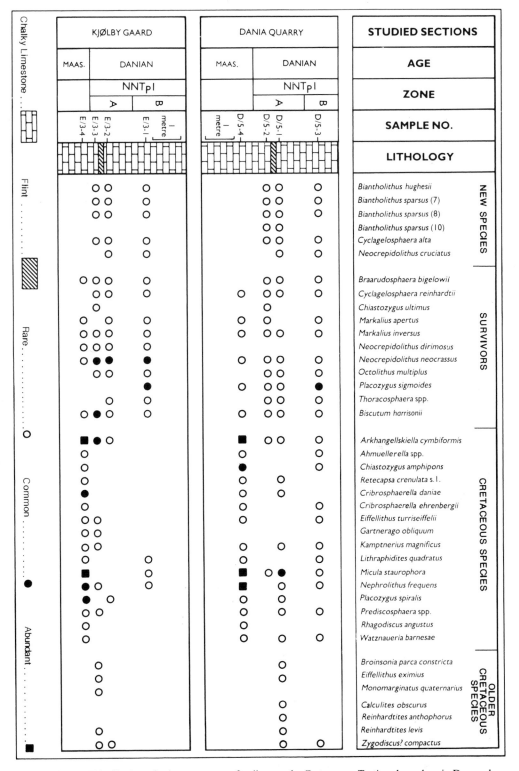

Fig. 12.11 — Distribution of calcareous nannofossils near the Cretaceous–Tertiary boundary in Denmark.

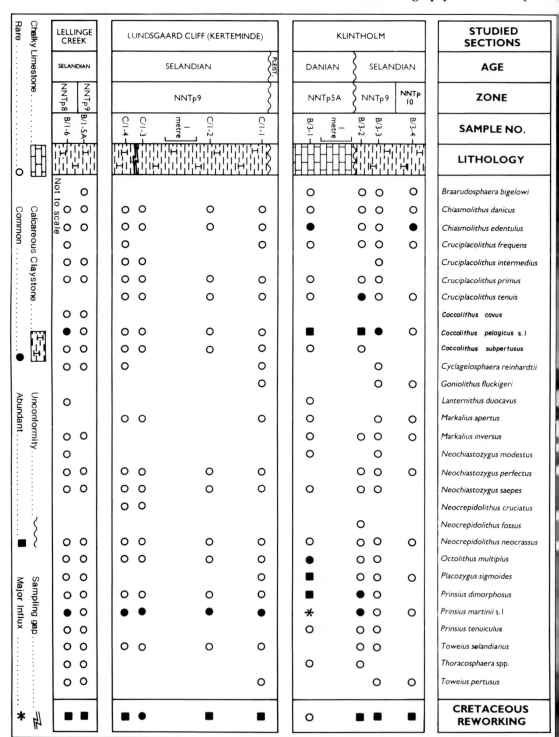

Fig. 12.12 — Distribution of calcareous nannofossils in the Danian and Selandian sediments in Denmark.

AGE			ZONES/SUBZONES		
M					
	LATE		NNT$_p$10	*Toweius pertusus*	
			NNT$_p$9	*Neochiastozygus saepes*	
60					
	PALAEOCENE		NNT$_p$8	*Lanthernithus duocavus*	
			NNT$_p$7	*Prinsius bisulcus*	NNT$_p$7 B	*Toweius selandianus*
					NNT$_p$7A	*Cruciplacolithus frequens*
			NNT$_p$6	*Neocrepidolithus neocrassus*	
			NNT$_p$5	*Markalius apertus*	NNT$_p$5 B	*Neochiastozygus perfectus*
					NNT$_p$5A	*Neochiastozygus imbriei*
			NNT$_p$4	*Coccolithus cavus*	NNT$_p$4D	*Chiasmolithus edentulus*
					NNT$_p$4C	*Prinsius martinii*
					NNT$_p$4B	*Neochiastozygus modestus*
					NNT$_p$4A	*Neocrepidolithus dirimosus*
			NNT$_p$3	*Prinsius tenuiculus*	
		EARLY	NNT$_p$2	*Octolithus multiplus*	NNT$_p$2G	*Coccolithus subpertusus*
					NNT$_p$2F	*Hornibrookina edwardsii*
					NNT$_p$2E	*Neocrepidolithus cruciatus*
					NNT$_p$2D	*Prinsius dimorphosus*
					NNT$_p$2C	*Coccolithus pelagicus*
65					NNT$_p$2B	*Cruciplacolithus intermedius*
					NNT$_p$2A	*Cruciplacolithus primus*
			NNT$_p$1	*Cyclagelosphaera alta*	NNT$_p$1B	*Placozygus sigmoides*
66.5					NNT$_p$1A	*Biantholithus sparsus*

Fig. 12.13 — Proposed Palaeocene nannofossil zonation for the North Sea (time in Myears after Haq *et al.* 1987).

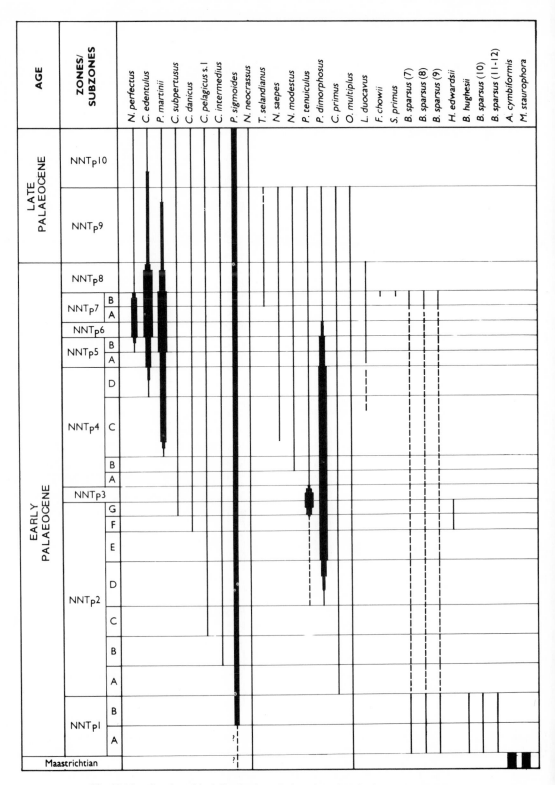

Fig. 12.14 — Stratigraphical distribution of selected markers in the North Sea area.

although it is not clear whether these occurrences are a result of sample contamination. *C. alta* is as good a marker for the definition of the lower boundary of Zone NNTp1 and may in fact prove to be more reliable since it occurs consistently in the studied area. When dealing with ditch-cuttings samples, however, it is best to use the last abundant occurrence of *A. cymbiformis* and *M. staurophora*.

The upper boundary of this zone is defined by the last occurrence of *B. hughesii*. However, the last reduction in abundance of *Biantholithus* species can also be used since, above this boundary, the genus is either absent or extremely rare.

Biantholithus sparsus Subzone (NNTp1A)

Definition: Interval from the last abundant occurrence of *A. cymbiformis* together with *M. staurophora* [this study] and/or the FO of *B. sparsus* [Romein 1979] to the first marked increase in abundance of *P. sigmoides* [Perch-Nielsen, 1979a].

Remarks: common to abundant reworking from Maastrichtian and Campanian sediments was observed within this zone. Reworking from Campanian deposits is either absent or extremely rare in the underlying Maastrichtian sediment in the North Sea area. Thus, ascertaining the source of reworking may help to identify the stratigraphic position of a studied interval. Forms such as *Neocrepidolithus neocrassus*, *Neocrepidolithus dirimosus*, *Markalius inversus*, *Markalius apertus*, *Braarudosphaera bigelowii*, *Biscutum blackii* and *Octolithus multiplus* are also present in the underlying Maastrichtian sediments and are considered to be species which survived the Cretaceous–Tertiary boundary event. *P. sigmoides* is extremely rare or absent in this subzone and the underlying Maastrichtian sediments. *B. sparsus* and *C. alta* have their first occurrence at the base of this subzone whereas *Neocrepidolithus cruciatus* appears to have its first appearance in the upper part of the subzone.

Placozygus sigmoides Subzone (NNTp1B)

Definition: interval from the first marked increase of *P. sigmoides* [Perch-Nielsen 1979a] to the LO of *B. hughesii* [this study] and/or the FO of *C. primus* [Romein 1979].

Remarks: typical Maastrichtian species decrease in numbers and become extremely rare towards the top of this subzone. Scarcity of reworked Cretaceous species is a characteristic feature of the interval between the top of this zone and the top of Zone NNTp5.

Octolithus multiplus Zone (NNTp2)

Definition: interval from the LO of *B. hughesii* [this study] and/or FO of *C. primus* [Romein 1979] to the LO of *Hornibrookina edwardsii* [this study].

Remarks: both species diversity and abundance increase towards the top of the zone. *Biantholithus* spp. are generally absent or extremely rare and no specimens of the genus *Futyania* were observed. Species of the latter are stratigraphically useful in low and mid latitude areas. Perch-Nielsen (1979a) recorded very rare specimens of *Futyania* in Subzone D2.

Neochiastozygus species are extremely rare in this zone but minute forms with symmetrical bars and sporadic occurrences of specimens with asymmetrical bars were observed. Species of *Thoracosphaera* occur in abundance within this zone but no discernible trends were observed in their stratigraphical distribution and abundance variations.

Cruciplacolithus primus Subzone (NNTp2A)

Definition: interval from the LO of *B. hughesii* [this study] and/or FO of *C. primus* [Romein 1979] to the FO of *C. intermedius* [Mohler and Hay in Hay *et al.* 1967].

Cruciplacolithus intermedius Subzone (NNTp2B)

Definition: interval from the FO of *C. intermedius* [Mohler and Hay in Hay *et al.* 1967] to the FO of *Coccolithus pelagicus* s.1. [this study].

Remarks: in the present study the subdivision of

Cruciplacolithus given by van Heck and Prins (1987) is followed.

Coccolithus pelagicus Subzone (NNTp2C)

Definition: interval from the FO of *C. pelagicus* s.1. [this study] to the FO of *Prinsius dimorphosus* [Perch-Nielsen 1979a].
Remarks: in the present study, elliptical forms previously assigned to *Coccolithus cavus* or *Ericsonia ovalis* have been included within *C. pelagicus*.

Prinsius dimorphosus Subzone (NNTp2D)

Definition: interval from the FO of *P. dimorphosus* [Perch-Nielsen 1979a] to the first major influx of *P. dimorphosus* [this study].
Remarks: forms very similar to *Prinsius tenuiculus* but possibly belonging to *Prinsius africanus* were observed in this subzone. Consistent separation of these forms is extremely difficult, however, and therefore, while recognising their status as separate species, *P. tenuiculus* also includes *P. africanus* in this study.

Neocrepidolithus cruciatus Subzone (NNTp2E)

Definition: interval from the first major influx of *P. dimorphosus* [this study] to the FO of *Chiasmolithus danicus* [Martini 1970] and/or *H. edwardsii* [this study].
Remarks: the definition of *C. danicus* provided by van Heck and Perch-Nielsen (1987) has been followed in the present study although their selection of the neotype is not considered valid (ICBN art. 7.4). Whether the original concept of the species put forward by Brotzen (1959) is adhered to in their study is not clear owing to the poor quality of Brotzen's illustrations.

The forms illustrated under *C. danicus* by van Heck and Perch-Nielsen (1987) should be re-named and re-described to avoid further confusion and the use of the name *C. danicus* abandoned.

P. dimorphosus dominates the nannofloral assemblages from this subzone until the top of

Zone NNTp4. *N. cruciatus* is frequently found in this subzone.

Hornibrookina edwardsii Subzone (NNTp2F)

Definition: interval from the FO of *C. danicus* [Martini 1970] and/or *H. edwardsii* [this study] to the FO of *C. subpertusus* and/or the first marked increase in *P. tenuiculus* [this study].
Remarks: extremely rare occurrences of *C. cavus* (only circular forms included) were recorded in this subzone although the individuals tended to be smaller than those found in younger sequences.

Coccolithus subpertusus Subzone (NNTp2G)

Definition: interval from the FO of *C. subpertusus* and/or the first marked increase of *P. tenuiculus* [this study] to the LO of *H. edwardsii* [this study].
Remarks: although *P. tenuiculus* is common to abundant within this subzone it is subordinate to larger numbers of *P. dimorphosus*. In some of the sections studied a slight increase in the abundance of *H. edwardsii* was observed in this subzone.

Prinsius teniculus Zone (NNTp3)

Definition: interval from the LO of *H. edwardsii* [this study] to the last common to abundant occurrence of *P. tenuiculus* [this study].
Remarks: *M. apertus* is present throughout the studied sections but occurs more consistently from this zone upwards. The occurrence of this form within this zone is slightly older than the range given by van Heck and Prins (1987). Extremely rare occurrences of *C. latipons* were observed within this zone. Although *P. tenuiculus* is common to abundant, the nannofloral assemblages are nevertheless dominated by *P. dimorphosus*. The great majority of specimens of *P. dimorphosus* are rounded.

Coccolithus cavus Zone (NNTp4)

Definition: interval from the last common to abundant occurrence of *P. tenuiculus* [this study] to the last major influx of *Prinsius dimorphosus* [this study].

Remarks: the most characteristic feature of this zone is the predominance of *P. dimorphosus* over *Prinsius martinii*, *C. cavus*, *C. pelagicus* and *C. danicus* which are also common to abundant in this zone.

Neocrepidolithus dirimosus Subzone (NNTp4A)

Definition: interval from the last common to abundant occurrence of *P. tenuiculus* [this study] to the FO of *Neochiastozygus modestus* [Perch-Nielsen 1979a].

Neochiastozygus modestus Subzone (NNTp4B)

Definition: interval from the FO of *N. modestus* [Perch-Nielsen 1979a] to the FO of *P. martinii* [Edwards 1971].
Remarks: below this subzone specimens of *P. dimorphosus* are rounded whereas above this subzone this species becomes subelliptical in shape. It is, however, difficult to ascertain precisely where this change in morphology takes place and it may have occurred gradually, within the limits of this subzone. Identification of the various morphotypes of *P. dimorphosus* assemblages helps greatly in the identification of reworking.

Prinsius martinii Subzone (NNTp4C)

Definition: interval from the FO of *P. martinii* [Edwards 1971] to the FO of *Chiasmolithus edentulus* [Perch Nielsen 1979a].
Remarks: the first occurrence of *Neochiastozygus saepes* is within this subzone. Because this species is sporadic in the lower part of its range in most of the sections examined it has not been used as a subzonal marker in the studied area. This variation in the range of *N. saepes* was also

observed by van Heck and Prins (1987) in the central North Sea. The first occurrence of *Neochiastozygus imbriei* and a slight increase in the abundance of *C. subpertusus* are also recorded in this subzone. *P. martinii* becomes abundant in the upper part of this subzone. Extremely rare occurrences of *Ellipsolithus macellus* were recorded in Subzone D8 by Perch-Nielsen (1979a) which is partly equivalent to the proposed Subzone NNTp4C. However, no evidence for the occurrence of this species as old as the latter subzone was observed in the course of the present study.

Chiasmolithus edentulus Subzone (NNTp4D)

Definition: interval from the FO of *Chiasmolithus edentulus* [Perch-Nielsen 1979] to the last major influx of *Prinsius dimorphosus* [this study].
Remarks: *P. martinii* is abundant within this subzone but remains less numerous than *P. dimorphosus*.

Markalius apertus Zone (NNTp5)

Definition: interval from the last major influx to the last abundant occurrence of *P. dimorphosus* [this study].
Remarks: *P. martinii* is more numerous than *P. dimorphosus* although the latter remains abundant. *C. edentulus* is also present in abundance but to a much lesser extent than the previous two species. The upper limit of this zone corresponds to the boundary between the Ekofisk and Maureen Formations. *M. apertus* is consistently observed in this zone.

Neochiastozygus imbriei Subzone (NNTp5A)

Definition: interval from the last major influx of *P. dimorphosus* [this study] to the FO of *Neochiastozygus perfectus* [Perch-Nielsen 1979a].
Remarks: extremely rare occurrences of *E. macellus* were observed as low as this subzone in this study. *N. imbriei* is rare but persistently found in this subzone.

Neochiastozygus perfectus Subzone (NNTp5B)

Definition: interval from the FO of *N. perfectus* [Perch-Nielsen 1979] to the last abundant occurrence of *P. dimorphosus* [this study].
Remarks: *N. perfectus* becomes common to abundant closely above its first occurrence at the base of this subzone.

Neocrepidolithus neocrassus Zone (NNTp6)

Definition: interval from the last abundant to the last common occurrence of *P. dimorphosus* [this study].
Remarks: *P. martinii* dominates the assemblages together with *C. edentulus*. A marked reduction in the abundance of *P. dimorphosus* is noted at the base of this zone while there is a corresponding increase in the numbers of *N. perfectus*, which becomes abundant.

Prinsius bisulcus Zone (NNTp7)

Definition: interval from the last common occurrence of *P. dimorphosus* [this study] to the last abundant occurrence of *N. perfectus* and/or the last major influx of *P. martinii* [this study].
Remarks: the nannofloral assemblages of this zone are dominated by *P. martinii* and to a lesser extent by *C. edentulus* and *N. perfectus*. *P. dimorphosus* is extremely rare. *P. bisulcus* is found common to abundant in this zone.

Cruciplacolithus frequens Subzone (NNTp7A)

Definition: interval from the last common occurrence of *P. dimorphosus* [this study] to the FO of *Toweius selandianus* [Perch-Nielsen 1979a].
Remarks: *C. frequens* is persistently observed in this subzone.

Toweius selandianus Subzone (NNTp7B)

Definition: interval from the FO of *T. selandianus* [Perch-Nielsen 1979a] to the last abundant occurrence of *N. perfectus* and/or the last major

influx of *P. martinii* [this study].
Remarks: the last occurrence of *Biantholithus* spp. is recorded in close proximity to the top of this subzone and can therefore be used to approximate its upper limit. In fact, apart from its presence in the older Zone NNTp1, *B. sparsus* s.l. is only found within this subzone. Extremely rare occurrences of *Fasciculithus chowii*, *Sphenolithus primus* and *E. macellus* were observed in the upper part of this subzone.

Lanternithus duocavus Zone (NNTp8)

Definition: interval from the last abundant occurrence of *N. perfectus* and/or the last major influx of *P. martinii* [this study] to the last occurrence of *Lanternithus duocavus* [this study].
Remarks: *P. martinii* is common to abundant throughout this zone whereas *C. edentulus* is predominant in the lower part of the zone but becomes less abundant in the upper part. The last common to abundant occurrence of *C. tenuis* is usually recorded in the lower part of this zone.

The top of this zone is tentatively taken to approximate the Lower–Upper Palaeocene boundary.

Neochiastozygus saepes Zone (NNTp9)

Definition: interval from the LO of *Lanternithus duocavus* [this study] to the LO of *Octolithus multiplus* and/or *N. saepes* [this study].
Remarks: *P. martinii* and *C. edentulus* usually are common in this zone whereas all other recorded species are rare.

Toweius pertusus Zone (NNTp10)

Definition: interval from the LO of *O. multiplus* and/or *N. saepes* [this study] to the LO of *N. perfectus* [this study].

Remarks: *C. edentulus* is rare to common in this zone whereas *P. martinii*, *C. frequens*, *C. intermedius*, *C. danicus* and *N. perfectus* are rare. Reworking of calcareous nannofossils is usually more extensive in this zone than in any of the underlying zones. It has, therefore, been very difficult to ascertain whether the consistent occurrence of *Neocrepidolithus* species is a result of reworking or whether the specimens are *in situ*. *T. pertusus* is commonly observed in this zone.

12.6 TAXONOMICAL NOTES

Biantholithus hughesii Varol, n.sp.
Plate 12.5, Figs. 26–29

Diagnosis: *Biantholithus* species with six elements.
Derivation of name: in honour of Dr. G. W. Hughes, micropalaeontologist.
Holotype: Plate 12.5, Figs. 26, 27.
Type level: Lower Palaeocene (Zone NNTp1).
Type locality: Kjølby Gaard, Denmark.
Dimensions of holotype: maximum diameter 9 μm
Remarks: *B. hughesii* is distinguished from *Biantholithus astralis* and *Biantholithus sparsus* by having only six elements. In the present study only single-cycled forms were observed; however, it is possible that well-preserved specimens may have double cycles.
Occurrence: *B. hughesii* was only observed in the North Sea area and is restricted to Zone NNTp1, Lower Palaeocene.

Biscutum harrisonii Varol, n.sp.
Plate 12.1, Fig. 1; Plate 12.4, Figs. 16–20

1979a *Biscutum* sp. Perch-Nielsen: Plate 3, Figs. 19, 20.
Diagnosis: large, elliptical placolith with single-cycled shield and a small closed central area which is strongly birefringent under polarised light.
Derivation of name: in honour of Mr. D. J. Harrison, micropalaeontologist.
Holotype: Plate 12.1, Fig. 1.
Type level: Lower Palaeocene (Zone NTp5).
Type locality: Kokaksu Section, Zonguldak, northern Turkey.
Dimensions of holotype: length 6.9 μm, width 6.1 μm.
Description: this large elliptical species has two single-cycled shields which are non-birefringent under polarised light. The central area is very small and strongly birefringent under polarised light.
Remarks: *B. harrisonii* is distinguished from other species of *Biscutum* by its large size and a comparatively small central area.
Occurrence: *B. harrisonii* occurs throughout the Palaeocene and has a wide geographical distribution in both low and high latitudes.

Neocrepidolithus rimosus (Bramlette and Sullivan) Varol, n. comb
Plate 12.4, Fig. 1.

Basionym: *Discolithus rimosus* Bramlette and Sullivan, 1961: p. 143, Plate 3, Figs. 12–13.

Fasciculithus chowii Varol, n.sp.
Plate 12.5, Figs. 11–13

Diagnosis: species of *Fasciculithus* having a parallel-sided or constricted column and a laterally reduced cone extending distally.
Derivation of name: in honour of Mr. C. Y. Chow, palynologist.
Holotype: Plate 12.5, Fig. 11.
Type level: Lower Palaeocene (Zone NTp7).
Type locality: Kokaksu Section, Zonguldak, northern Turkey.
Dimensions of holotype: maximum height 6.7 μm, maximum width 6.5 μm.
Description: the contact between column and cone is always straight. The column is usually constricted and strongly concave proximally while the cone extends distally to a height almost equal to half of the column but is reduced laterally. Its diameter is always smaller than that of the column.
Remarks: *F. chowii* differs from *Fasciculithus billii* and *Fasciculithus ulii* by having a column with a flat distal side and a comparatively high cone. The distal sides of *F. billii* and *F. ulii* are strongly concave and the cone is weakly developed or absent. *F. chowii* is distinguished from *Fasciculithus pileatus* and *Fasciculithus janii* by its laterally reduced cone.

Occurrence: *F. chowii* is one of the oldest species of the genus recovered and is widely distributed in both low and high latitudes. It is so far found restricted to Zone NTp7, Lower Palaeocene.

Genus: *Futyania* Varol, n.gen.

Type species: *Toweius petalosus* Ellis and Lohmann, 1973.

Diagnosis: small, circular to elliptical placoliths with two equal-size shields and relatively large petaloid distal elements.

Derivation of name: in honour of Dr. A. R. I. Futyan, petroleum geologist.

Remarks: this small, circular to elliptical genus has two equal-size shields, which are non-birefringent under cross-polarised light, and petaloid distal elements which are strongly birefringent under cross-polarised light. These characteristics allow *Futyania* to be easily distinguished from any other Tertiary placolith.

Futyania attewellii Varol, n.gen. n.sp.
Plate 12.1, Fig. 8; Plate 12.2, Figs. 1–8a

?1979 *Prinsius dimorphosus* (Perch-Nielsen 1969) Perch-Nielsen, 1979; Romein, Plate 3, Fig. 6.
1981c *Toweius petalosus* Ellis and Lohmann, 1973; Perch-Nielsen, Plate 2, Figs. 2, 3, 4, non Plate 2, Figs. 9, 10, 14.

Diagnosis: small strictly circular placolith having equal-size shields and distally extended, petaloid plates.

Derivation of name: in honour of Mr. R. A. K. Attewell, micropalaeontologist.

Holotype: Plate 12.2, Fig. 7.

Type level: Lower Palaeocene (Zone NTp3).

Type locality: Kokaksu Section, Zonguldak, northern Turkey.

Dimensions of holotype: diameter 1.7 μm.

Remarks: this small species is distinguished from *Futyania petalosa* by having strictly circular shields and having a much smaller, usually closed, central area. The central area of *F. petalosa* is occupied by irregularly arranged elements possibly representing the proximal ends of distal extensions. In the central area of *F. attewellii* no elements of this nature were observed until now. Under cross-polarised light

PLATE 12.1

Bar = 1 μm. All scanning electron micrographs from the Kokaksu Section, Zonguldak, northern Turkey.

Plate 12.1, Fig. 1. *Biscutum harrisonii* n.sp..
Zone NTpl, Lower Palaeocene, Kokaksu Section, Zonguldak, northern Turkey, holotype.

Plate 12.1, Fig. 2. *Cruciplacolithus primus.*
Zone NTpl. Lower Palaeocene.

Plate 12.1, Fig. 3. *Toweius pertusus.*
Zone NTp14, Upper Palaeocene.

Plate 12.1, Fig. 4. *Prinsius bisulcus*
Zone NTp10, Upper Palaeocene.

Plate 12.1, Figs. 5–7, 9, 10, 13, 16. *Futyania petalosa* n.comb.
Zone NTpl, Lower Palaeocene.

Plate 12.1, Fig. 8. *Futyania attewellii* n.gen., n.sp.
Zone NTp2, Lower Palaeocene.

Plate 12.1, Figs. 11, 12, 15. *Neobiscutum parvulum* n.comb.
Zone NTp1, Lower Palaeocene.

Plate 12.1, Fig. 14. *Neobiscutum romeinii* n.comb.
Zone NTp1, Lower Palaeocene.

Pl. 12.1] **Palaeocene calcareous nannofossil biostratigraphy** 299

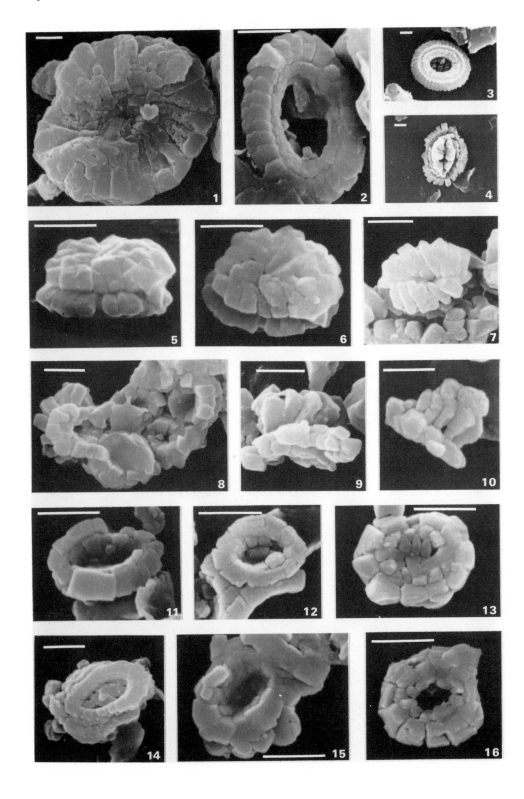

the shields are non-birefringent whereas the petal-like elements are strongly birefringent and appear ragged. *F. petalosa* appears much smoother under cross-polarised light.

Occurrence: *F. attewellii* is common to abundant in Lower Palaeocene sediments of mid and low latitudes (Zones NTp1 to NTp3). *F. attewellii* has not been observed in the North Sea area.

Futyania petalosa (Ellis and Lohmann, 1973)
Varol, n.comb.
Plate 12.1, Figs. 5–7, 9–10, 13, 16; Plate 12.2, Figs. 24, 25

Basionym: *Toweius petalosus* Ellis and Lohmann, 1973; p. 107, Plate 1, Figs. 1–11.

Heliolithus aktasii Varol, n.sp.
Plate 12.5, Figs. 21–25

Diagnosis: a small species of *Heliolithus* having a distal cycle and a column which are almost equal in diameter.

Derivation of name: in honour of Dr. G. Aktas, sedimentologist.

Holotype: Plate 12.5, Figs. 24, 25.

Type level: Upper Palaeocene (Zone NTp12).

Type locality: Kokaksu Section, Zonguldak, northern Turkey.

Dimensions of holotype: maximum diameter 7.1 μm, maximum height 7.0 μm.

Remarks: *H. aktasii* has no median cycle which makes it easy to distinguish from *Heliolithus cantabriae* and *Bomolithus elegans*. It differs from *Heliolithus riedelii* by having an inflated distal cycle instead of flaring outwards and distally as in the latter species. Both species have a serrated outer rim in plan view. *H. aktasii* is strongly birefringent under polarised light.

PLATE 12.2

Bar = 10 μm. All cross-polarised micrographs (except Fig. 31, phase contrast).

Plate 12.2, Figs. 1–8a *Futyania attewellii* n.gen., n.sp.
Zone NTp3, Lower Palaeocene, Kokaksu Section, Zonguldak, northern Turkey. Fig. 7, holotype.

Plate 12.2, Figs. 8b–10, 11b–15. *Prinsius dimorphosus*. Fig. 8b, Zone NTp3, Lower Palaeocene, Kokaksu Section, Zonguldak, northern Turkey. Figs. 9, 10, Zone NNTp3, Lower Palaeocene, North Sea area (Quadrant 1). Figs. 11b–15. Zone NNTp4, Lower Palaeocene, North Sea area (Quadrant 30).

Plate 12.2, Fig. 11a. *Coccolithus pelagicus*.
Zone NNTp4, Lower Palaeocene, North Sea area (Quadrant 30).

Plate 12.2, Figs. 16–18. *Prinsius martinii*. Figs. 16, 17, Zone NNTp5A, Lower Palaeocene, Klintholm, Denmark. Fig. 18, Zone NTp10, Upper Palaeocene, Kokaksu Section, Zonguldak, northern Turkey.

Plate 12.2, Figs. 19–20. *Toweius selandianus*.
Zone NNTp8, Lower Palaeocene, Lellinge Creek, Denmark.

Plate 12.2, Figs. 21–23. *Prinsius tenuiculus*.
Zone NNTp3, Lower Palaeocene, North Sea area (Quadrant 21).

Plate 12.2, Figs. 24, 25 *Futyania petalosa* n.comb.
Zone NTp1, Lower Palaeocene, Kokaksu Section, Zonguldak, northern Turkey.

Plate 12.2, Figs. 26–30. *Neobiscutum parvulum* n.comb.
Zone NTp1, Lower Palaeocene, Kokaksu Section, Zonguldak, northern Turkey.

Plate 12.2, Figs. 31–35. *Munarinus emrei* n.sp.
Zone NTp12, Upper Palaeocene, Kokaksu Section, Zonguldak, northern Turkey. Figs. 31, 32, holotype.

Pl. 12.2] **Palaeocene calcareous nannofossil biostratigraphy** 301

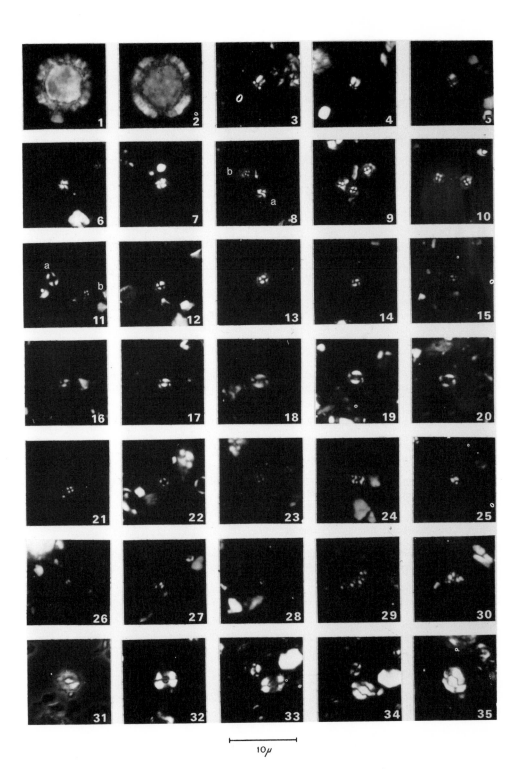

10μ

PLATE 12.3

Bar = 10 μm. All cross-polarised micrographs.

Plate 12.3, Figs. 1, 2. *Cruciplacolithus frequens.* Fig. 1, Zone NTp18, Upper Palaeocene, Irian Jaya, Indonesia. Fig. 2, Zone NTp17, Upper Palaeocene, DSDP Leg 40, Site 363.

Plate 12.3, Figs. 3–5. *Cruciplacolithus tenuis.* Fig. 3, Zone NTp8, Lower Palaeocene, India. Fig. 4, Zone NTp18, Upper Palaeocene, Irian Jaya, Indonesia. Fig. 5, Zone NNTp9, Upper Palaeocene, Klintholm, Denmark.

Plate 12.3, Figs. 6–8. *Chiasmolithus danicus.* Figs. 6, 7, Zone NNTp8, Lower Palaeocene, Lellinge Creek, Denmark. Fig. 8, Zone NTp7, Lower Palaeocene, Kokaksu Section, Zonguldak, northern Turkey.

Plate 12.3, Fig. 9a. *Coccoldithus pelagicus.*
Zone NTp8, Lower Palaeocene, Kokaksu Section, Zonguldak, northern Turkey.

Plate 12.3, Figs. 9b, 10. *Cruciplacolithus subrotundus.*
Zone NTp8, Lower Palaeocene, Kokaksu Section, Zonguldak, northern Turkey.

Plate 12.3, Figs. 11–13. *Cruciplacolithus edwardsii.*
Zone NTp6, Lower Palaeocene, Kokaksu Section, Zonguldak, northern Turkey.

Plate 12.3, Figs. 14,15. *Cruciplacolithus intermedius.*
Zone NTp8, Lower Palaeocene, Kokaksu Section, Zonguldak, northern Turkey.

Plate 12.3, Figs. 16–18. *Cruciplacolithus primus.* Figs. 16, 17, Zone NTp3, Lower Palaeocene, Kokaksu Section, Zonguldak, northern Turkey. Fig. 18, Zone NNTp5A, Lower Palaeocene, Klintholm, Denmark.

Plate 12.3, Fig. 19. *Cruciplacolithus latipons.*
Zone NTp7, Lower Palaeocene, Kokaksu Section, Zonguldak, northern Turkey.

Plate 12.3, Fig. 20. *Campylosphaera eodela.*
Zone NTp17, Upper Palaeocene, DSDP Leg 40, Site 363.

Plate 12.3, Figs. 21, 22. *Neochiastozygus perfectus.* Fig. 21, Zone NNTp10, Upper Palaeocene, Klintholm, Denmark. Fig. 22, Zone NNTp9, Upper Palaeocene, Lellinge Creek, Denmark.

Plate 12.3, Fig. 23. *Neochiastozygus modestus.*
Zone NNTp5A, Lower Palaeocene, Klintholm, Denmark.

Plate 12.3, Figs. 24, 25. *Neochiastozygus imbriei.* Fig. 24, Zone NTp6, Lower Palaeocene, Irian Jaya, Indonesia. Fig. 25, Zone NTP6, Lower Palaeocene, Kokaksu Section, Zonguldak, northern Turkey.

Plate 12.3, Figs. 26–28. *Neochiastozygus saepes.*
Zone NTp8, Lower Palaeocene, Kokaksu Section, Zonguldak, northern Turkey.

Plate 12.3, Figs. 29, 30. *Neochiastozygus eosaepes.*
Zone NTp6, Lower Palaeocene, Kokaksu Section, Zonguldak, northern Turkey.

Plate 12.3, Figs. 31, 32. *Hornibrookina teuriensis.*
Zone NTp10, Upper Palaeocene, DSDP Leg 40, Site 363.

Plate 12.3, Figs. 33, 34. *Hornibrookina edwardsii.* Fig. 33, Zone NNTp2G, Lower Palaeocene, North Sea area (Quadrant 30). Fig. 34, Zone NTp3, Lower Palaeocene, DSDP Leg 40, Site 363.

Plate 12.3, Fig. 35. *Hornibrookina australis.*
Zone NTp12, Upper Palaeocene, Kokaksu Section, Zonguldak, northern Turkey.

Pl. 12.3]   **Palaeocene calcareous nannofossil biostratigraphy**   303

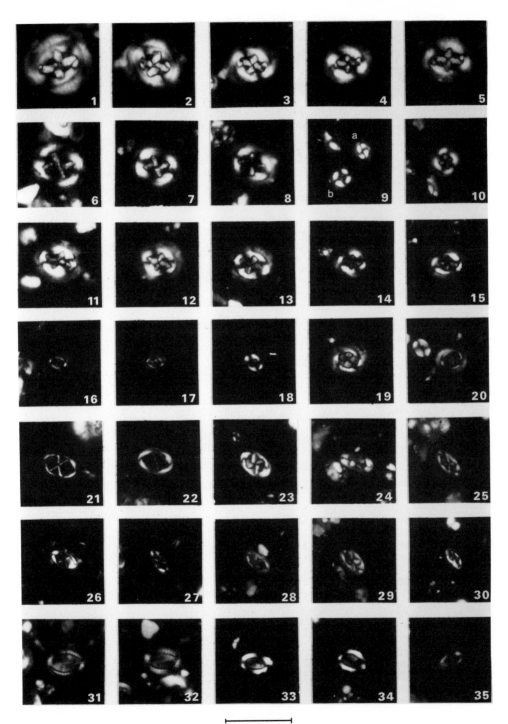

10μ

PLATE 12.4

Bar = 10 μm. All cross-polarised micrographs (except Fig. 31, phase contrast).

Plate 12.4, Fig.1. *Neocrepidolithus rimosus* n. comb.
Zone NTp17, Upper Palaeocene, DSDP Leg 40, Site 363.

Plate 12.4, Figs. 2, 3. *Neocrepidolithus fossus.*
Zone NTp1, Lower Palaeocene, Kokaksu Section, Zonguldak, northern Turkey.

Plate 12.4, Figs. 4, 5. *Neocrepidolithus cruciatus.* Fig. 4, Zone NTp5, Lower Palaeocene, Kokaksu Section, Zonguldak, northern Turkey. Fig. 5, Zone NNTp2, Lower Palaeocene, North Sea area (Quadrant 1).

Plate 12.4, Figs. 6, 7. *Neocrepidolithus dirimosus.*
Zone NNTp1, Lower Palaeocene, Kjølby Gaard, Denmark.

Plate 12.4, Fig. 8. *Neocrepidolithus neocrassus.*
Zone NNTp1, Lower Palaeocene, Dania Quarry, Denmark.

Plate 12.4, Fig. 9. *Octolithus multiplus.*
Zone NNTp5A, Lower Palaeocene, Klintholm, Denmark.

Plate 12.4, Fig. 10. *Lanternithus* sp.I.
Zone NTp5, Lower Palaeocene, Kokaksu Section, Zonguldak, northern Turkey.

Plate 12.4, Fig. 11. *Cyclagelosphaera reinhardtii.*
Zone NNTp1, Lower Palaeocene, Dania Quarry, Denmark.

Plate 12.4, Figs. 12–15. *Cyclagelosphaera alta.*
Zone NNTp1, Lower Palaeocene, Kjølby Gaard, Denmark.

Plate 12.4, Figs. 16–20. *Biscutum harrisonii* n.sp.
Zone NNTp1, Lower Palaeocene, Kjølby Gaard, Denmark.

Plate 12.4, Figs. 21, 22. *Zygodiscus clausus.*
Zone NTp13, Upper Palaeocene, Kokaksu Section, Zonguldak, northern Turkey.

Plate 12.4, Fig. 23a *Placozygus sigmoides.*
Zone NTp1, Lower Palaeocene, Kokaksu Section, Zonguldak, northern Turkey.

Plate 12.4, Fig. 23b *Markalius inversus.*
Zone NTp1, Lower Palaeocene, Kokaksu Section, Zonguldak, northern Turkey.

Plate 12.4, Fig. 24. *Markalius apertus.*
Zone NTp2, Lower Palaeocene, Kokaksu Section, Zonguldak, northern Turkey.

Plate 12.4, Fig. 25. *Coccolithus subpertusus.*
Zone NTp17, Upper Palaeocene, DSDP Leg 40, Site 363.

Plate 12.4, Figs. 26, 27. *Sphenolithus anarrhopus.*
Zone NTp14, Upper Palaeocene, Kokaksu Section, Zonguldak, northern Turkey.

Plate 12.4, Fig. 28. *Scapholithus apertus.*
Zone NTp13, Upper Palaeocene, Kokaksu Section, Zonguldak, northern Turkey.

Plate 12.4, Fig. 29. *Ellipsolithus macellus.*
Zone NTp18, Upper Palaeocene, DSDP Leg 40, Site 363.

Plate 12.4, Fig. 30. *Coccolithus cavus.*
Zone NNTp8, Lower Palaeocene, Lellinge Creek, Denmark.

Plate 12.4, Figs. 31–35. *Multipartis ponticus* n.sp.
Zone NTp12, Upper Palaeocene, Kokaksu Section, Zonguldak, Northern Turkey. Figs. 31–32, Holotype.

Pl. 12.4]   **Palaeocene calcareous nannofossil biostratigraphy**   305

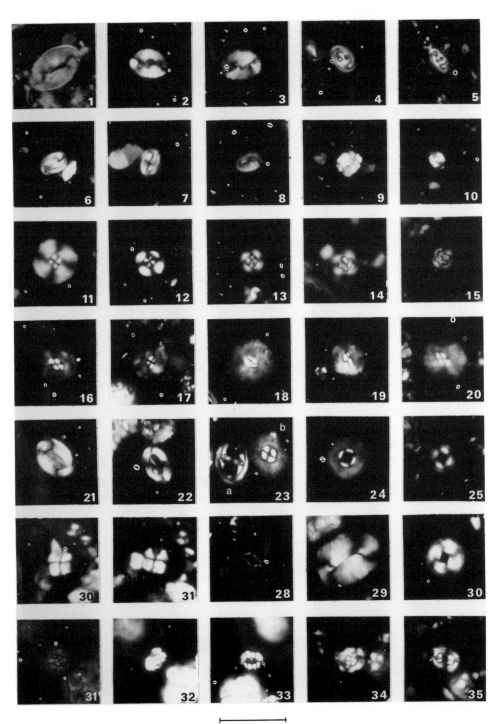

10μ

PLATE 12.5

Bar = 10 μm. All cross-polarised micrographs (except Fig. 26, phase contrast).

Plate 12.5, Figs. 1, 2. *Fasciculithus janii.*
Zone NTp10, Palaeocene, Kokaksu Section, Zonguldak, northern Turkey.

Plate 12.5, Figs. 3, 4. *Fasciculithus pileatus.*
Zone NTp10, Upper Palaeocene, Kokaksu Section, Zonguldak, northern Turkey.

Plate 12.5, Fig. 5. *Fasciculithus involutus.*
Zone NTp20, Upper Palaeocene, India.

Plate 12.5, Figs. 6, 7. *Fasciculithus billii.*
Zone NTP10, Upper Palaeocene, Kokaksu Section, Zonguldak, northern Turkey.

Plate 12.5, Figs. 8, 9. *Fasciculithus tympaniformis.*
Zone NTp13, Upper Palaeocene, DSDP Leg 40, Site 363.

Plate 12.5, Fig. 10. *Fasciculithus ulii.*
Zone NTp10, Upper Palaeocene, Kokaksu Section, Zonguldak, northern Turkey.

Plate 12.5, Figs. 11–13. *Fasciculithus chowii* n.sp. Figs. 11, 12, Zone NTp7, Lower Palaeocene, Kokaksu Sections, Zonguldak, northern Turkey Fig. 11, holotype. Fig. 13, Zone, NNTp7, Lower Palaeocene, North Sea area (Block 21).

Plate 12.5, Figs. 14, 15. *Heliolithus riedelii.*
Zone NTp12, Upper Palaeocene, Kokaksu Section, Zonguldak, northern Turkey.

Plate 12.5, Fig. 16. *Fasciculithus hayi.*
Zone NTp19, Upper Palaeocene, Irian Jaya, Indonesia.

Plate 12.5, Fig.17. *Lanternithus duocavus.*
Zone NNTp8, Lower Palaeocene, Lellinge Creek, Denmark.

Plate 12.5, Fig. 18. *Lanternithus* sp.II.
Zone NNTp4, Lower Palaeocene, North Sea area (Quadrant 1).

Plate 12.5, Figs. 19, 20. *Lanternithus jawzii* n.sp.
Zone NTp5, Lower Palaeocene, Kokaksu Section, Zonguldak, Northern Turkey. Fig. 19, holotype.

Plate 12.5, Figs. 21–25. *Heliolithus aktasii* n.sp.
Zone NTp12, Upper Palaeocene, Kokaksu Section, Zonguldak, Northern Turkey. Figs. 24, 25, holotype (same specimens).

Plate 12.5, Figs. 26–29. *Biantholithus hughesii* n.sp. Figs. 26–27, 29, Zone NNTp1, Lower Palaeocene, Kjølby Gaard, Denmark Figs. 26, 27, holotype. Fig. 28, Zone NNTp1, Lower Palaeocene, Dania Quarry, Denmark.

Plate 12.5, Fig. 30. *Biantholithus sparsus* (7 armed).
Zone NNTp1, Lower Palaeocene, Kjølby Gaard, Denmark.

Plate 12.5, Fig. 31. *Biantholithus sparsus* (8 armed).
Zone NNTp1, Lower Palaeocene, Kjølby Gaard, Denmark.

Plate 12.5, Fig. 32. *Biantholithus sparsus* (9 armed).
Zone NTp5, Lower Palaeocene, Kokaksu Section, Zonguldak, northern Turkey.

Plate 12.5, Fig. 33. *Biantholithus sparsus* (11 armed).
Zone NTp2, Lower Palaeocene, Irian Jaya, Indonesia.

Plate 12.5, Fig. 34. *Bomolithus elegans.*
Zone NTp13, Upper Palaeocene, Kokaksu Section, Zonguldak, northern Turkey.

Plate 12.5, Fig. 35. *Discoaster drieveri.*
Zone NTp12, Upper Palaeocene, Kokaksu Section, Zonguldak, northern Turkey.

Pl. 12.5]　Palaeocene calcareous nannofossil biostratigraphy　307

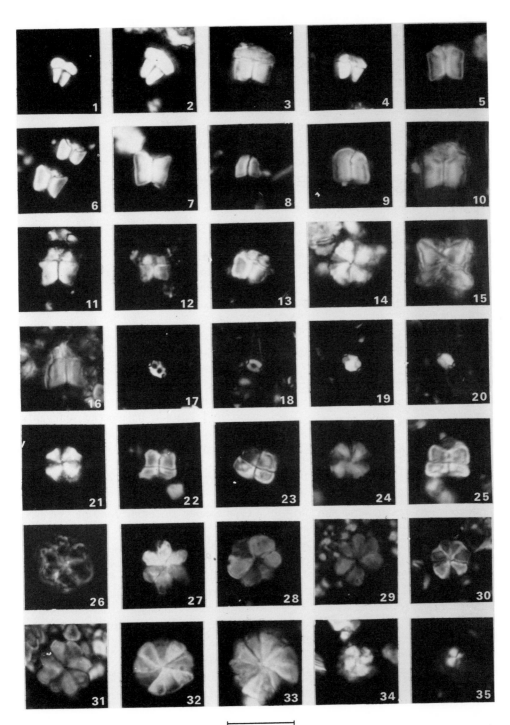

10µ

Occurrence: observed in mid and low latitude Upper Palaeocene sediments from various parts of the world including Turkey, Irian Jaya, India, DSDP Leg 40, Site 363.

Lanternithus jawzii Varol, n.sp.
Plate 12.5, Figs. 19, 20

1981b *Semiholithus*? sp. Perch-Nielsen: Plate 3, Fig. 7.
Diagnosis: small holococcolith having four large imperforated elements and two small elements at both ends of the long axis of species.
Derivation of name: in honour of Mr. A. Jawzi, petroleum geologist.
Holotype: Plate 12.5, Fig. 19.
Type level: Lower Palaeocene (Zone NTp5).
Type locality: Kokaksu Section, Zonguldak, northern Turkey.
Dimensions of holotype: maximum length 3.7 μm, maximum width 2.5 μm.
Remarks: *L. jawzii* is distinguished from *Lanternithus duocavus* by having imperforated plates.
Occurrence: *L. jawzii* occurs in low latitudes worldwide and is also rarely observed in the North Sea area.

Multipartis ponticus Varol, n.sp.
Plate 12.4, Figs. 31–35

Diagnosis: elliptical holococcolith having 8 to 12 wedge-shaped elements in the outer rim and 4 to 6 irregular elements in the central area which are arranged in two rows, along the long axis of holococcolith.
Derivation of name: *Ponticus* is Latin for 'Black Sea'.
Holotype: Plate 12.4, Figs. 31, 32.
Type level: Upper Palaeocene (Zone NTp12).
Type locality: Kokaksu Section, Zonguldak, northern Turkey.
Dimensions of holotype: maximum length 4.2 μm, maximum width 2.9 μm.
Remarks: this species is easily distinguished from other Palaeocene holococcoliths by having multiple elements in its central area.

Occurrence: *M. ponticus* was observed in Lower and Upper Palaeocene sediments of Turkey, Libya and Irian Jaya but was not observed in the North Sea area.

Munarinus emrei Varol, n.sp.
Plate 12.2, Figs. 31–35

Diagnosis: holococcolith having four to six elements in its outer rim and a large element in its central area.
Derivation of name: in honour of Dr. M. F. Emre, sedimentologist.
Holotype: Plate 12.2, figs. 31, 32.
Type level: Upper Palaeocene (NTp12).
Type locality: Kokaksu Section, Zonguldak, northern Turkey.
Dimensions of holotype: maximum length 4.1 μm, maximum width 2.8 μm.
Remarks: *M. emrei* is distinguished from *Munarinus marszalekii* by having four to six elements in its outer rim whereas the latter species has only two elements. *M. emrei* is similar to *Munarinus lesliae* but differs by having fewer elements in its rim and a larger central element.
Occurrence: *M. emrei* is widely distributed in low and mid latitude Palaeocene sediments from areas including Turkey, Irian Jaya and Libya. This species is not observed in the North Sea area.

Genus: *Neobiscutum* Varol, n.gen.

Type species: *Biscutum? romeinii* Perch-Nielsen, 1981.
Diagnosis: small, elliptical placoliths with two single-cycled shields and a central area which is either open or occupied by a reticulate structure.
Remarks: as in *Biscutum* the shields are non-birefringent under cross-polarised light. However, the former genus has a relatively small, infilled central area which is strongly birefringent, whereas *Neobiscutum* has either an open central area or one which is covered with a reticulate structure.

Neobiscutum parvulum (Romein 1979) Varol, n.comb.

Plate 12.1, Figs. 11, 12, 15; Plate 12.2, Figs. 26–30

Basionym: *Biscutum parvulum* Romein, 1979; pp. 96–97, Plate 1, Fig. 10; Plate 2, Figs. 1, 2.
Remarks: the central area is strongly birefringent under cross-polarised light. It has a small opening in the central area in some specimens but is closed in the others.

Neobiscutum romeinii (Perch-Nielsen 1981)
Varol, n.comb
Plate 12.1 Fig. 14

Basionym: *Biscutum?* *romeinii* Perch-Nielsen, 1981a: pp. 834–835, Plate 2, Figs. 7, 11–14.

12.7 ACKNOWLEDGEMENTS

I would like to thank Mr. D. Harrison for his critical reading of the manuscript, and his suggestions and comments and Mr. M. Jakubowski and Dr. M. H. Girgis for their review of the manuscript. Thanks are also due to Alice Leow and D. Kenworthy for drafting the diagrams, Tony Owen for developing the photographs, Miss J. Smith for typing the manuscript, all from Robertson Group, and to Mr. S. Davis and the University of Bangor for the use of a scanning electron microscope.

12.8 REFERENCES

Bramlette, M. N. and Sullivan, F. R. 1961. Coccolithophorids and related nannoplankton of the Early Tertiary in California. *Micropaleontology*, 7, 129–174.

Brotzen, F. 1959. On *Tylocidaris* species (Echinoidea) and the stratigraphy of the Danian of Sweden. *Årsbok. Sver. Geol. Unders.*, 54, 1–81.

Deegan, C. E. and Scull, B. J. 1977. A standard lithostratigraphic nomenclature for the central and northern North Sea. *Rep. Inst. Geol. Sci. U.K.*, 77/25, 1–33.

Desor, M. 1846. Sur le terrain danien nouvel étage de la craie. Bull. Soc. Géol. France, 4, 179–182.

Edwards, A. R. 1971. A calcareous nannoplankton zonation of the New Zealand Paleogene. In A. Farinacci (ed.), *Proc. II Plankt. Conf., Roma, 1970*, 1, Tecnoscienza, Rome, 381–419.

Ellis, C. H. and Lohman, W. H. 1973. *Toweius petalosus* new species, a Paleocene calcareous nannofossil from Alabama. *Tulane Stud. Geol. Paleontol.*, 10, 107–110.

Gry, H. 1935. Petrology of the Paleocene Sedimentary Rocks of Denmark. *Danm. Geol. Unders. IIrk.*, 18, 1–102.

Haq, B. U., Hardenbol, J. and Vail, P. R. 1987. Chronology of fluctuating sealevels since the Triassic. *Science*, 235, 1156–1167.

Hay, W. W. 1964. The use of the electron microscope in the study of fossils. *Annu. Rep. Smithsonian Inst.*, 1963, 409–415.

Hay, W. W. and Mohler, H. P. 1967. Calcareous nannoplankton from early Tertiary rocks at Pont Labau, France, and Paleocene–Eocene correlations. *J. Paleontol.*, 41, 1505–1541.

Hay, W. W., Mohler, H. P., Roth, P. H., Schmidt, R. R. and Boudreaux, J. E. 1967. Calcareous nannoplankton zonation of the Cenozoic of the Gulf Coast and Caribbean–Antillean area, and transoceanic correlation. *Trans. Gulf Coast Assoc. Geol. Soc.*, 17, 428–480.

Heck, S. E. van and Perch-Nielsen, K. 1987. Validation of *Chiasmolithus danicus*. *Abh. Geol. B.-A.*, 39, 279–283.

Heck, S. E. van and Prins, B. 1987. A refined Nannoplankton Zonation for the Danian of the Central North Sea. *Abh. Geol. B.-A.*, 39, 285–303.

Martini, E. 1970. Standard Palaeogene calcareous nannoplankton zonation. *Nature*, 226, 560–561.

Martini, E. 1971. Standard Tertiary and Quaternary calcareous nannoplankton zonation. In A. Farinacci (ed.), *Proc. II Plankt. Conf., Roma, 1970*, 2, Tecnoscienza, Rome, 739–785.

Okada, H. and Bukry, D. 1980. Supplementary modification and introduction of code numbers to the low-latitude coccolith biostratigraphic zonation (Bukry 1973, 1975). *Mar. Micropaleont.*, 5, 321–325.

Perch-Nielsen, K. 1968. *Naninfula*, genre nouveau des nannofossiles calcaires. *C.P. Acad. Sci. Paris*, 267, 2298–2300.

Perch-Nielsen, K. 1969. Die Coccolithen einiger dänischer Maastrichtien-und Danienlokalitäten. *Bull. Geol. Soc. Denmark*, 19, 51–66.

Perch-Nielsen, K. 1979a. Calcareous nannofossil zonation at the Cretaceous/Tertiary boundary in Denmark. In T. Birkelund and R. G. Bromley (Eds.), *Cretaceous–Tertiary Boundary Events*, 1, *The Maastrichtian and Danian of Denmark*, University of Copenhagen, Copenhagen, 115–135.

Perch-Nielsen, K. 1979b. Calcareous nannofossils in Cretaceous–Tertiary boundary section in Denmark. In T. Birkelund and R. G. Bromley (Eds.), *Cretaceous–Tertiary Boundary Events*, 2, *Proceedings*, University of Copenhagen, Copenhagen, 120–126.

Perch-Nielsen, K. 1979c. Calcareous nannofossils from the Cretaceous between the North Sea and the Mediterranean. In *Aspek de Kreide Europas. IUGS Ser. A.*, 6, 223–272.

Perch-Nielsen, K. 1981a. New Maastrichtian and Paleocene calcareous nannofossils from Africa, Denmark, the USA and the Atlantic, and some Paleocene lineages. *Eclogae Geol. Helv.*, 74, 831–863.

Perch-Nielsen, K. 1981b. Les coccolithes du Paléocène près de El Kef, Tunisie et leurs ancêtres. *Cah. Micropaléontol.*, 1981(3), 7–23.

Perch-Nielsen, K. 1981c. Les nannofossiles calcaires à la limite Crétacé–Tertiaire près de El Kef, Tunisie. *Cah. Micropaléontol.*, **1981**(3), 25–37.

Perch-Nielsen, K. 1985. Cenozoic calcareous nannofossils. In H. M. Bolli, J. B. Saunders and K. Perch-Nielsen (Eds.), *Plankton Stratigraphy*, Cambridge University Press, Cambridge, 422–454.

Perch-Nielsen, K. and Hansen, J. M. 1981. Selandian. In C. Pomerol (Ed.), Stratotypes of Palaeogene Stages. *Mem. Hors. Séries (2) Bull. d'Inf. Geol. Bassin Paris*, 215–230.

Radomski, A. 1967. Some stratigraphic units based on nannoplankton in the Polish outer Carpathians. *Bull. Ins. Geol.*, **5**, 385–393.

Radomski, A. 1968. Calcareous nannoplankton zones in Palaeogene of the western Polish Carpathians. *Rocz. Pol.* *Tow. Geol.*, **38**, 545–605.

Romein, A. J. T. 1979. Lineages in early Paleogene calcareous nannoplankton. *Utrecht. Micropaleont. Bull.*, **22**, 1–231.

Rosenkrantz, A. 1924. De Kobenhavnske og deres Placering i den Danske Lagraeke. *Medd. Dansk Geol. Foren.*, **6**, 3–39.

Sullivan, F. R. 1964. Lower Tertiary nannoplankton from the California Coast Ranges. I. Paleocene. *Univ. Calif. Publ. Geol. Sci.*, **44**, 163–227.

Varol, O. 1983. Late Cretaceous–Paleocene calcareous nannofossils from the Kokaksu Section (Zonguldak, Northern Turkey). *N. Jb. Geol. Paläont, Abh.*, **166**, 431–460.

13

Standard Palaeocene–Eocene calcareous nannoplankton zonation of Turkey

Vedia Toker

In the investigation of the calcareous nannoplankton in Turkey, seven Palaeocene and eleven Eocene biostratigraphical zones have been distinguished. The material studied comes from the land sections of Haymana, Adiyaman, Şarkişla, Zonguldak, Kaman, Thrace and Antalya. Lithologically these are represented by marls, shales, sandstones and limestones.

1200 samples from these regions were examined for their nannoplankton content. 99 nannoplankton species were identified in the Palaeocene–Eocene interval. In most of the sections, nannoplankton zones were correlated with the planktonic foraminifera zones to achieve a greater stratigraphical refinement. The following zones were distinguished.

In the Palaeocene: the *Cruciplacolithus tenuis, Chiasmolithus danicus, Ellipsolithus macellus, Fasciculithus tympaniformis, Heliolithus kleinpellii, Discoaster mohleri* and *Discoaster multiradiatus* Zones.

In the Eocene: the *Tribrachiatus contortus, Discoaster binodosus, Tribrachiatus orthostylus, Discoaster lodoensis, Discoaster sublodoensis, Nannotetrina fulgens, Discoaster tanii nodifer, Discoaster saipanensis, Discoaster oamaruensis, Isthmolithus recurvus* and *Sphenolithus pseudoradians* Zones.

13.1 INTRODUCTION

This paper contains brief descriptions and discussions of the zones that can be distinguished in Turkey by the use of nannoplankton. It consists of a summary of published and unpublished information and includes the result of original work which the author carried out in different areas in Turkey (Toker 1977, 1980, 1982, 1987, Toker and Erkan 1985). The paper compiles all the information now available for the standard calcareous nannoplankton zonation of Turkey.

The Palaeocene–Eocene samples were collected for this study from the regions Haymana (south-western Ankara), Adiyaman (south-eastern Turkey), Zonguldak (northern Turkey), Kaman (Central Anatolia), Thrace (north-western Turkey), Şarkişla (north-eastern Turkey), and Antalya (southern Turkey) (Fig. 13.1). The Palaeocene–Eocene sediments are well developed in the stratigraphical succession of Turkey. These sediments are represented by sandstones, shales, marls and limestones.

In Anatolia the Palaeocene–Eocene sequence is about 1000 m thick. The Palaeocene series consists of an alternation of sandstones, marls and limestones. In this sequence rich nannoplankton and planktonic foraminifera assemblages have been recorded (Toker 1977). Eocene sediments are generally represented by limestones with abundant larger foraminifera such as *Nummulites, Alveolina, Assilina* and *Discocyclina*. These limestones grade east and upwards into marls with abundant planktonic faunas and floras.

311

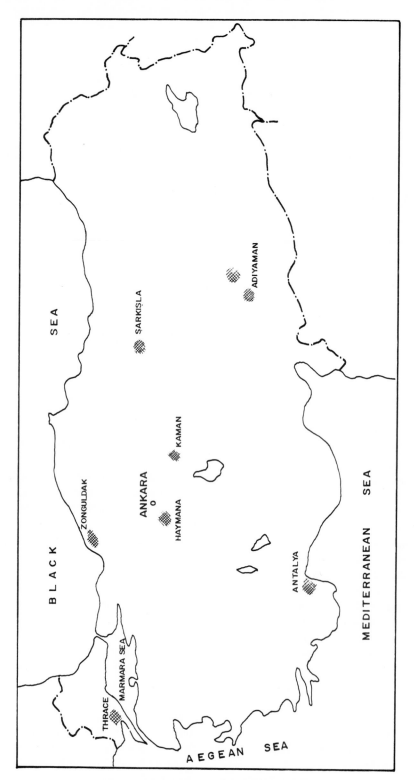

Fig. 13.1 — Map of Turkey showing the localities mentioned in this study.

In the Thrace and Zonguldak regions the Palaeocene–Eocene sequence is between 400 and 1770 m thick. This thick series consists of shallow marine conglomerates, sandstones and shales with poor planktonic foraminifera and moderately rich nannoplankton assemblages.

In the Antalya and Adiyaman regions in southern Turkey the whole marine series is of sublittoral origin and the sediments consist of marls and limestones which contain abundant planktonic foraminifera and nannoplankton.

1200 samples were taken for micropalaeontological analysis. Smear slides were prepared with the commonly accepted technique introduced by Bramlette and Sullivan (1961) and Hay (1961, 1965) and for more detailed observations the electron microscope was used. Nannoplankton species were observed with a light microscope between crossed-nicols with $10 \times$ and $12 \times$ oculars and $100 \times$ objectives with oil immersion.

13.2 BIOSTRATIGRAPHICAL ZONATION

The first Tertiary nannoplankton zones were introduced by Brönnimann and Stradner (1960). Since then calcareous nannoplankton zonations have been presented by Hay *et al.* (1967), Bramlette and Wilcoxon (1967), Martini and Worsley (1970), Martini (1970), Ellis (1975), Bukry (1981), Perch-Nielsen (1985), and van Heck and Prins (1987). A combination of the standard nannoplankton zonations of Martini (1970) and that of Hay *et al.* (1967) has been successfully applied in Turkey (Table 13.1)

Table 13.2 shows the proposed subdivision with indications of the age and geographical distribution of the zones in Turkey. Table 13.3 shows the stratigraphical distribution of 99 nannoplankton species that were observed in different localities.

The biostratigraphy in Turkey as presented in this paper is based on nannoplankton and planktonic foraminifera. The nannoplankton zonation of Hay *et al.* (1967) and of Martini (1970) and the planktonic foraminifera zonations of Bolli (1957), Stainforth *et al.* (1975) and Toumarkine and Luterbacher (1985) are

used as the framework for most subdivisions.

The Maastrichtian–Danian boundary has been drawn at the top of the *Micula murus* and *Abathomphalus mayaroensis* Zones in Turkey. Planktonic foraminifera showed an abrupt extinction of the Globotruncanidae at the end of the Maastrichtian in the Haymana and Adiyaman regions. This is followed by the first appearance of the Globigerinidae. The *Globigerina eugibina* Zone which is recognised in many other localities in the world was not found in the basal Danian of Turkey (Table 13.4). No lithological change was observed between the Upper Maastrichtian and basal Danian sediments. Instead of the *G. eugibina* Zone, Dizer and Meriç (1981) have recognised the *Globigerina daubjergensis* Zone with rich planktonic foraminifera in the sediments of the shallow marine facies which can be observed in the entire northern belt of Turkey. They consider these two zones equivalent. The absence of the *G. eugibina* Zone is most probably the result of unsuitable ecological factors. The *Biantholithus sparsus* Zone can be correlated with the *G. daubjergensis* Zone.

The early Danian is represented by the *Cruciplacolithus tenuis* Zone in three localities. The *Markalius inversus* Zone is missing, but the successions do not show any hiatus at this level as the basal Danian in the Haymana and Adiyaman regions is represented by sandstones and sandy limestones in which only benthonic foraminifera have been recognised. The same situation occurs with the *Heliolithus riedelii* Zone. The zonal marker has not yet been found in Turkey, owing to ecological factors.

The Palaeocene–Eocene boundary has been drawn at the top of the *Discoaster multiradiatus* and the *Globorotalia velascoensis* Zones.

The Eocene–Oligocene boundary is placed at the top of the *Sphenolithus pseudoradians* Zone. *Discoaster saipanensis* and *Discoaster barbadiensis* disappear at this level.

(a) *Cruciplacolithus tenuis* Zone
Definition: interval from the first occurrence of

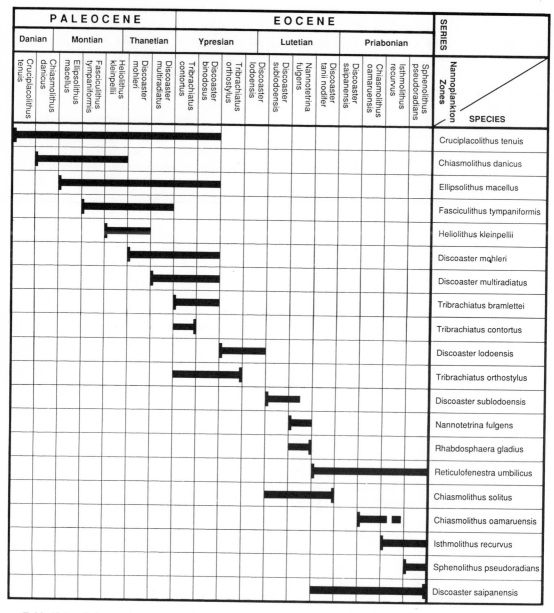

Table 13.1 — Palaeocene–Eocene zonal nannoplankton markers and proposed subdivision with indications of age.

Toker - 1987 Standard zones of Turkey	Toker - 1987 NW Antalya	Toker-Erkan - 1985 Thrace	Toker - 1982 Kaman	Toker-Meriç - 1986 W.NW Adiyaman	Varol - 1983 Zonguldak	Gökten-Toker - 1981 Şarkişla	Aköz - 1981 SW Adiyaman	Toker - 1977-1980 SW Ankara	NP	Standard Zones. Martini 1971	SERIES	
Sphenolithus pseudoradians	Sphenolithus pseudoradians	Sphenolithus pseudoradians							NP20	Sphenolithus pseudoradians	Priabonian	EOCENE
Isthmolithus recurvus	Isthmolithus recurvus	Isthmolithus recurvus							NP19	Isthmolithus recurvus	Priabonian	EOCENE
Chiasmolithus oamaruensis	Chiasmolithus oamaruensis	Chiasmolithus oamaruensis							NP18	Chiasmolithus oamaruensis	Priabonian	EOCENE
Discoaster saipanensis	Discoaster saipanensis	Discoaster saipanensis	Discoaster saipanensis						NP17	Discoaster saipanensis	Priabonian	EOCENE
Discoaster tanii nodifer		Discoaster tanii nodifer	Discoaster tanii nodifer						NP16	Discoaster tanii nodifer	Lutetian	EOCENE
Nannotetrina fulgens			Nannotetrina fulgens						NP15	Chiphragmalithus alatus	Lutetian	EOCENE
Discoaster sublodoensis			Discoaster sublodoensis				Discoaster sublodoensis	Discoaster sublodoensis	NP14	Discoaster sublodoensis	Lutetian	EOCENE
Discoaster lodoensis							Discoaster lodoensis	Discoaster lodoensis	NP13	Discoaster lodoensis	Ypresian	EOCENE
Tribrachiatus orthostylus							Tribrachiatus orthostylus	Tribrachiatus orthostylus	NP12	Tribrachiatus orthostylus	Ypresian	EOCENE
Discoaster binodosus								Discoaster binodosus	NP11	Discoaster binodosus	Ypresian	EOCENE
Tribrachiatus contortus				Tribrachiatus contortus				Tribrachiatus contortus	NP10	Tribrachiatus contortus	Ypresian	EOCENE
Discoaster multiradiatus				Discoaster multiradiatus		Discoaster multiradiatus	Discoaster multiradiatus	Discoaster multiradiatus	NP9	Discoaster multiradiatus	Thanetian	PALEOCENE
Discoaster mohleri					Discoaster mohleri	Discoaster gemmeus	Discoaster gemmeus	Discoaster gemmeus	NP8	Heliolithus riedelii	Thanetian	PALEOCENE
									NP7	Discoaster gemmeus	Thanetian	PALEOCENE
Heliolithus kleinpellii					Heliolithus kleinpellii	Heliolithus kleinpellii	Heliolithus kleinpellii	Heliolithus kleinpellii	NP6	Heliolithus kleinpellii	Montian	PALEOCENE
Fasciculithus tympaniformis				Fasciculithus tympaniformis	Fasciculithus tympaniformis	Fasciculithus tympaniformis	Fasciculithus tympaniformis	Fasciculithus tympaniformis	NP5	Fasciculithus tympaniformis	Montian	PALEOCENE
Ellipsolithus macellus				Ellipsolithus macellus		Ellipsolithus macellus	Ellipsolithus macellus	Ellipsolithus macellus	NP4	Ellipsolithus macellus	Montian	PALEOCENE
Chiasmolithus danicus				Chiasmolithus danicus	Cruciplacolithus tenuis	Chiasmolithus danicus	Chiasmolithus danicus	Chiasmolithus danicus	NP3	Chiasmolithus danicus	Danian	PALEOCENE
Cruciplacolithus tenuis				Cruciplacolithus tenuis			Cruciplacolithus tenuis	Cruciplacolithus tenuis	NP2	Cruciplacolithus tenuis	Danian	PALEOCENE
					Biantholithus sparsus				NP1	Markalius inversus	Danian	PALEOCENE

Table 13.2 — Correlation of the Palaeocene–Eocene nannoplankton zones in the sections of Turkey.

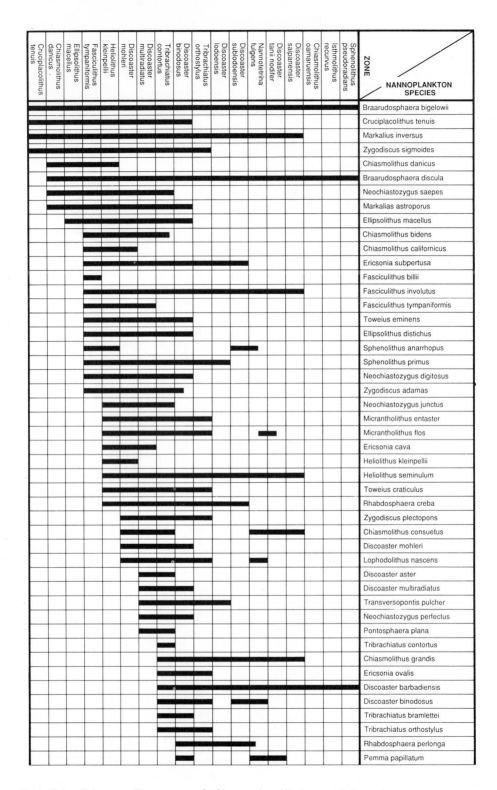

Table 13.3 — Palaeocene–Eocene nannoplankton species of Turkey and their stratigraphical distribution.

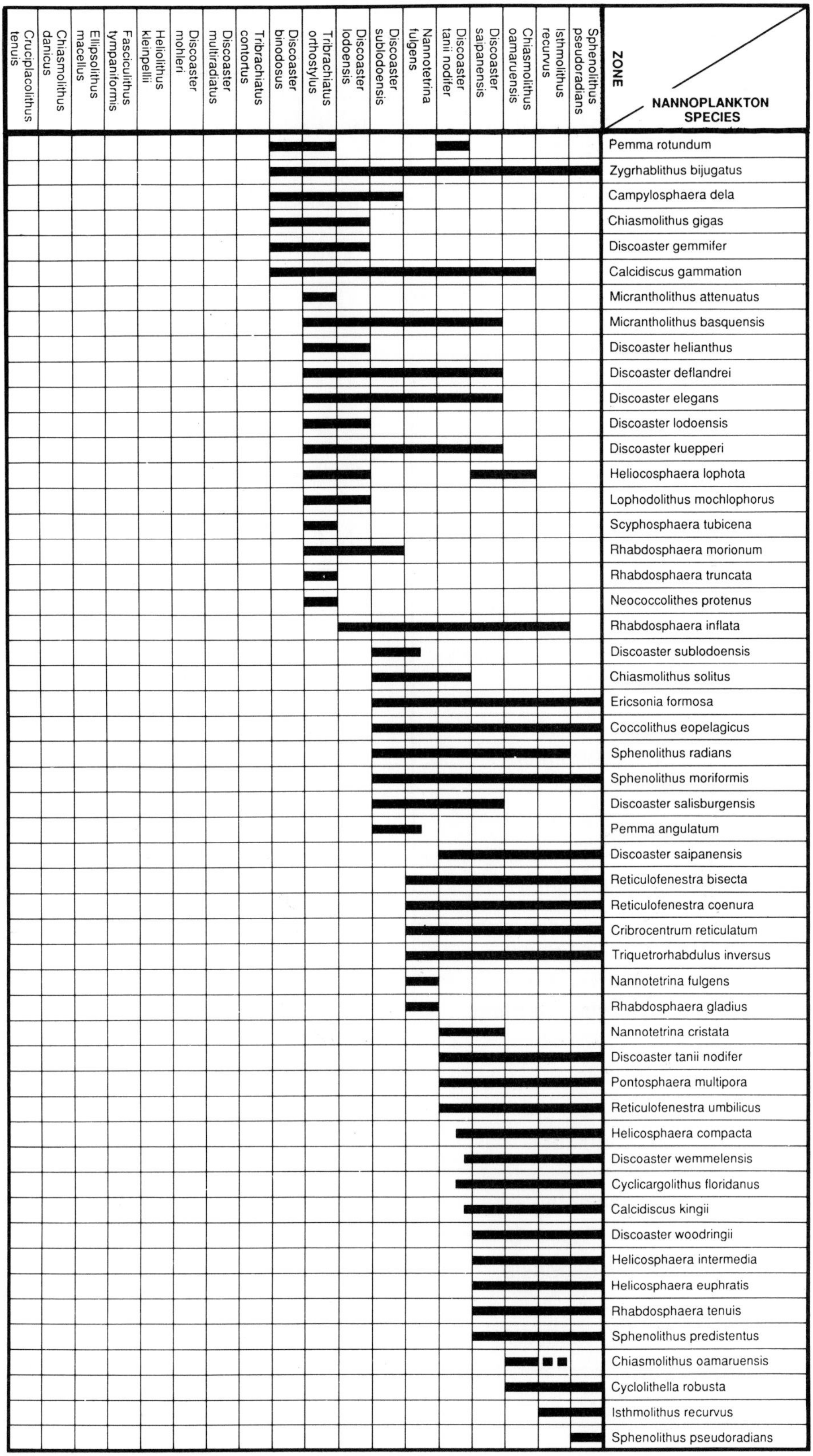

Column headers (stratigraphic zones, left to right):
Cruciplacolithus tenuis · Chiasmolithus danicus · Ellipsolithus macellus · Fasciculithus tympaniformis · Heliolithus kleinpellii · Discoaster mohleri · Discoaster multiradiatus · Tribrachiatus contortus · Discoaster binodosus · Tribrachiatus orthostylus · Discoaster lodoensis · Discoaster sublodoensis · Nannotetrina fulgens · Discoaster tanii nodifer · Discoaster saipanensis · Chiasmolithus oamaruensis · Isthmolithus recurvus · Sphenolithus pseudoradians

ZONE

NANNOPLANKTON SPECIES

- Pemma rotundum
- Zygrhablithus bijugatus
- Campylosphaera dela
- Chiasmolithus gigas
- Discoaster gemmifer
- Calcidiscus gammation
- Micrantholithus attenuatus
- Micrantholithus basquensis
- Discoaster helianthus
- Discoaster deflandrei
- Discoaster elegans
- Discoaster lodoensis
- Discoaster kuepperi
- Heliocosphaera lophota
- Lophodolithus mochlophorus
- Scyphosphaera tubicena
- Rhabdosphaera morionum
- Rhabdosphaera truncata
- Neococcolithes protenus
- Rhabdosphaera inflata
- Discoaster sublodoensis
- Chiasmolithus solitus
- Ericsonia formosa
- Coccolithus eopelagicus
- Sphenolithus radians
- Sphenolithus moriformis
- Discoaster salisburgensis
- Pemma angulatum
- Discoaster saipanensis
- Reticulofenestra bisecta
- Reticulofenestra coenura
- Cribrocentrum reticulatum
- Triquetrorhabdulus inversus
- Nannotetrina fulgens
- Rhabdosphaera gladius
- Nannotetrina cristata
- Discoaster tanii nodifer
- Pontosphaera multipora
- Reticulofenestra umbilicus
- Helicosphaera compacta
- Discoaster wemmelensis
- Cyclicargolithus floridanus
- Calcidiscus kingii
- Discoaster woodringii
- Helicosphaera intermedia
- Helicosphaera euphratis
- Rhabdosphaera tenuis
- Sphenolithus predistentus
- Chiasmolithus oamaruensis
- Cyclolithella robusta
- Isthmolithus recurvus
- Sphenolithus pseudoradians

P.FORAMINIFERA ZONE	NANNOPLANKTON ZONE
G. cerroazulensis	S. pseudoradians
	I. recurvus
G. semiinvoluta	C. oamaruensis
T. rohri	D. saipanensis
– – –?–?–?–?– – –	
G. subconglobata	D. tanii nodifer
	N. fulgens
G. pentacamerata	D. sublodoensis
G. aragonensis	D. lodoensis
G. formosa	T. orthostylus
	D. binodosus
G. subbotinae	T. contortus
G. velascoensis	D. multiradiatus
G. pseudomenardii	D. mohleri
	H. kleinpellii
G.pusilla	F. tympaniformis
G. angulata	
G. uncinata	E. macellus
G. trinidadensis	C. danicus
G. pseudobulloides	C. tenuis

Table 13.4 — Correlation of the nannoplankton and planktonic foraminifera zones.

Cruciplacolithus tenuis to the first occurrence of *Chiasmolithus danicus*.
Authors: Mohler and Hay in Hay *et al.* 1967 emend. Martini 1970.
Age: Early Danian.
Localities: Haymana, Adiyaman, Zonguldak.
Common species: *C. tenuis, Zygodiscus sigmoides, Markalius inversus.*
Remarks: this zone characterises the lowest part of the Palaeocene succession in the Haymana, Adiyaman and Zonguldak areas, where it was easily recognised. The *Biantholithus sparsus* Zone has been described by Varol (1983) in the basal part of the Palaeocene of the Zonguldak area. The *Markalius inversus* Zone is not recognised in these localities because of the unfavourable lithology.

(b) *Chiasmolithus danicus* Zone

Definition: interval from the first occurrence of *C. danicus* to the first occurrence of *Ellipsolithus macellus.*
Author: Martini 1970.
Age: Late Danian.
Localities: Haymana, Adiyaman, Şarkişla.
Common species: *C. danicus, C. tenuis, Braarudosphaera discula, Neochiastozygus saepes.*
Remarks: the zone was not distinguished in the Zonguldak area by Varol (1983) because he considered it more practicable to combine Martini's *C. tenuis, C. danicus* and *E. macellus* Zones into the single *Cruciplacolithus tenuis* Zone of Mohler and Hay in Hay *et al.* (1967).

(c) *Ellipsolithus macellus* Zone

Definition: interval from the first occurrence of *E. macellus* to the first occurrence of *Fasciculithus tympaniformis.*
Author: Martini 1970.
Age: Early Montian.
Localities: Haymana, Adiyaman, Şarkişla, Zonguldak.
Common species: the assemblage of the *Chiasmolithus danicus* Zone and *E. macellus.*
Remarks: the zonal marker *E. macellus* is easily recognised in Turkey. It has also been found in

the Zonguldak area by Varol (1983). Although the *E. macellus* Zone has not been distinguished by him, it can be used in this region.

(d) *Fasciculithus tympaniformis* Zone

Definition: interval from the first occurrence of *F. tympaniformis* to the first occurrence of *Heliolithus kleinpellii.*
Authors: Mohler and Hay in Hay *et al.* 1967.
Age: Middle Montian.
Localities: Haymana, Adiyaman, Şarkişla, Zonguldak.
Common species: the assemblage of the *Ellipsolithus macellus* Zone and *F. tympaniformis, Chiasmolithus bidens, Chiasmolithus californicus, Ericsonia subpertusa, Fasciculithus billii, Ellipsolithus distichus, Sphenolithus anarrhopus, Sphenolithus primus.*
Remarks: this zone was easily recognised in the four localities in Turkey. Several species have their first appearance in this zone.

(e) *Heliolithus kleinpellii* Zone

Definition: interval from the first occurrence of *H. kleinpellii* to the first occurrence of *Discoaster mohleri* (previously *Discoaster gemmeus*).
Authors: Mohler and Hay in Hay *et al.* 1967.
Age: Late Montian–Early Thanetian.
Localities: Haymana, Adiyaman, Şarkişla, Zonguldak.
Common species: the assemblage of the *Fasciculithus tympaniformis* Zone and *Micrantholithus flos, Micrantholithus entaster, Ericsonia cava, H. kleinpellii, Helicosphaera seminulum, Rhabdosphaera crebra, Neochiastozygus junctus.*
Remarks: this zone was easily recognised in the four localities as the zonal marker is widely distributed in low and high latitudes.

(f) *Discoaster mohleri* Zone

Definition: interval from the first occurrence of *D. mohleri* to the first occurrence of *Discoaster multiradiatus.*
Authors: Hay 1964, Mohler and Hay in Hay *et al.* 1967, emend. Toker 1980.
Age: Early Thanetian.

Localities: Haymana, Adiyaman, Şarkişla, Zonguldak.

Common species: those of the *Heliolithus kleinpellii* Zone and *D. mohleri, Chiasmolithus consuetus, Lophodolithus nascens, Zygodiscus plectopons*, minus *C. danicus* and *S. anarrhopus*.

Remarks: in the original definition the zone was called the *D. gemmeus* Zone. Its lower boundary is indicated by the first occurrence of *D. mohleri*, of which *D. gemmeus* is considered to be a junior synonym. The upper boundary was defined by the first occurrence of *Heliolithus riedelii*. This species has not been found in Turkey, or in the north-western Pacific Ocean (Bukry 1975), the North Atlantic (Perch-Nielsen 1972) or France (Hay *et al.* 1967). For this reason the upper boundary of the *Discoaster mohleri* Zone is here defined by the first occurrence of *D. multiradiatus*, thereby combining the *D. gemmeus* Zone and the *D. riedelii* Zone of Hay (1964). Measured stratigraphical sections of the Haymana and Adiyaman regions do not show any hiatus in the succession of planktonic foraminifera.

(g) *Discoaster multiradiatus* Zone

Definition: interval from the first occurrence of *D. multiradiatus* to the first occurrence of *Tribrachiatus bramlettei*.

Authors: Bramlette and Sullivan 1961, emend. Hay *et al.* 1967.

Age: Late Thanetian.

Localities: Haymana, Adiyaman, Şarkişla.

Common species: the assemblage of the *Discoaster mohleri* Zone and *D. multiradiatus, Transversopontis pulcher, Neochiastozygus perfectus*, but without *C. californicus* and *H. kleinpellii*.

Remarks: the zonal marker was widely distributed in these localities. The Palaeocene–Eocene boundary based on nannoplankton is drawn at the top of the *Discoaster multiradiatus* Zone.

(h) *Tribrachiatus contortus* Zone

Definition: interval from the first occurrence of *T.*

T. bramlettei to the last occurrence of *Tribrachiatus contortus*.

Authors: Hay 1964 and Martini 1970, 1971.

Age: Early Ypresian.

Localities: Haymana, Adiyaman.

Common species: those of the *Discoaster multiradiatus* Zone, plus *T. contortus, T. bramlettei, Chiasmolithus grandis, Tribrachiatus orthostylus, Discoaster barbadiensis, Discoaster binodosus, Ericsonia ovalis*, minus *N. saepes* and *C. bidens*.

Remarks: the nannoplankton diversity increases in the lower part of the Ypresian succession in Turkey. This zone was not recognised in the southwest Adiyaman region, but the stratigraphically recorded sections do not show any hiatus. As the absence of *T. contortus* probably has an ecological cause, the author extended the *Tribrachiatus orthostylus* Zone down to this level (Table 13.2).

(i) *Discoaster binodosus* Zone

Definition: Interval from the last occurrence of *T. T. contortus* to the first occurrence of *Discoaster lodoensis*.

Authors: Mohler and Hay in Hay *et al.* 1967.

Age: Middle Ypresian.

Localities: Haymana.

Common species: the assemblage of the *Tribrachiatus contortus* Zone and *Discoaster gemmifer, Rhabdosphaera perlonga*, but without *T. contortus*.

(j) *Tribrachiatus orthostylus* Zone

Definition: Interval from the first occurrence of *D. lodoensis* to the last occurrence of *T. orthostylus* (= *Marthasterites tribrachiatus*).

Authors: Brönnimann and Stradner 1960.

Age: Late Ypresian.

Localities: Haymana, Adiyaman.

Common species: those of the *Discoaster binodosus* Zone, plus *Discoaster deflandrei, Discoaster elegans, Discoaster lodoensis, Discoasteroides kuepperi, Helicosphaera lophota, Lophodolithus mochlophorus, Scyphosphaera tubicena*.

Remarks: several species, especially of the genus

Discoaster, appeared in this zone for the first time.

(k) *Discoaster lodoensis* Zone

Definition: interval from the last occurrence of *T. orthostylus* to the first occurrence of *Discoaster sublodoensis*.

Authors: Brönnimann and Stradner 1960.

Age: Late Ypresian to Early Lutetian.

Localities: Haymana, Adiyaman.

Common species: those of the *Tribrachiatus orthostylus* Zone, plus *Rhabdosphaera inflata*.

Remarks: this zone was easily recognised in the two regions.

(l) *Discoaster sublodoensis* Zone

Definition: interval from the first occurrence of *D. sublodoensis* to the first occurrence of *Nannotetrina fulgens*.

Author: Hay 1964.

Age: Middle Lutetian.

Localities: Haymana, Adiyaman, Kaman.

Common species: the assemblage of the *Discoaster lodoensis* Zone and *D. sublodoensis, Ericsonia formosa, Coccolithus eopelagicus, Chiasmolithus solitus, Sphenolithus radians, Sphenolithus moriformis*.

Remarks: in the original definition the zone was called the *Chiphragmalithus quadratus* Zone by Hay (1964). Martini (1971) used *Chiphragmalithus alatus* to define the upper boundary of this zone. *C. quadratus* and *C. alatus* are considered synonyms of *Nannotetrina fulgens*. Several nannoplankton species appeared in this zone for the first time.

(m) *Nannotetrina fulgens* Zone

Definition: interval from the first occurrence of *N. fulgens* to the last occurrence of *Rhabdosphaera gladius* or the first occurrence of *Reticulofenestra umbilicus*.

Authors: Hay in Hay *et al.* 1967, emend. Martini 1970, emend. Bukry 1973.

Age: Middle Lutetian.

Locality: Kaman.

Common species: the assemblage of the *Discoaster sublodoensis* Zone and *N. fulgens,*

Reticulofenestra coenura, Cribrocentrum reticulatum, R. gladius.

Remarks: the lower boundary of the *Nannotetrina fulgens* Zone is drawn at the first occurrence of *Nannotetrina*. Although the top of the zone is defined by the last occurrence of *R. gladius*, this species is generally rare or absent. Therefore the first appearance of *R. umbilicus* was used to recognise the top of this zone in the Kaman region.

(n) *Discoaster tanii nodifer* Zone

Definition: interval from the last occurrence of *R. gladius* or the first occurrence of *R. umbilicus* to to the last occurrence of *C. solitus*.

Authors: Hay in Hay *et al.* 1967, emend. Martini 1970, emend. Bukry 1973.

Age: Late Lutetian.

Localities: Kaman, Thrace.

Common species: those of the *Nannotetrina fulgens* Zone, plus *Discoaster tanii* ssp. *nodifer, R. umbilicus, Helicosphaera compacta, Discoaster wemmelensis, Cyclicargolithus floridanus, Calcidiscus kingii*.

(o) *Discoaster saipanensis* Zone

Definition: interval from the last occurrence of *C. solitus* to the first occurrence of *Chiasmolithus oamaruensis*.

Author: Martini 1970.

Age: Early Priabonian.

Localities: Kaman, Thrace, Antalya.

Common species: the assemblage of the *Discoaster tanii nodifer* Zone and *Discoaster woodringii, Helicosphaera intermedia, Helicosphaera euphratis, Rhabdosphaera tenuis, Sphenolithus predistentus*.

Remarks: several species appear in this zone for the first time.

(p) *Chiasmolithus oamaruensis* Zone

Definition: interval from the first occurrence of *C. oamaruensis* to the first occurrence of *Isthmolithus recurvus*.

Author: Martini 1970.

Age: Middle Priabonian.

Localities: Thrace, Antalya.

Common species: Those of the *Discoaster saipanensis* Zone, plus *C. oamaruensis, Cyclolithella robusta*.
Remarks: only two new species appeared in this zone, one of them the zonal marker. The diversity of the nannoplankton assemblage is low in the upper part of the Eocene sections in Turkey.

(q) *Isthmolithus recurvus* Zone
Definition: Interval from the first occurrence of *I. recurvus* to the first occurrence of *Sphenolithus pseudoradians*.
Author: Martini 1970.
Age: Late Priabonian.
Localities: Thrace, Antalya.
Common species: the assemblage of the *Chiasmolithus oamaruensis* Zone and *I. recurvus*.

(r) *Sphenolithus pseudoradians* Zone
Definition: interval from the first occurrence of *S. pseudoradians* to the last occurrence of *D. saipanensis*.
Author: Martini 1970.

Age: Late Priabonian.
Localities: Thrace, Antalya.
Common species: those of the *Isthmolithus recurvus* Zone plus *S. pseudoradians*.
Remarks: the Eocene–Oligocene boundary is drawn at the top of the *Sphenolithus pseudoradians* Zone in Turkey. *D. saipanensis* and *D. barbadiensis* both disappear at the top of this zone, which makes the Eocene–Oligocene boundary easy to recognise.

13.3 CONCLUSIONS
(1) The standard Palaeocene–Eocene zonation of Turkey is presented for the first time. .
(2) The Cretaceous–Tertiary boundary has been drawn at the top of the *Micula murus* Zone.
(3) The Palaeocene–Eocene boundary has been defined as the boundary between the *Discoaster multiradiatus* and *Tribrachiatus contortus* Zones.
(4) The Ypresian–Lutetian boundary has been drawn in the middle of the *Discoaster lodoensis* Zone.

PLATE 13.1

Plate 13.1, Fig. 1. *Chiasmolithus danicus*. Scanning electron micrograph; Haymana/1301, ×4000.

Plate 13.1, Fig. 2. *Discoaster binodosus*. Transmitted light; Haymana/2404, ×3000.

Plate 13.1, Fig. 3. *Discoaster multiradiatus*. Transmitted light; Haymana/1407, ×3000.

Plate 13.1, Figs. 4a, 4b. *Tribrachiatus orthostylus*. Transmitted light; Haymana/1510, ×2000.

Plate 13.1, Figs. 5a and 5b. *Heliolithus kleinpellii*. Fig. 5a, transmitted light. Fig. 5b, crossed-nicols. Haymana/1602, ×3000.

Plate 13.1, Fig. 6. *Ellipsolithus macellus*. Crossed-nicols; Haymana/1540, ×4000.

Plate 13.1, Figs. 7a, 7b. *Cruciplacolithus tenuis*. Fig. 7a, crossed-nicols. Fig. 7b, transmitted light. Haymana/2512, ×2500.

Plate 13.1, Fig. 8. *Tribrachiatus contortus*. Transmitted light; Haymana/1508, ×2500.

Plate 13.1, Fig. 9. *Discoaster lodoensis*. Transmitted light; Haymana/2205, ×2000.

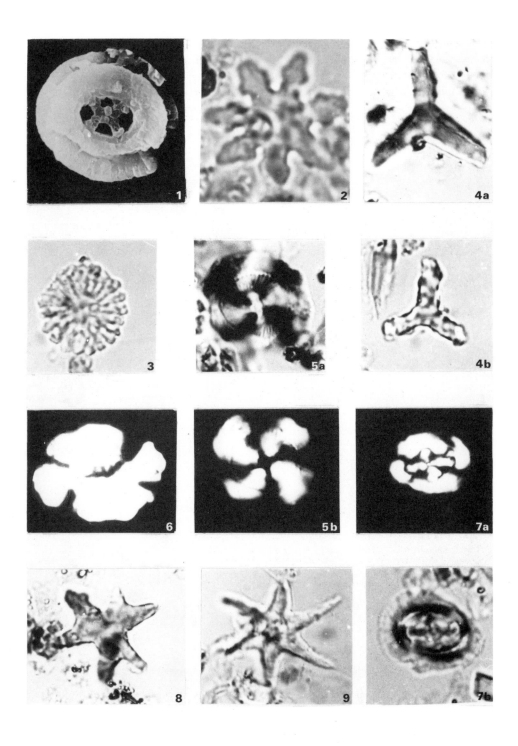

(5) The Lutetian–Priabonian boundary has been drawn at the boundary between the *Discoaster tanii nodifer* and the *Discoaster saipanensis* Zones.

(6) The Eocene–Oligocene boundary has been defined at the top of the *Sphenolithus pseudoradians* Zone.

13.4 ACKNOWLEDGEMENTS

I wish to thank Drs. Shirley van Heck of Shell EXPRO UK London for valuable comments and help with the correction of the manuscript and the English. Thanks are also due to Dr. Alan R. Lord of the Micropalaeontology Department of University College London for help with the correction and final drawing of the text figures.

PLATE 13.2

Plate 13.2, Fig. 1. *Chiasmolithus grandis*. Crossed-nicols; Thrace/GC.8.96, × 2000.

Plate 13.2, Fig. 2. *Discoaster saipanensis*. Scanning electron micrograph; Antalya/D.26, × 4500.

Plate 13.2, Fig. 3. *Discoaster tanii* ssp. *nodifer*. Scanning electron micrograph; Antalya/K.25, × 4000.

Plate 13.2, Fig. 4. *Discoaster barbadiensis*. Scanning electron micrograph; Antalya/K.10, × 4000.

Plate 13.2, Fig. 5. *Nannotetrina fulgens*. Transmitted light; Kaman/303, × 2500.

Plate 13.2, Fig. 6. *Ericsonia formosa*. Fig. 6a, transmitted light. Fig. 6b, crossed-nicols. Kaman/303, × 2500.

Plate 13.2, Fig. 7. *Sphenolithus pseudoradians*. Fig. 7a, crossed-nicols, 0°. Fig. 7b, crossed-nicols, 45°. Thrace/ENB.0.14, × 2000.

Plate 13.2, Fig. 8. *Isthmolithus recurvus*. Crossed-nicols; Thrace/GC.8.102, × 2000.

13.4 REFERENCES

Aköz, O. 1981. Biostratigraphic investigation of the Karababa type stratigraphic section based on nannoplankton. *Thesis*, Ankara University (unpublished).

Bolli, H. M. 1957. The genera *Globigerina* and *Globorotalia* in the Palaeocene–Lower Eocene Lizard Springs Formation of Trinidad, B.W.I.U.S. *Nat. Mus. Bull.*, **215**, 51–81.

Bolli, H. M. 1966. Zonation of Cretaceous to Pliocene marine sediments based on planktonic foraminifera. *Boletino Informativo Asociacion Venezolana de Geologia, Mineria y Petroleo*, **9**, 3–32.

Bramlette, M. N. and Sullivan, F. R. 1961. Coccolithophorids and related nannoplankton of the Early Tertiary in California. *Micropal.*, **7**, 129–174.

Bramlette, M. N. and Wilcoxon, J. A. 1967. Middle Tertiary calcareous nannoplankton of the Cipero Section, Trinidad W. I. *Tulane Stud. Geol. Paleontol.*, **5**, 93–131.

Brönnimann, P. and Stradner, H. 1960. Die Foraminiferen- und Discoasteridenzonen von Kuba und ihre interkontinentale Korrelation. *Erdöl-Z.*, **76**, 364–369.

Bukry, D. 1973. Low latitude coccolith biostratigraphic zonation. *Init. Rep. DSDP*, **15**, 685–703.

Bukry, D. 1975. Coccolith and silicoflagellate stratigraphy, Northwestern Pacific Ocean Leg 32. *Init. Rep. DSDP*, **32**, 677–711.

Bukry, D. 1981. Cenozoic coccoliths from the Deep Sea Drilling Project. *Spec. Publ. SEPM Tulsa*, **32**, 335–353.

Dizer, A. and Meriç, E. 1981. Upper Cretaceous–Palaeocene biostratigraphy of the Northwest Anatolia. *Bull. Min. Res. Explor. Inst. Turkey*, **96**, 149–163.

Ellis, C. H. 1975. Calcareous nannofossil biostratigraphy Leg 31. *Init. Rep. DSDP*, **31**, 655–676.

Gökten, E. and Toker, V. 1981. Stratigraphy and geological evolution of Şarkişla (Sivas). In E. Gökten *Geology of the Sivas Region. Bull. Geol. Soc. Turkey*, **26**, 167–177.

Hay, W. W. 1961. Note on the preparation of samples for discoasterids. *J. Paleontol.*, **35**, 873.

Hay, W. W. 1964. Utilisation stratigraphique des discoastéridés pour la zonation de Paléocène et l'Eocène inférieur. *Mém. Bur. Rech. Géol. Minières*, **28**, 885–889.

Hay, W. W. 1965. Calcareous nannofossils. In B. Kummel and D. Raup (Eds.), *Handbook of Paleontological Techniques*, Freeman, San Francisco, 3–7.

Hay, W. W., Mohler, H. P. and Wade, M. E. 1966. Calcareous nannofossils from Nal'chik (NW Caucasus). *Eclogae Geol. Helv.*, **59**, 379–399.

Hay, W. W., Mohler, H. P., Roth, P. H., Schmidt, R. R. and Boudreaux, J. E. 1967. Calcareous nannoplankton zonation of the Cenozoic of the Gulf Coast and Caribbean–Antillean area, and transoceanic correlation. *Trans. Gulf Coast Assoc. Geol. Soc.*, **17**, 428–480.

Heck, S. E. van and Prins, B. 1987. A refined nannoplankton zonation for the Danian of the central North Sea. *Abh. Geol. B.-A.*, **39**, 285–303.

Martini, E. 1970. Standard Paleogene calcareous nannoplankton zonation. *Nature*, **226**, 560–561.

Martini, E. 1971. Standard Tertiary and Quaternary calcareous nannoplankton zonation. In A. Farinacci (Ed.), *Proc. II Plankt. Conf. Roma, 1970*, **2**, Tecnoscienza, Rome, 739–785.

Martini, E. and Worsley, T. 1970. Standard Neogene calcareous nannoplankton zonation. *Nature*, **225**, 289–290.

Mohler, H. P. and Hay, W. W. 1967. Calcareous nannoplankton from early Tertiary rocks at Pont Labau, France, and Paleocene–Eocene correlations. *J. Paleontol.*, **41**, 1505–1541.

Perch-Nielsen, K. 1972. Remarks on Late Cretaceous to Pleistocene coccoliths from the North Atlantic. *Init. Rep. DSDP*, **12**, 1003–1069.

Perch-Nielsen, K. 1985. Cenozoic calcareous nannofossils. In H. M. Bolli, J. B. Saunders and K. Perch-Nielsen (Eds.), *Plankton Stratigraphy*, Cambridge University Press, Cambridge, 427–554.

Stainforth, R. M., Lamb, J. L., Luterbacher, H., Beard, J. H. and Jeffords, R. M. 1975. Cenozoic planktonic foraminiferal zonation and characteristics of index forms. *Univ. Kansas Paleontol. Contrib. Article*, **62**, 1–425.

Toker, V. 1977. Biostratigraphic investigation of Haymana region (SW Ankara) based on Nannoplankton and Planktonic foraminifera. *Thesis*, Ankara University (unpublished).

Toker, V. 1980. Upper Cretaceous–Lower Tertiary Nannoplankton biostratigraphy of the Haymana region (SW Ankara). *Bull. Geol. Soc. Turkey*, **23**, 165–178.

Toker, V. 1982. Calcareous nannoplankton in the Eocene formation at the Kaman region. *Comm. Fac. Sciences Univ. Ankara, Suppl.*, **2**, 1–33.

Toker, V. 1987. Upper Eocene–Lower Oligocene calcareous nannoplankton biostratigraphy of the Korkuteli region (NW Ankara). *Bull. Geol. Soc. Turkey*, **30**, 19–34.

Toker, V. and Erkan, E. 1985. Nannoplankton biostratigraphy of Eocene formation in the Gelibolu peninsula. *Bull. Min. Res. Explor. Inst. Turkey*, **101**, 25–44.

Toker, V. and Meriç, E. 1988. Planktonic foraminifera and nannoplankton biostratigraphy of the Adiyaman Region. *Bull. Geol. Soc. Turkey*, **31**, (in press).

Toumarkine, M. and Luterbacher, H. P. 1985. Palaeocene and Eocene planktic foraminifera. In H. M. Bolli, J. B. Saunders and K. Perch-Nielsen (Eds.), *Plankton Stratigraphy*, Cambridge University Press, Cambridge, 87–154.

Varol, O. 1983. Late Cretaceous–Paleocene calcareous nannofossils from the Kokaksu sections (N Turkey). *N. Jb. Geol. Paleont. Abh.*, **3**, 431–460.

14

A morphometric analysis of the *Arkhangelskiella* group and its stratigraphical and palaeoenvironmental importance

Magdy H. Girgis

A detailed morphometric analysis of the *Arkhangelskiella* group in the Maastrichtian–Danian sequence at two outcrop sections in the Gulf of Suez region, Egypt, is made. It reveals a gradual increase in the mean length (*M*) during most of the Maastrichtian. This feature was also observed in the latest Campanian–Maastrichtian sequences in a number of well sections in the Middle East, North and West Africa areas. Synthesis of the outcrop and well sections leads to the recognition of four nannofossil zonal units based on the mean length and other size characteristics of the group. These size characteristics have facilitated the recognition of the zonal units and may be used as alternative indicators of stratigraphical age in the absence of conventional index species. In the North Sea area, the size characteristics of the group appear to be both generally different and less useful stratigraphically than those elsewhere.

In the outcrop sections, the group displays a decrease in the *M* values in the latest Maastrichtian which are associated with an increase in the abundance of the calcareous dinoflagellate *Thoracosphaera operculata*. This feature is thought to reflect stressful environments for nannoplankton prior to the Cretaceous–Tertiary (K–T) boundary events. The survival of the 'Cretaceous' nannoplankton into the earliest Danian, suggested in the literature, is difficult to assess in this study owing to the wide sample interval employed around the K–T boundary. However, the overall size characteristics of the *Arkhangelskiella* group in the earliest Danian studied here suggest reworking of specimens of this group, from sediments of Maastrichtian age, rather than survival.

14.1 INTRODUCTION AND PREVIOUS WORK

The importance of the size variation of nannofossils was first pointed out by Stradner (1963) who stated '. . . variation in the size of nannofossils can be used to a certain extent for paleogic (palaeoenvironment) and stratigraphic purposes'. Many studies on size or morphometry were carried out on Cenozoic nannofossils, while only a few were made on the Mesozoic material, mainly on the genus *Broinsonia* (Verbeek 1977, Crux 1982).

Perch-Nielsen (1985a) reported a general increase in the mean length of *Arkhangelskiella cymbiformis* (? = *Arkhangelskiella* group here) in the sediments of Late Campanian age to the Maastrichtian Zone CC25 of Sissingh (1977). In the same year Girgis (1985) made a detailed size (length) analysis of the *Arkhangelskiella* group from two Egyptian sections which reveals a gradual persistent increase in the mean length throughout most of the Maastrichtian. This paper presents this analysis and focuses on its stratigraphical and palaeoenvironmental significance and its applications in other areas.

14.2 TAXONOMIC CONCEPT AND MORPHOLOGICAL VARIATIONS

Terminology adopted here is that proposed by Gartner (1968). In the literature a number of forms (or species) belonging to the genus *Arkhangelskiella* have been distinguished based on the number of perforations in the central plate, the number of tiers making up the rim (or marginal structure), the relative width of the distal tier and the number of cycles in this tier (one or two). In the present material, none of these characters, especially the number of pores and tiers, can be consistently recognised. This, together with the lack of a clear stratigraphical value of all known Late Campanian–Maastrichtian members of this group such as *A. specillata* and *A. cymbiformis* in Sissingh (1977) (see Perch-Nielsen 1985a, p. 354), has prompted research into size stratigraphy of the group. The *Arkhangelskiella* group under investigation includes forms with a broad or narrow single cycle distal tier, 4–12 pores per quadrant and three or four tiers in the rim. It also includes a form with numerous pores (about 16–20 per quadrant) and a narrow distal tier with two cycles (the inner cycle is pitted or dimpled), a feature which is well developed in the genus *Broinsonia*. This form was only identified under the scanning electron microscope from the two outcrop sections examined. This form is probably the same member of the *Arkhangelskiella* group 2 of Lauer (1975) which evolved around the Santonian–Campanian boundary.

14.3 MATERIAL

The two outcrop sections (A and B) are situated around St. Paul's Monastery, Gulf of Suez region, Egypt (Fig. 14.1a). Section A is a cliff on a prominent hill which is about 1 km to the south-west of the Monastery. Section B is a steep cliff situated about 3 km to the west of the Monastery. Size analysis was carried out only on the Maastrichtian–early Danian parts of these sections. The lithology, position of samples and zones recognised are given in Fig. 14.1b. Ditch cuttings samples from a number of Late Campanian–Maastrichtian well sections in the Middle East, North and West Africa and the North Sea areas were also examined. The lithology of these sections varies, including chalk, limestone, grey calcareous mudstone and dark brown shale.

14.4 ZONATION SCHEME

A zonation scheme (Fig. 14.2) was established for the Late Campanian–Maastrichtian sequence in Egypt and the Middle East and serves as reference for the size characteristics of the *Arkhangelskiella* group. The scheme is a combination, with slight modifications, of the zonal system of Sissingh (1977) and 'events' of Perch-Nielsen (1979). A full discussion of the zonal scheme is not intended here but a few remarks on the selection of some zonal markers are made.

Some of the markers in published schemes are not used in the present zonation owing to (1) scarcity (such as last occurrence (LO) of *Eiffellithus eximius, Reinhardtites anthophorus* and *Broinsonia parca* ssp. *constricta*, (2) often being difficult to distinguish (*R. anthophorus*) and (3) the nature of the material examined. The Late Campanian samples examined were mainly ditch cuttings in which first occurrences are not reliable datums, and thus the first occurrence (FO) of *Reinhardtites levis* is excluded. Last consistent occurrences of *E. eximius* and *B. parca* ssp. *constricta* were found to be more reliable than their absolute LO, and thus were used as zonal and subzonal markers in the present scheme.

In the outcrop sections, only four Danian samples contain sufficient specimens of the *Arkhangelskiella* group for size analysis. The

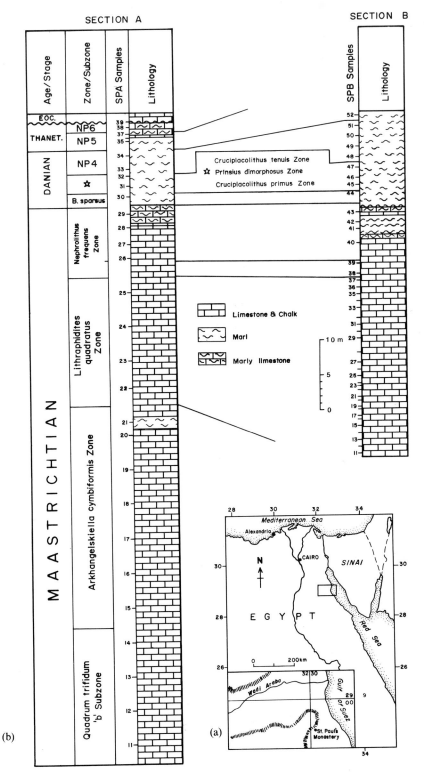

Fig. 14.1 — (a) Location map for St. Paul's Monastery. (b) Sections A and B, around St. Paul's Monastery; lithology, position of samples and biozonation.

lowest Danian samples SPA 30 and SPB 44 are assigned to the *Biantholithus sparsus* Zone, while samples SPB 45 and 46 are assigned to the *Prinsius dimorphosus* Zone as defined by Romein (1979). These two zones correspond to the lower part of the NP1 Zone and lower part of the NP3 Zone respectively of Martini (1971), according to Perch-Nielsen (1985b). The presence of the latest Maastrichtian *Nephrolithus frequens* 'b' Subzone (= range of *Micula prinsii*) and the earliest Danian *B. sparsus* Zone, together with the absence of any signs of break in sedimentation, suggests that the sequence across the Maastrichtian–Danian boundary is more or less continuous.

14.5 METHODS OF INVESTIGATION

(a) Sample Preparation

Centrifuge preparations as described by Taylor and Hamilton (1982) were made from the outcrop samples, mainly chalk–limestone. Smear and modified centrifuge techniques were applied to ditch cuttings. The organic-rich shales from West Africa were prepared using sieving, centrifuging and heavy liquid separation techniques as described in Girgis (1986). The length distribution of the specimens observed in smear preparation is almost identical to that in other preparations, i.e. the non-smear preparations have not biased the size characteristics determined for the *Arkhangelskiella* group.

(b) Size Analysis

The length or maximum diameter of a specimen was the criterion chosen for measurements. The length of 50 specimens in most Maastrichtian samples (12 samples in the *Lithraphidites quadratus* Zone of Section B were omitted owing to the already sufficient data) and in four Danian samples from both sections was measured and the results are presented in histograms (Figs. 14.3 and 14.4). The mean length (*M*) and coefficient of variation (CV) of the specimens were calculated and plotted for each section (Figs. 14.5 and 14.7). Measurements were made on

well-preserved specimens as far as possible. Some heavily overgrown specimens, mainly from section B, could not be avoided; the overgrowth of these specimens may cause an increase of 0.5 to 1.0 μm to their actual lengths.

14.6 DISCUSSION AND RESULTS

(a) Stratigraphical aspects

In the following discussion, the mean length (*M*) and coefficient of variation (CV) values of the highest Maastrichtian samples in both sections (SPA29 and SPB43) as well as sample SPB41 are ignored owing to environmental factors as will be discussed later.

(i) Mean length (M)

A general increase in the *M* values (7.45 to 12.6 μm) during the Maastrichtian is evident. This increase is less distinct in section B because it represents a much shorter time span.

When the highest Maastrichtian samples are ignored, the variations of the mean length (*M*) within each of the four Maastrichtian zones of section A do not overlap, while they are slightly overlapping in the two zones of section B (Fig. 14.6). This is probably due to the closer sample interval employed in section B (Fig. 14.1b). The increases of *M* values in each of the *L. quadratus* and *N. frequens* Zones in section A are almost identical to those in the equivalent zones of section B. However, the highest and lowest *M* values in these two zones in section A are lower than those in section B. The magnitude of these differences is relatively small (0.3–0.8 μm), which may be attributed to the fact that the specimens measured in section B are generally more overgrown, and possibly slightly younger.

(ii) Coefficient of variation (CV)

The coefficient of variation in this size analysis is the dispersion of the 50 measured specimens away from the average length, expressed as a percentage of the mean value. It is, therefore, a useful parameter to identify the size characteristics of the *Arkhangelskiella* group.

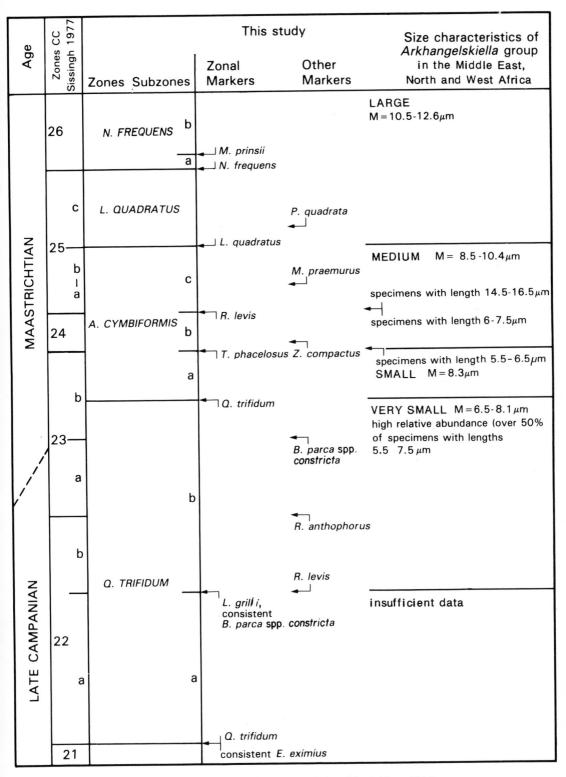

Fig. 14.2 — Zonation schemes and size characteristics of the *Arkhangelskiella* group.

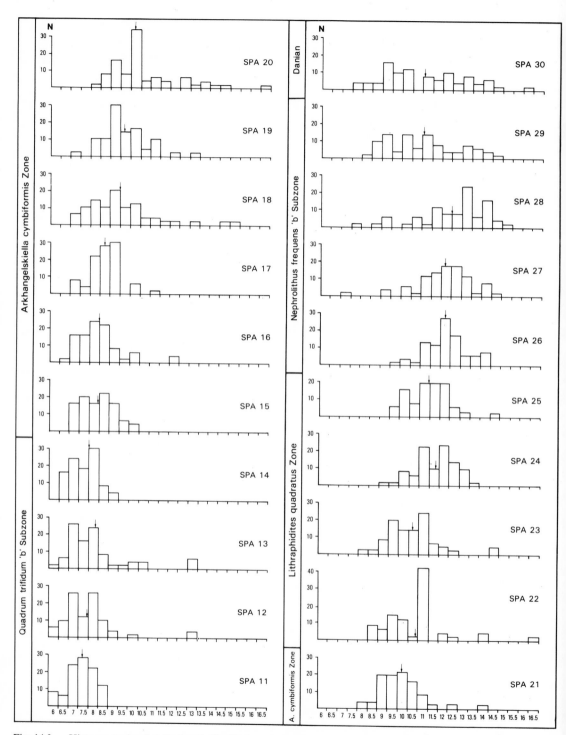

Fig. 14.3 — Histograms showing the length distribution of *Arkhangelskiella* group (number of specimens $N = 50$) in the Maastrichtian–Danian interval in Section A. Arrows denote mean length.

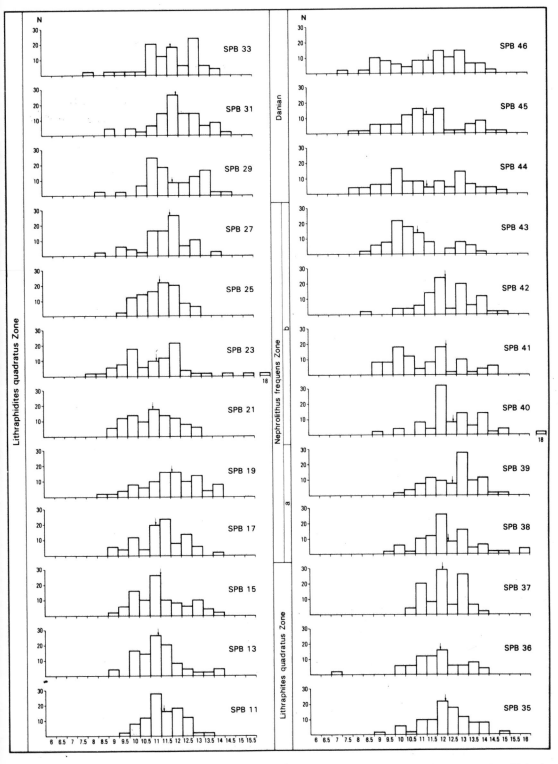

Fig. 14.4 — Histograms showing the length distribution of *Arkhangelskiella* group (number of specimens *N* = 50) in the Maastrichtian–Danian interval in Section B. Arrows denote mean length.

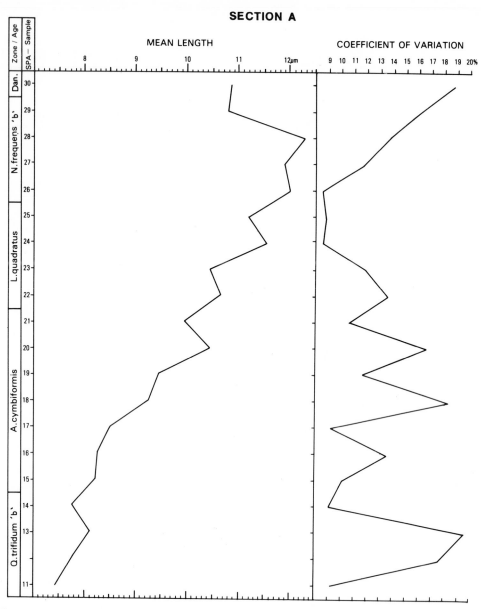

Fig. 14.5 — Mean length *M* and coefficient of variation (CV) of *Arkhangelskiella* group (50 specimens per sample) in Section A.

Zone	MEAN LENGTH			
	8	9	10	11	12µm
N.frequens				
L.quadratus				
A.cymbiformis				
Q.trifidum				

Fig. 14.6 — The range of the mean length *M* in the Maastrichtian Zones of Section A (thin bars) and Section B (thick bars); excluding samples SPA 29 and SPB 43.

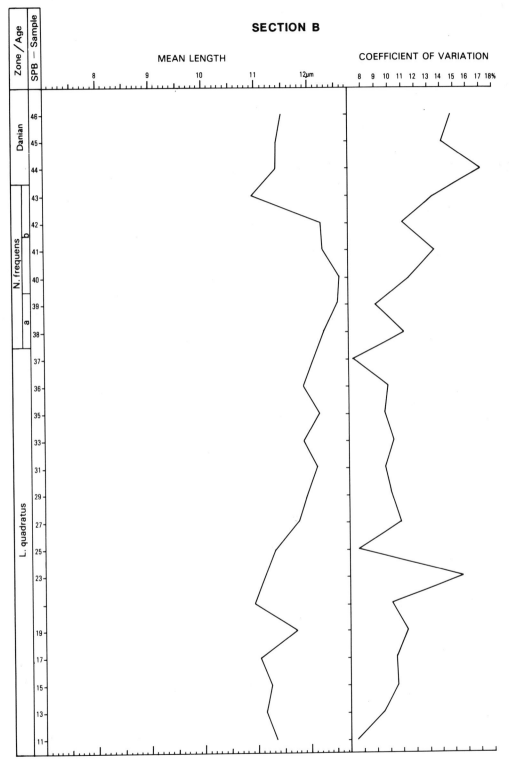

Fig. 14.7 — Mean length *M* and coefficient of variation (CV) of *Arkhangelskiella* group (50 specimens per sample) in Section B.

The CV values are low (7.3–11.6%), except in four samples in section A (SPA 12, 13, 18 and 20) and in one sample in section B (SPB 23) where the values are considerably higher (16–19.5%). The introduction of exceptionally large forms relative to the mean length, especially in sample SPB 23, causes these peak values.

(iii) Size characteristics and conventional datums

Fig. 14.2 shows the size characteristics of the *Arkhangelskiella* group based on synthesis of information from both the outcrop and well sections under investigation. In the uppermost Campanian–Maastrichtian, four units can be identified by a combination of the mean length and the extinction, appearance and abundance of extreme (in size) members of the group. In the *Quadrum trifidum* 'a' Subzone, no detailed size analysis was made, owing to low frequency of the *Arkhangelskiella* group. The group, however, is tentatively characterised by M values between 7.5 and 9.5 μm, rare occurrence of specimens with length 5.5–6.5 μm and lack of 'giant' specimens (>14.5 μm) as observed in younger Maastrichtian sediments.

Campanian–Maastrichtian boundary: in the literature, this boundary is drawn within the range of the marker species *Q. trifidum*; two datums, namely the LO of *R. anthophorus* and *B. parca* ssp. *constricta*, are close to the boundary and may be useful in its approximate determination. *R. anthophorus* is somewhat difficult to distinguish consistently from *R. levis* and other related forms owing to poor preservation of the material examined. In the Middle East and Egypt, the LO of *B. parca* ssp. *constricta* is difficult to pin-point owing to its scarcity towards the upper limit of its stratigraphical range. The size characteristics of the *Arkhangelskiella* group may, therefore, facilitate the recognition of the boundary, especially in poorly preserved assemblages where conventional marker species are lacking or beyond recognition. Around the Campanian–Maastrichtian boundary, the group is characterised by a mean length of 6.5–8.1 μm

and the dominance of specimens with length 5.5–7.5 μm.

It is difficult to pin-point the LO of *Tranolithus phacelosus* and *R. levis* (employed to subdivide the *A. cymbiformis* Zone into three subzones 'a', 'b' and 'c') in the outcrop section examined. Based on regional experience in Middle East and Egyptian well sections, these two datums were found to coincide approximately with the LO of specimens with a length of 5.5–6.5 μm and 6–7.5 μm respectively. These specimens display a high relative abundance in the *Q. trifidum* 'b' Subzone (Fig. 14.3). A 'giant' form (14.5–16.5 μm) appears at the base of the *A. cymbiformis* 'c' Subzone (LO of *R. levis*) and continues until the end of the Maastrichtian. Future studies may warrant the erection of new species or amending the existing ones to accommodate these forms identified by their length.

(iv) Geographical significance

Size characteristics in these units have been observed in wide geographic areas covering the Middle East and North and West Africa, and are expected to be recognisable in similar latitudes world-wide. They have been found useful to aid the recognition of the zonal units and were employed as age indicators in the absence of conventional marker species. Some of these markers (such as *N. frequens* and *B. parca* ssp. *constricta*) are sparse in the areas under investigation or are susceptible to diagenetic alteration (*M. prinsii*) and/or difficult to distinguish consistently (*R. anthophorus*). The *Arkhangelskiella* group is more resistant to diagenetic alteration and has a wider geographic distribution than some conventional markers. This, together with the fact that it usually occurs in sufficient numbers for the determination of its size characteristics, makes the group a reliable alternative marker.

Only a few well sections from the North Sea area were examined for the size characters of the *Arkhangelskiella* group. These characters were found to be different, except in the general trend of the increase in M values, and appear to be less useful stratigraphically than in the other areas under investigation. No detailed size analysis has been made, and thus only tentative remarks can be made. In the interval between the Coniacian FO of *Micula staurophora* to the Maastrichtian LO of *R. levis*, the mean length (*M*) of the group varies from 8.5 to 11.3 μm, while in the late Maastrichtian *N. frequens* Zone the *M* values are distinctively higher, usually within the range from 10.5–14.5 μm. In the lower part of the *N. frequens* Zone at a northerly located section, *M* values are slightly lower (around 10 μm). This anomaly can be explained by the earlier evolution of *N. frequens* in this site than in other parts of the North Sea area. As in the Middle East and North and West Africa, the specimens with length 5.5–6.5 μm have their last occurrence around the extinction of *T. phacelosus*.

(b) Environmental aspects around the Cretaceous–Tertiary boundary

(i) Cretaceous–Tertiary (K–T) boundary

A discussion on the boundary is beyond the scope of this study and only basic background information is given. The K–T (or Maastrichtian-Danian) boundary is known to be marked by a major crisis for marine biota, especially the calcareous plankton (nannoplankton and foraminifera). In the literature, the mass extinction of nannoplankton around the boundary is well documented and is thought to be either catastrophic (sudden) or gradual. Recently the survival of the nannoplankton in the earliest Danian was suggested (Perch-Nielsen *et al.* 1982), as will be discussed later. In the two Egyptian outcrop sections, the sequence around the K–T boundary is thought to be more or less complete, and the boundary is placed below both a massive occurrence of the calcareous dinoflagellate *Thoracosphaera operculata* and a drop in the absolute abundance of nannofossils (Girgis 1985). In the following size analysis of the *Arkhangelskiella* group around the K–T boundary, only the outcrop sections are considered, as the cuttings samples from the well sections are not suitable for such analysis.

(ii) Latest Maastrichtian

High proportions of specimens with length 8.5–11.5 μm appeared in the upper part of the *Arkhangelskiella cymbiformis* Zone and disappeared near the top of the *Lithraphidites quadratus* Zone (Fig. 14.3). The reappearance of this high frequency is recorded in the 'middle' part of the *Nephrolithus frequens* Zone — seen only in section B — at sample SPB 41, which is clearly reflected in its higher CV value. The apparent absence of this feature in section A is probably due to a sample gap. This high frequency is well established in the youngest Maastrichtian samples (SPA 29 and SPB 43), especially in sample SPB 43, which accounts for increase in CV and considerable decrease in *M* values.

The re-occurrence of high proportions of these specimens (lengths 8.5–11.5 μm) in the upper part of the *N. frequens* Zone may be explained by (1) reworking from older strata which contains specimens of this size, (2) contamination via bioturbation from the immediately overlying Danian sediments which contain specimens of similar size distribution, and (3) environmental factors prior to or related to K–T boundary event(s).

Reworking from older strata is disregarded because (1) there are no obviously reworked specimens (of species whose extinctions are well below *N. frequens* Zone such as *T. phacelosus*) in one of the samples (SPA 29) and (2) the abundance of reworked specimens (rare to few) in the other two samples (SPB 41 and 43) is not greater than that in other samples of section B. Contamination through bioturbation in the youngest Maastrichtian samples (SPA 29 and SPB 43) is possible although it is difficult to assess owing to lack of proper field observations. However, the presence of a high frequency of these 'small' specimens well below the Danian sediments in sample SPB 41 (at least 2.5 m below the Danian) dispels the possibility of a Danian source.

It is noted that the three samples which contain a high proportion of 'small' specimens also contain a higher frequency of the calcareous dinoflagellate *Thoracosphaera operculata* than any other Maastrichtian samples. Blooms of this species are well known to be associated with the unfavourable environments for plankton around the K–T boundary events. It is suggested, therefore, that the reappearance of high proportions of 'small' specimens of the *Arkhangelskiella* group is caused by stressful environments for nannoplankton prior to their mass extinction at the K–T boundary.

(iii) Danian

The Cretaceous nannofossils in the basal Danian sediments (Zone NP1) were suggested to have actually survived the K–T boundary events and to have continued to reproduce in the earliest Tertiary oceans for some tens of thousands of years after the boundary events (Perch-Nielsen *et al.* 1982). This conclusion is based mainly on isotope stratigraphical evidence, supported by the similarity in size, morphology and composition of nannofossils from directly above and below the boundary.

The lowest Danian samples examined here are probably not directly above (i.e. within a few centimetres of) the K–T boundary because of the wide sample interval employed across the boundary in this study (2.5 m and 3 m in sections A and B respectively). However, even if the present lowest Danian samples examined were immediately above the K–T boundary and if it is true that the 'Cretaceous' nannoplankton survived into the earliest Danian, then at least some specimens measured in the lowest Danian samples (assigned to lower part of Zone NP1) would be the products of 'surviving' organisms, while the specimens in the younger samples (lower part of NP3 Zone) are reworked. There is no clear difference in the size distribution of the *Arkhangelskiella* group in all Danian samples studied. The *M* values are slightly higher than in the youngest Maastrichtian samples and CV values are higher than those in the Maastrichtian except in a few samples which contain reworked or considerably larger specimens than average, as explained above. These size characteristics clearly do not reflect 'surviving' organisms in

normal open marine conditions as in the Maastrichtian time, where most specimens show smaller size variations.

The size (length) distribution can be attributed to (1) stressful environments or (2) reworking from the Maastrichtian and/or older sediments. All palaeontological evidence indicates stressful environments, especially in the open marine surface water during the earliest Danian (Thierstein 1982). The re-introduction of the 'small' specimens, in association with the increase in the abundance of *Thoracosphaera operculata* during the latest Maastrichtian, is thought to be caused by unfavourable conditions. These conditions are far less severe than those in the earliest Danian, which caused the mass extinction of nannoplankton, and a higher proportion of 'small' specimens would, therefore, be expected in the earliest Danian. Such higher proportions are not seen except vaguely in sample SPA 30, which can hardly be considered as evidence. Future studies on closely spaced samples are essential to reveal the absence or presence of signs of living in stressful environments. At this stage of research reworking from Upper Cretaceous sediments is the most plausible explanation for the occurrence of the *Arkhangelskiella* group in the Danian samples examined. The *M* value of the group and the absence of specimens with length 5.5–6.5 μm indicates reworking from Maastrichtian sediments of age ranging from the *A. cymbiformis* 'b' Subzone to the *N. frequens* Zone.

14.7 CONCLUDING REMARKS

Based on morphometric analysis, new stratigraphical applications of the *Arkhangelskiella* group are evident. The analysis also provides valuable data for palaeoenvironmental interpretations. Future detailed size analyses of the group are needed from (1) high latitude areas to evaluate its stratigraphical significance, and (2) closely spaced samples around the K–T boundary which may shed more light on the environments and possible survival of the group in the earliest Danian. A byproduct of this study is the identification of a 'continuous' sequence around the K–T boundary in the Gulf of Suez region, a new source of information on an important geological event.

14.8 ACKNOWLEDGEMENTS

The work on the outcrop sections is part of a Ph.D. project undertaken at University College London. Professor T. Barnard and Dr. A. R. Lord supervised the project. I am grateful to the management of the Geological Survey of Egypt for providing field facilities during collecting the outcrop samples. The remaining work presented in this paper was done at the Robertson Group plc. who gave financial support for publication. Mr. D. R. Clowser (Robertson Group) kindly checked the English. Dr. K. Perch-Nielsen and Miss S. van Heck (London) critically read the manuscript.

14.9 REFERENCES

Crux, J. A. 1982. Upper Cretaceous (Cenomanian to Campanian) calcareous nannofossils. In A. R. Lord (Ed.), *A Stratigraphical Index of Calcareous Nannofossils* Ellis Horwood, Chichester, 81–135.

Gartner, S. 1968. Coccoliths and related calcareous nannofossils from Upper Cretaceous deposits of Texas and Arkansas. *Univ. Kansas Paleontol. Contrib., Artic.*, **48**, 1–56.

Girgis, M. H. 1985. Upper Cretaceous–Lower Tertiary calcareous nannofossils from Egypt and Spain. *Thesis*, University College London (unpublished).

Girgis, M. H. 1986. A new preparation technique for calcareous nannofossils from organic-rich argillaceous sediments. *INA Newsletter*, **8**, 93–95.

Lauer, G. 1975. Evolutionary trends in the *Arkhangelskiellacae* (calcareous nannoplankton) of the Upper Cretaceous of central Oman, S. E. Arabia. *Arch. Sci. Genève*, **28**, 259–262.

Martini, E. 1971. Standard Tertiary and Quaternary calcareous nannoplankton zonation. In A. Farinacci (Ed.). Tecnoscienza, Rome. *Proc. II Plank. Conf., Roma, 1970* **2**,, 739–785.

Perch-Nielsen, K. 1979. Calcareous nannofossils from the Cretaceous between the North Sea and the Mediterranean. In *Aspekte der Kreide Europas. IUGS Series A*, **6**, 223–272.

Perch-Nielsen, K. 1985a. Mesozoic calcareous nannofossils. In H. M. Bolli, J. B. Saunders and K. Perch-Nielsen (Eds.), *Plankton Stratigraphy*, Cambridge University Press, Cambridge, 329–426.

Perch-Nielsen, K. 1985b. Cenozoic calcareous nannofossils. In H. M. Bolli, J. B. Saunders and K. Perch-Nielsen (Eds.), *Plankton Stratigraphy*, Cambridge University Press, Cambridge, 427–554.

Perch-Nielsen, K., McKenzie, J. and He, Q. 1982. Biostratigraphy and isotope stratigraphy and the 'catastrophic' extinction of calcareous nannoplankton at the Cretaceous/Tertiary boundary. *Geol. Soc. Amer., Special Paper*, **190**, 353–371.

Romein, A. J. T. 1979. Evolutionary lineages in Early Paleogene calcareous nannoplankton. *Utrecht Micropaleont. Bull.*, **22**, 1–231.

Sissingh, W. 1977. Biostratigraphy of Cretaceous calcareous nannoplankton. *Geol. Mijnbouw*, **65**, 37–65.

Stradner, H. 1963. New contribution to Mesozoic stratigraphy by means of nannofossils. *Proc. World Petrol. Congr., Sect. 1*, Paper 4 (preprint), 167–184.

Taylor, R. J. and Hamilton, G. B. 1982. Techniques. In A. R. Lord (Ed.), *A Stratigraphical Index of Calcareous Nannofossils*, Ellis Horwood, Chichester, 11–15.

Thierstein, H. R. 1982. Terminal Cretaceous plankton extinctions: A critical assessment. *Geol. Soc. Amer., Special Paper*, **190**, 385–399.

Verbeek, J. W. 1977. Calcareous nannoplankton biostratigraphy of Middle and Upper Cretaceous deposits in Tunisia, southern Spain and France. *Utrecht Micropaleont. Bull.*, **16**, 1–157.

Index of Taxa

Index

General Index